WITHDRAWN
UTSA LIBRARIES

Challenges in the Management of New Technologies

Management of Technology

Series Editor: Tarek Khalil *(University of Miami, USA)*

Published

Vol. 1 Challenges in the Management of New Technologies
edited by Marianne Hörlesberger, Mohamed El-Nawawi & Tarek Khalil

Management of Technology – VOL. 1

Challenges in the Management of New Technologies

edited by

Marianne Hörlesberger
Austrian Research Centers GmbH-ARC, Austria

Mohamed El-Nawawi
UNIDO, Austria

Tarek Khalil
University of Miami, USA

World Scientific

NEW JERSEY · LONDON · SINGAPORE · BEIJING · SHANGHAI · HONG KONG · TAIPEI · CHENNAI

Published by

World Scientific Publishing Co. Pte. Ltd.
5 Toh Tuck Link, Singapore 596224
USA office: 27 Warren Street, Suite 401-402, Hackensack, NJ 07601
UK office: 57 Shelton Street, Covent Garden, London WC2H 9HE

Library of Congress Cataloging-in-Publication Data
Challenges in the management of new technologies / edited by Marianne Hörlesberger, Mohamed El-Nawawi & Tarek Khalil.
   p. cm. -- (Management of technology ; v 1)
   ISBN-13 978-981-270-855-7
   ISBN-10 981-270-855-3
   1. Technological innovations--Management--Congresses. 2. Technological innovations--Economic aspects--Congresses. I. Hörlesberger, Marianne. II. Nawawi, Mohamed A. III. Khalil, Tarek M.

HD45.C445 2007
658.5'14--dc22

                                                         2007030396

British Library Cataloguing-in-Publication Data
A catalogue record for this book is available from the British Library.

Copyright © 2007 by World Scientific Publishing Co. Pte. Ltd.

All rights reserved. This book, or parts thereof, may not be reproduced in any form or by any means, electronic or mechanical, including photocopying, recording or any information storage and retrieval system now known or to be invented, without written permission from the Publisher.

For photocopying of material in this volume, please pay a copying fee through the Copyright Clearance Center, Inc., 222 Rosewood Drive, Danvers, MA 01923, USA. In this case permission to photocopy is not required from the publisher.

Printed in Singapore by World Scientific Printers (S) Pte Ltd

Contents

Preface		xi
SECTION I: MANAGING NEW TECHNOLOGIES		1
Chapter 1	An Exploratory Analysis of TSS Firms: Insights from the Italian Nanotech Industry *Vittorio Chiesa, Alfredo De Massis, and Federico Frattini*	3
Chapter 2	Knowledge Creation Dynamics and Financial Governance: Crisis of Growth in Biotech Firms *Anne-Laure Saives, Mehran Ebrahimi, Robert H. Desmarteau, and Catherine Garnier*	17
Chapter 3	Partnerships between Technology-Based Start-Ups and Established Firms: Case Studies from the Cambridge (U.K.) High-Tech Business Cluster *Tim Minshall, Rob Valli, Pete Fraser, and David Probert*	33
Chapter 4	Impacts of RFID on Warehouse Management in the Retail Industry *Louis-A. Lefebvre, Elisabeth Lefebvre, Samuel Fosso Wamba, and Harold Boeck*	47
Chapter 5	Factors Driving the Broadband Internet Growth in the OECD Countries *Petri Kero*	59

Contents

Chapter 6	What Comes After "New-to-the-World" Product Success for a Small Firm? Utilize MOT Analysis and Implementation for Innovative Products and Competitive Lead *Myra Urness*	69
Chapter 7	Demand Forecast and Adoption Factors for Portable Internet Service *Moon-Koo Kim and Jong-Hyun Park*	87
Chapter 8	The Use of Scenario Analysis in Studying Emerging Technologies — Case Grid Computing *Mika Lankila, Liisa-Maija Sainio, and Jukka-Pekka Bergman*	101
Chapter 9	Computer Services as Innovation Agents for Local Production Areas *Enrico Scarso and Ettore Bolisani*	119
Chapter 10	Towards Digital Integration: Platform Thinking in the Fashion Business *Finn Kehlet Schou*	135
Chapter 11	Value-Centric e-Government: The Case of Dubai Municipality *Habib Talhami and Mohammed Arif*	145
Chapter 12	E-Business and the Company Strategy: The Case of the Celta at GM Brazil *Sílvia Novaes Zilber*	163

SECTION II: BUSINESS ORGANIZATION — 181

Chapter 13	Business Model Design and Evolution *Michael Weiss and Daniel Amyot*	183
Chapter 14	The Entrepreneur Has "Sold Out": An Exploratory Study of the Sale of High-Tech Companies to Off-Shore Buyers *Sally Davenport*	195
Chapter 15	Technology Acquisition Through Convergence: The Role of Dynamic Capabilities *Fredrik Hacklin, Christian Marxt, and Martin Inganäs*	211

Chapter 16	The Process for Aligning Project Management and Business Strategy: An Empirical Study *Sabin Srivannaboon and Dragan Z. Milosevic*	227
Chapter 17	Risk-Sharing Partnerships with Suppliers: The Case of Embraer *Paulo Figueiredo, Silveira Gutenberg, and Roberto Sbragia*	241
Chapter 18	The Limits of Business Development *Mats Larsson*	263
Chapter 19	Knowledge Management Across Borders: Empirical Evidence of the Current Status and Practices of Knowledge Management in Multinational Corporations *Helmut Kasper and Florian Kohlbacher*	279
SECTION III: TECHNOLOGY AND INNOVATION MANAGEMENT		**295**
Chapter 20	The Influence of Organizational Maturity in the Project Performance: A Research in the Information Technology Sector *Renato de Oliveira Moraes and Isak Kruglianskas*	297
Chapter 21	Organization for Product Development: The Successful Case of Embraer *Paulo Tromboni do Nascimento, Eduardo Vasconcellos, and Paulo César Lucas*	311
Chapter 22	A Study of R&D Outsourcing in the Manufacturing Industry: A Case Study of PS2 *Fujio Niwa and Midori Kato*	323
Chapter 23	A Strategic Science and Technology Planning and Development Process Model *Grace Bochenek, Carey Iler, Bruce Brendle, Timothy Kotnour, and James Ragusa*	341
Chapter 24	Institutional MOT: Co-Evolutionary Dynamism of Innovation and Institution *Chihiro Watanabe*	355

viii   Contents

Chapter 25	Firms with Adaptability Lead a Way to Innovative Development *Akihisa Yamada and Chihiro Watanabe*	367
Chapter 26	Toward Project Strategy Typologies: Cases in Pharmaceutical Industry *Peerasit Patanakul, Aaron J. Shenhar, Dragan Z. Milosevic, and William Guth*	381
Chapter 27	Technology Know-How Protection: Promote Innovators, Discourage Imitators *Christoph Neemann and Günther Schuh*	397
Chapter 28	Comparison of Technology Forecasting Methods for Multi-National Enterprises: The Case for a Decision-Focused Scenario Approach *Oliver Yu*	409

SECTION IV: STANDARDS AND EVALUATIONAL METHODS   425

Chapter 29	Standards of Quality and Quality of Standards for Telecommunications and Information Technologies *Mostafa Hashem Sherif, Kai Jakobs, and Tineke M. Egyedi*	427
Chapter 30	A Methodology to Measure the Innovation Process Capacity in Enterprises *José Ramón Corona-Armentats, Laure Morel Guimaraes, and Vincent Boly*	449
Chapter 31	Auditing Technology Development Projects *Dominique R. Jolly*	465
Chapter 32	Complexity, Cost Reduction and Productivity Improvement Through History *Mats Larsson and Carl-Henric Nilsson*	485

SECTION V: SUSTAINABILITY   497

Chapter 33	Involvement of Small Manufacturing Firms in Organic Production Systems *Stéphane Talbot*	499

Chapter 34	Innovation, Competitiveness and Sustainability: Study on a Plywood Industry Cluster *Sieglinde Kindl da Cunha and João Carlos da Cunha*	515
Chapter 35	Sustainable Development in Companies: An International Survey *Daniela Ebner and Rupert J. Baumgartner*	535
Chapter 36	Enhancement of Environmental Performance Through Total Productive Maintenance *Rupert J. Baumgartner*	553
Chapter 37	Integrating Sustainable Business Management into Daily Business via Generic Management *Rupert J. Baumgartner*	563
SECTION VI: SOCIAL AND EDUCATIONAL ASPECTS IN MOT		575
Chapter 38	The New Company-School Relationship in the Knowledge Age *Luís Henrique Piovezan and Afonso Carlos Corrêa Fleury*	577
Chapter 39	Always Connected: How Young Brazilians Use Short Message Services (SMS) *Marie Agnes Chauvel*	589
Chapter 40	The Individual in Management of Technology *Marianne Hörlesberger*	601
Author Index		613

Preface

New developments in bio- and nanotechnologies and also in information and communication technologies have shaped the research environment in the last decade. Highly educated experts in R&D departments work together with scientists and researchers at universities and research institutes for creating new technologies. Transnational companies which have acquired various firms in different countries deal with the management of their diverse R&D strategies and cultures. Different disciplines have to be brought together in developing new technologies. Researchers and educators increasingly collaborate across borders throughout the globe. The knowledge-based economy permeates across companies, universities, research institutes, and countries. Managing technology in this new environment presents real challenges.

Some aspects of these challenges have been reflected in varied contributions to the 14th International Conference on Management of Technology convened during the period of May 22-26, 2005 in Vienna, Austria. The conference subtitled "Productivity Enhancement for Social Advance: The Role of Management of Technology" was hosted by United Nations Industrial Development Organization (UNIDO) and organized by Mohamed El-Nawawi and Marianne Hörlesberger in Vienna together with Tarek Khalil and his team in Miami.

Remarkable contributions came from researchers working for UNIDO in its different offices around the world. They focused more on the industrial and technology developments in developing countries. When we started working on this book, UNIDO requested to publish a specific book dealing with these aspects. Hence, the emphasis of this book is on the general contributions to the conference in other challenging technology management issues.

The book is organized in six sections, with papers in each section revolving around specific themes: managing new technologies; business organisation; technology and innovation management; standards and evaluation methods; sustainability; and social and educational aspects in MOT.

IAMOT expresses its appreciation to the many organisations, companies and the government in Austria that supported the 2005 IAMOT conference, especially UNIDO; the Federal Ministry of Transport, Innovation and Technology; the Federal Ministry of Economics and Labour of the Republic of Austria; and the Federal Ministry for Education, Science and Culture.

The editors would also like to express their appreciation for the individual contributors to this volume.

Marianne Hörlesberger and Tarek Khalil
Austria Research Centers GmbH - ARC, Vienna, Austria
and University of Miami, Florida, USA

SECTION I

MANAGING NEW TECHNOLOGIES

CHAPTER 1

AN EXPLORATORY ANALYSIS OF TSS FIRMS: INSIGHTS FROM THE ITALIAN NANOTECH INDUSTRY

Vittorio Chiesa, Alfredo De Massis, and Federico Frattini

Politecnico di Milano, Italy

Private service firms have largely diffused in the last years as external sources of technology that are accessed by innovative companies for supporting their innovation process. This paper in particular focuses on Technical and Scientific Service (TSS) firms that basically sell technical and scientific knowledge. Examples of TSS are contract Research and Development, laboratory testing, technology consulting and product development. The purpose is to offer a preliminary insight into the business models adopted by TSS firms and analyze the resulting strategic, managerial and organizational choices. This investigation is based upon the results of an empirical study on Italian nanotech companies.

1. Introduction

In the last years, technological innovation has become increasingly complex and expensive, whereas markets have turned out to be dramatically fast-changing. For innovating successfully, firms need a wide set of competencies and resources they can hardly develop by themselves. As a result, they have been increasingly relying on other companies, universities, research institutes, start ups, with the aim of creating "networks" of relationships through which knowledge and technological assets are exchanged (Amidon Rogers, 1996). Among the possible actors of these "creation nets" (Brown and Hagel III, 2006), private service firms have largely diffused. These companies develop and sell technology-intensive services to clients that use them for improving their innovation process. Examples of these Technical and Scientific Services (or TSS) are: (i) product design, engineering, testing, rapid and virtual prototyping; (ii) Contract Research & Development (CRO); (iii) software instruments supporting the R&D process; (iv) technological consulting, brokering and training. The market for TSS is rapidly growing and offers great opportunities for the future (Arora *et al.*, 2001; Howells, 1999). Literature on TSS has addressed the following major issues: the form of the relationships between the TSS organization and its clients, e.g. R&D contracts (Haour, 1992), partnering (Bruce *et al.*, 1995; Millson *et al.*, 1996); the role of TSS firms as partners in technological collaborations (Chatterji and Manuel, 1993; Chesbrough and Teece, 1996); the role that TSS play in different phases of the innovation process (Bruce *et al.*, 1995; Millson *et al.*, 1996; Turpin and Garret, 1996); the impact of TSS on national or local economies (Windrum and Tomlinson, 1999; Mansfield and Lee, 1996). What is almost totally missing is the attempt to study the phenomenon from the perspective of the company that develops and sells the TSS, in order to understand the peculiar strategic, organizational and managerial problems it is called to face. In order to make a step further in this direction, this paper focuses on a specific cluster of TSS companies (i.e., those operating in the nanotechnology industry), with the purpose of describing the business models they apply. Many authors have provided definitions for the term "business model" (e.g., Chesbrough and Rosenbloom, 2002), although a largely accepted one has not emerged yet (Shafer *et al.*, 2005); here we define "business model" as the set of answers to the following questions: What does the company offer to its customers? How is that offer conveyed to them? Specifically, how does the TSS firm structure its marketing approach and manage the interaction with its clients? The choice of the nanotechnology industry as the scope of the investigation was suggested by the growing attention paid, in the last seven to eight years, to nanotechnology, both in academic and industrial environments. The expectations that have surrounded nanotechnology are witnessed by the growth of public funds devoted to research into the field; in 2004 they raised the amount of

4.6 billion US Dollars worldwide, with an increase of about 700% with respect to 1997 (www.luxresearchinc.com, 2004). However, the novelty of the nanotech market and the pervasiveness of the underlying technology, together with a lack of common definitions, make it very difficult to clearly distinguish between nanotech and non-nanotech firms. This is why it was necessary to develop an empirical framework that would serve as an instrument for identifying the players of the nanotech market and supporting the investigation of their business models.

2. Research Objectives and Methodology

The first objective of this paper is to develop an empirical framework that clearly defines the boundaries of the nanotech market and help identify its players. All the Italian companies (together with the most important European ones) that are currently labeled by scholars and practitioners as "nanotech", were extensively surveyed. They were interviewed in order to understand the nanotechnology domain(s) and the applications they have developed or applied. This investigation suggested the basic classification criteria to be used for clustering nanotech companies; they represent the dimensions of the empirical framework. Moreover, the empirical analysis served the purpose of identifying those Italian nanotech companies that actually offer technology-intensive services (TSS firms) to be studied in the second step of the research. The second objective of the paper is to offer preliminary insights into the business models of nanotech TSS companies. To this aim, a multiple case study on five Italian nanotech TSS firms was undertaken. The selected companies were homogeneously distributed in the empirical framework, so that a certain degree of theoretical replication is allowed.

3. A Framework for Nanotechnology Firms Classification

The objective of this section is to illustrate the classification framework that can be used for bringing order into the "world" of nanotech applications; it has been elaborated on the basis of the information collected through the semi-structured interviews with the Italian nanotech firms and some non-Italian ones, and the available public and private documents and reports. It was finally tested through a panel study that involved experts in nanotechnologies. A list of the studied companies is reported in Table 1.

Table 1. List of studied companies.

Italian Firms	Non-Italian Firms	
A.P.E. Research	Ntera Ltd	Bayer AG
Crf	Carbo Microelectronics Corp.	Sony Corporation
Csm	Mitsui & Co. Ltd	Motorola Inc
Geal	DuPont	DSM
Kedrion	Xerox Corporation Ltd	Basf
Microcoat	Intel Corporation	Toshiba Corporation
Moma	Elan Corporation	Toyota Motor Corporation
Olivetti I-jet	Aveka Group	ABB Group
Organic Spintronics	Lucent Technologies	ExxonMobil Corporation
Prometon	BAE Systems	Canon Inc
Saes Getters	Fujitsu Group	General Motor
Siad	Infineon Technologies AG	Agfa-Gevaert N.V.
Sorin Biomedica Cardio	NEC Corporation	Schering AG

The analysis has shown the possibility to classify nanotech companies on three levels, connected in a hierarchical structure. At the **first level**, nanotech firms are classified into "**nanotechnology categories**", i.e. the macro-areas through which a company can access the market for nanotech applications. Each category aggregates applications that are homogeneous in terms of the type of use they are destined to. Obviously, it is possible for a firm to access contemporarily more than a single macro-area (e.g., IBM, NEC, General Electric, and Intel). There are four main categories at this level:

1. Nanomaterials. The properties of a solid material depend upon its microstructure, i.e. the atomic structure, the shape and dimension of the material itself in one, two or three dimensions. Conventional materials have grains with a dimension that can vary between a few

microns and a few millimeters, while the nanostructured materials are made of grains with dimensions of 1-100 nanometers. The simple material has a structure that is repeated homogeneously along one, two or three axis (characteristic that a nanomaterial may not have), and therefore shows homogeneous features in terms of chemical, physical and mechanical properties. This dramatically limits the possible applications of a traditional material, with respect to a nanostructured one.
2. <u>Nanostructures</u>. These complex systems are composed of different parts (inorganic or organic) assembled together; at least one of them has dimensions of 1-100 nanometers, and they are typically made of nanostructured materials. They are assembled and made interacting in order to create structures with macro-dimensions. They normally show innovative chemical, electrical, mechanical and optical properties and are capable of autonomously functioning.
3. <u>Nanotools</u>. This nanotechnology category contains: (i) instruments capable of operating with an atomic precision in order to manipulate and measure materials on a nanometric scale; and, (ii) interface software that are capable of 3D modeling materials at the atomic scale and of conducting very accurate simulations on the behavior of nanomaterials and nanostructures.
4. <u>Nanoprocesses</u>. Within this category are mainly included synthesis methods for obtaining nanomaterials and nanostructures.

At the **second level**, categories are divided into "**nanotechnology sub-categories**", each aggregating nanotech applications homogeneous in terms of the scientific or technological domain they belong to. Finally, the **third level** of the model includes the "**nanotechnology applications**" that are available on the market or are actually being developed. For instance, major nanotech applications in the sub-category of "nanotubes" are field emission displays, optic biosensors, nanoneedles, nanotubes antennas, probes for scanning tunneling microscopes, electron guns.

The nanotechnology categories and the sub-categories they are composed of are represented in Figures 1, 2, 3, 4.

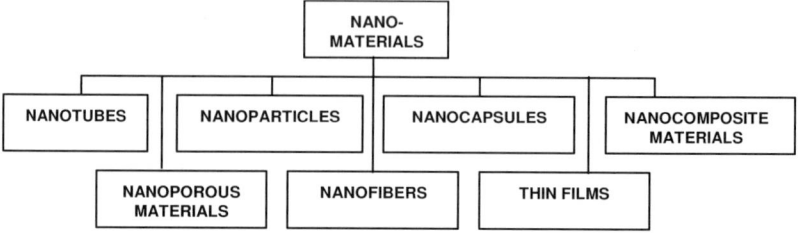

Figure 1. The articulation of "nanomaterials" category into sub-categories.

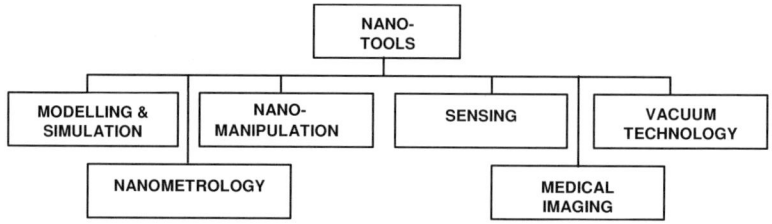

Figure 2. The articulation of "nanotools" category into sub-categories.

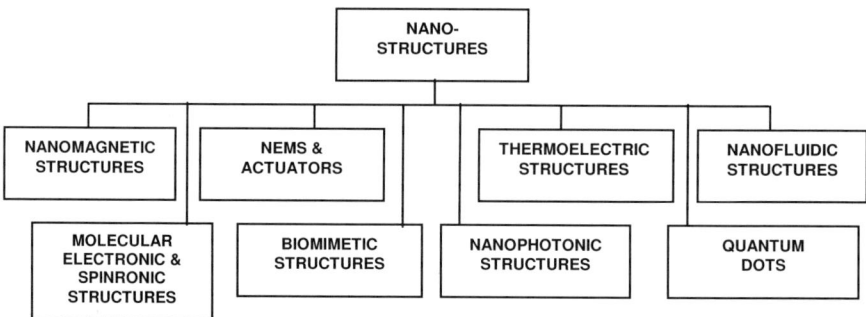

Figure 3. The articulation of "nanostructures" category into sub-categories.

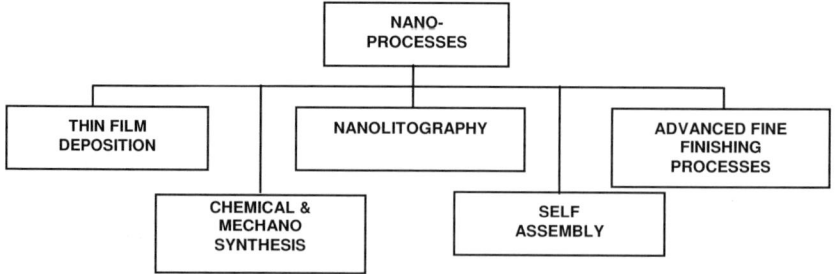

Figure 4. The articulation of "nanoprocesses" category into sub-categories.

This framework has been tested through a panel study involving experts in physics and, especially, nanotechnologies, and the analysis has confirmed its completeness. Therefore it is possible to state that the model has a general validity and can be effectively used for classifying every type of nanotech firm. Moreover, it can be applied for supporting macroeconomic investigations into nanotech markets that represent a fundamental starting point for the definition of public funding projects or other types of governmental initiatives. Similarly, it represents an important basis for technology forecasting activities that have a great relevance because of the novelty of the nanotech environment. The generality of the framework makes it possible to apply it as a reference model for discriminating between nanotech and non-nanotech companies. We label as "nanotech" a company that develops, sells or incorporates in its production processes one or more nanotech applications that can be included at least in one of the sub-categories encompassed by the model.

4. Emerging Business Models among Nanotech TSS Companies

Companies listed in Table 1 adopted two different types of business models: (i) the "**A-type**" business model, i.e. the one pursued by those nanotech companies offering a service (a process, a tool, an intermediate finding that needs to be further developed and included into an innovative product) through which their scientific competencies in a particular nanotechnology category(ies) are transferred to the client firm, and (ii) the "**B-type**" business model, i.e. the one adopted by those nanotech companies internally developing, acquiring or licensing in from other companies nanotechnological processes, tools, or intermediate findings to be used for innovating their processes or products.

Companies belonging to the first group can be labeled as TSS firms. The aim of this section of the paper is to study the cases of five of these firms in order to analyze in greater detail the business model they apply. This will give the opportunity to point out the major characteristics of the business models of TSS firms operating in different nanotechnology categories, and enlighten the managerial and organizational choices the five considered firms have made. The five companies that have been analyzed in this way are: Saes Getters, Pometon, A.P.E. Research, Olivetti I-jet, Sorin Biomedica Cardio.

4.1. *The empirical results*

Table 2 shows the classification of the five analyzed TSS companies in the framework presented in Section 3.

Table 2. The framework classification of studied companies.

	Nanotechnology Category	Nanotechnology Sub-category	Applications
Saes Getters	Nanoprocesses	Thin film deposition	Sputtering
	Nanomaterials	Thin film	Absorbers
Pometon	Nanomaterials	Nanoparticles	Nanopowders
A.P.E. Research	Nanotools	Nanomanipulation	Scanning probe microscope
		Modelling & simulation	Software for real time imaging
Olivetti I-jet	Nanostructures	MEMS & actuators	Ink jet technology
	Nanoprocesses	Thin film deposition	Sputtering
Sorin Biomedica Cardio	Nanostructures	Biomimetic structures	Implantable devices

If the activities undertaken by the companies are carefully considered, it clearly emerges that the business models of three companies (Saes Getters, Pometon and A.P.E. Research) are wholly focused on the sale of TSS ("A-type" nanotech companies). On the other hand, Olivetti I-jet and Sorin Biomedica Cardio are mainly "B-type" nanotech firms that expanded their business model in order to include, even if as a marginal activity, the development and sale of TSS.

The empirical evidence on nanotech companies adopting an "A-type" business model, i.e. developing and selling, at least as a marginal part of their activities, TSS, suggests the possibility of classifying them into three main categories:

A1. Nanotech TSS firms developing intermediate findings (e.g., nanopowders or nanotubes) to be sold, licensed out or partnered with the client company, that carries out the remaining tasks of the development process and includes them into its innovative products. This type of TSS can be labeled as "Work In Progress (WIP)

Innovation", and is typical of firms operating in the nanotechnology category called "nanomaterials". An example of this type of nanotech TSS firm is Pometon.

A2. Nanotech TSS companies carrying out, for the client firm, a particular process (e.g., magnetic abrasive finishing or thin film deposition) that requires excellent competencies in a specific scientific domain. It is possible to call this kind of TSS "Process Activity", which is common among firms operating in the "nanoprocesses" category. Examples of firms offering this kind of service are Saes Getters, Olivetti I-jet and Sorin Biomedica Cardio.

A3. Nanotech TSS firms developing and selling instrumentation operating at the nanoscale (e.g., machinery for nanolithography, Scanning Probe Microscopes or software for real-time imaging) that is used by the client firm for supporting its innovation processes (typically, its basic and applied research activities). This type of TSS can be referred to as "Technologies to develop technology" and is offered mainly by firms operating in the nanotechnology category called "nanotools" (e.g., A.P.E. Research).

Olivetti I-jet and Sorin Biomedica Cardio possess exclusive competencies in nanostructures. However, these skills are not exploited in order to offer a TSS, but to innovate their products. The TSS sales activity of these companies leverages competencies in the "nanoprocesses" category. The emerged empirical evidence stands therefore for a lower diffusion of nanotech TSS firms working in nanostructures.

The managerial and organizational decisions of the studied companies are summarized in Table 3.

First, it is possible to point out that all companies adopt a structured marketing approach, based upon a direct contact with clients, the participation at professional fairs and the use of the Web site as a window on the firm's projects and services. The only case of non-structured marketing approach is that of Sorin Biomedica Cardio, but it can be explained taking into consideration that the adoption of an "A-type" business model by this firm is still at an embryonic stage. In fact, the interviewees have recognized the importance of developing a

Table 3. Schematic description of the analyzed business model's variables.

	Marketing Approach	Degree of Standardization	Commercial Relationship Management Model	Phase(s) of Interaction
Saes Getters	• Structured; • Direct approach; Professional fairs; Web site	Customized output	• Structured; • Checkpoints; Regular meetings; Reference person	R&D
Pometon	• Structured; • Direct approach; Professional fairs; Web site	Mainly standardized output	•Not structured; • Reference person	Service delivery
A.P.E. Research	• Structured; • Direct approach; Professional fairs; Web site	Highly customized output	• Structured; • Checkpoints; Regular meetings; Reference person	R&D; Service delivery
Olivetti I-jet	• Structured; • Direct approach; Professional fairs; Web site	Customized output	• Very structured; • Checkpoints with reports; Regular meetings; Reference person; Co-development teams	R&D; Service delivery
Sorin Biomedica Cardio	• Not structured; • Professional fairs	Standardized output	•Not structured; • Reference person	Service delivery

structured marketing approach for the future. The contact of new potential clients seems therefore to be a critical aspect for the analyzed firms, and this is a typical attribute of TSS companies (Chiesa et al., 2004).

Considering the other three perspectives from which the five companies have been analyzed, two alternative models seem to be applied by Italian nanotech TSS firms in the management of the service sale. First, the offering of a **standardized output** through a non-structured commercial relationship management model in which the interaction with the client company is merely limited to the phase of service delivery and, second, the offering of a **customized output** which, on the other hand, requires an intense interaction since the early stages of the service development and a sort of co-design with the client company. This second case entails a greater complexity in the management of commercial relationships that have also a longer duration. All the companies offering a customized output, indeed, have to implement formal and structured approaches consisting in: (i) the definition of checkpoints with the client where it is informed, through different types of reports, about the project advancement; (ii) the establishment of regular meetings with the client; and, (iii) the identification of internal and client's reference persons in charge of managing the commercial relationships.

The empirical evidence has therefore allowed classifying the business models of TSS players operating in nanotechnology into six different typologies. They are summarized in Table 4.

Table 4. The taxonomy of nanotech TSS companies' business models.

		Degree of Standardization	
		Standardized output	Customized output
Type of Output	WIP innovation (A1)		
	Process activity (A2)		
	Technologies to develop technology (A3)		

This matrix seems to be a viable instrument for further analyzing nanotech TSS companies from a managerial perspective. Of course, some of the quadrants of the matrix will be less likely populated than others. For example, a firm adopting an "A1-type" business model is expected to offer a standardized output while a firm adopting an "A2-type" or an "A3-type" business model is likely to offer an output characterized by a high degree of customization. This is reasonable considering that in the latter cases the type of output provided has to be strictly modified to function with the client's needs and requirements.

5. Conclusions

In the last seven to eight years, interest in nanotechnology has steadily grown, both in academic and industrial environments, and it has proved to be an emerging and rapidly-growing field for the diffusion of TSS. The paper offered a preliminary insight into the business models adopted by TSS firms operating in nanotechnology. First, an empirical framework that is useful for identifying the players of the nanotech market and supporting the investigation of the business models they apply was developed. The analysis of the companies studied in order to create the aforementioned framework made it possible to identify two basic types of business model: (i) the sale of a Technical and Scientific Service (TSS) that supports the innovation process of the client company ("A-type" business model), and (ii) the use of nanotechnologies, internally developed or externally acquired, for innovating the offered products, processes or services ("B-type" business model). The second part of the paper focused exclusively on Italian nanotech TSS companies, i.e. those adopting an "A-type" business model. It reported the results of the multiple case study conducted on five of the previously identified nanotech TSS firms, and provided an insight into the managerial and organizational choices related to their business models. Finally, a taxonomy matrix of nanotech TSS companies' business models was developed. It can support further analyses of nanotech TSS firms from a managerial perspective. The research described here had an exploratory dimension as well. It has suggested some possible directions for deepening the knowledge of the business models adopted by nanotech TSS players. Further aspects that could be investigated are: (i) the organizational structures, i.e. the way the nanotech TSS company organizes its resources (mainly human) in order to develop and provide the services; (ii) the mechanisms implemented to protect the nanotech TSS company's intellectual capital; (iii) the applied managerial techniques, like the performance evaluation and reward systems, or the management control system; and (iv) the technical and scientific service's pricing model, that would require to identify the methodologies used to evaluate the intangible assets and the R&D activities of the company.

References

Amidon Rogers, D.L. (1996). The Challenge of Fifth Generation R&D. Research & Technology Management, July-August, 33-41.
Arora, A., A. Fosfuri and A. Gambardella (2001). Markets for Technology. The MIT
Brown, J. S. and J. Hagel III (2006). Creation nets: getting most from open innovation, The McKinsey Quarterly, 2, 40-51.
Bruce, M., F. Leverick, D. Littler and D. Wilson (1995). Success factors for collaborative product development: a study of suppliers of information and communication technology. *R&D Management*, 25, 33-44.

Chatterji, D. and T.A. Manuel (1993). Benefiting from external sources of technology. Research & Technology Management, November-December, 21-26.

Chesbrough, H. and R. S. Rosenbloom (2002). The role of the business model in capturing value from innovation: evidence from Xerox Corporation's technology spin-off companies. *Industrial and Corporate Change,* 11, 3, pp.529-555.

Chesbrough, H. and D.J. Teece (1996). When Is Virtual Virtuous? Organizing for Innovation. Harvard Business Review, January-February, pp. 65-73.

Chiesa, V., R. Manzini and E. Pizzurno (2004). "The externalisation of R&D activities and the growing market of product development services". *R&D Management,* 34, 1, pp. 65-76.

Haour, G. (1992). Stretching the knowledge-base of the enterprise through contract research. *R&D Management,* 22, 177-182.

Howells, J. (1999). Research and technology outsourcing. Technology Analysis & Strategic Management, 11, 17-29.

Mansfield, E. and J-Y. Lee (1996). The modern university: contributor to industrial innovation and recipient of industrial support. *Research Policy,* 25, 1047-1058.

Millson, M.R., S.P. Raj and D. Wilemon (1996). Strategic Partnering for developing new products. Research-Technology Management, May-June, 41-49.

Shafer, S.M., H.J. Smith and J.C. Linder (2005). The power of business models, *Business Horizons,* 48, 199-207.

Turpin, T. and S. Garret (1996). Bricoleurs and Boundary Riders: Managing Basic Research and Innovation Knowledge Networks. *R&D Management,* 26, 267-282.

Windrum, P. and M. Tomlinson (1999). Knowledge-intensive services and international competitiveness: a four country comparison. *Technology Analysis and Strategic Management,* 11, 391-405.

The main web sites consulted in the research are: www.luxresearchinc.com, 2004; www.nanoinvestornews.com, 2004; www.investinitaly.com, 2004.

CHAPTER 2

KNOWLEDGE CREATION DYNAMICS AND FINANCIAL GOVERNANCE: CRISIS OF GROWTH IN BIOTECH FIRMS

Anne-Laure Saives, Mehran Ebrahimi,
Robert H. Desmarteau, and Catherine Garnier

University of Quebec at Montreal, Canada

The business models of dedicated biotech enterprises currently focus on the management of inventions with the aim of entering a virtuous growth cycle based on judicious directing of R&D projects (towards the market), the choice of intellectual property to be protected and traded, as well as managing financial options. We here deepen our understanding of the technological and organizational development cycle of these firms. We base ourselves on a series of semi-structured interviews with over 110 biotechnology firms within Quebec, one of the largest bio-cluster in Canada. This exploratory field study leads us to major paradoxical observations between organizational knowledge creation within biotechnology firms and the type of financial governance that is present. It ultimately shows three different modes of development within those firms: that is, pre-entrepreneurial, entrepreneurial and managerial, each staking out the progress of biotechnology firms, and also equally marked by two transformational ruptures (teleological and creativity gaps).

1. Introduction

For more than a decade, the biopharmaceutical sector has been experiencing an expansion of new technologies, i.e. a "third wave" of biotechnologies[1], whereby the latter are succeeded by new ones at an ever increasing rate. Combinatorial chemistry techniques and High Throughput Screening (HTS) of molecules have facilitated the rationalizing of the traditional pharmaceutical discovery process. More recently, bio-technologies resulting from advances made in computer technologies and genetics has reshuffled the cards of innovation between new players (such as bio-technology firms for example). According to the conditions of the knowledge based approach (KBV), the business models of these companies focus on managing invention supported by a judicious choice of intellectual property to be protected and traded, along with the managing of various financial alternatives for growth. In this paper, within the context of a speculative economy, we will try to deepen our understanding of the technological and organisational development cycle of these firms. Using the hypothesis that techno-scientific knowledge creation is the driving force of value creation within the bio-pharmaceutical industry, we query the possible conciliation between the creation of techno-scientific knowledge and the predominating financial governance within the contemporary bio-industrial system.

By first reviewing the concept of the development or technological life cycle across a brief literature review on innovation management, knowledge management and strategic analysis as applied to the biotechnology sector, we formulate our research questions around the issue of understanding the reasoning behind "high technology" organizational transformations; that is, techno-scientific knowledge creating firms, during the various stages of development as encountered in biotechnology companies. We then present our findings from our empirical field of study, that is, the biotechnology system of Quebec (being the largest geographically concentrated area of biotechnology companies within Canada). As a last step, we emphasize above all, on the paradox that is highlighted by these observations, and subsequently propose a renewed framework of understanding of the "development modes" that succeed one another within these specific companies.

---

[1] New biotechnologies refer to the third generation of biotechnologies (i.e.: pharmacogenomics for the development of products for the treatment of complex genetic diseases and personalized and predictive Medicine).

2. Theoretical Framework and Research Questions

2.1. *Biopharmaceutical knowledge and innovation creation*

The abundant literature on the innovation economy within the biotechnology sectors suggest that bio-industries, especially the dominant biopharmaceutical industry (Hamdouch and Depret, 2003a, 2003b; Hamdouch and Perrochon, 2000), no longer functions around just a few large integrated companies, but rather, is characterized by the diversity, the evolving character and the entangled modes of interactions (horizontal, vertical and transversal) linking the different types of agents that today make-up the industrial bio-pharmaceutical system (Parolini, 1999). In particular, the reasoning of "pre-emption" (Hamdouch and Perrochon, 2000, p. 42) adopted by the "large pharmaceutical companies" consists of renewing their product pipelines by acceding as soon as possible to the knowledge incarnated in both patents and promising candidate drugs from biotechnology firms, whereby the latter dispose of distinctive technological competencies that qualify them to the rank of partners with pharmaceutical companies within an arrangement of co-operation that approaches coopetition.

Within high-technology sectors based on technological innovation ("technology-based" and "science-driven", Saives *et al.*, 2006), and in terms of the knowledge-based approach (Grant, 1996), the strategic resource of biotechnology firms is the capacity to create, absorb (Cohen and Levinthal, 1990) and monitor knowledge within networks of collaboration and innovation (Powell, 1998; Salman and Saives, 2005). These firms have integrated the reticular and partnering dimensions of value creation (Chesbrough, 2003), with the idea that the dynamics of "open" innovation is bi-directional; and whereby strategy must no longer be understood simply as a portfolio of activities, resources and competencies but also as a portfolio of relationships (Venkatraman *et al.*, 2002). Within this perspective, a key resource of the firm is also *its position within a network of expertise* from which it will achieve economies of scale, scope and "economies of expertise".

Equally numerous are the authors within the field "knowledge management" who have adopted this reticular vision of knowledge creation in the new economy founded on knowledge; an economy of relationships necessary to accede to and grow the expertise which is at the source of value creation. Hence, the creation and circulation of knowledge primarilly involves social processes permitting the articulation of explicit and codified knowledge, along with tacit knowledge (Nonaka and Takeuchi, 1995). Within organisational strategies, it becomes primordial to create and foster contexts for socialisation so as to reinforce the cycle of tacit, conceptual and operational knowledge around common metaphors. To the extent where the organisation is considered as a place of socialisation combining tacit and explicit knowledge, it becomes necessary to re-situate the individual within the centre of organisational preoccupations.

Thus, there is an important qualitative jump between the individual and organisational finality with respect to the invention (techno-scientific knowledge creation) and innovation phase (oriented towards commercialisation of knowledge) driven by this new pharmaceutical industrialisation system. If techno-scientific knowledge is the source of competitiveness amongst bio-pharmaceutical firms of tomorrow, how is such knowledge generated on the one hand, and activated on the other? What transformations occur and operate between intangible intellectual resources such as patents/techno-scientific/technological knowledge creation, and specific financial assets of bio-technology firms?

2.2. Development cycle of bio-technology firms

With the advent of the new scientific paradigm founded on genomics (Lacetera, 2001; Pisano, 2002), literature addressing the management of innovation, especially within the bio-technology sectors, often treats the bio-pharmaceutical system around two homogeneous groups (Lacetera, 2001), namely, large pharmaceutical corporations and companies dedicated to bio-technology. This literature does not enter within the black box of their respective organisations and describes even less the transformation which occurs within their organisational practices. Hence, the discontinuous metamorphosis of these organisations, let alone the characteristics of development phases as defined in 1972 by Greiner

(1998) or Churchill and Lewis (1983), or the conditions for the creation of organisational bases as defined by Godener (2002), is rarely explored.

The "metamorphosis" models of development (Godener, 2002) in particular, are built on various criteria[2] and can involve up to five stages of development. The first stage is usually characterized by a very *informal* organization: i.e. a low level of specialization managed by its entrepreneur. At the following stage, the formal structure of the organization starts to emerge: *task specialization* as well as the *centralization of decisions* within a *functional* structure increase. The third stage involves *delegation*. The formal structure of the firm is clearly *hierarchized*, but the decision modalities evolve towards a more participative model. The two following stages (*coordination* and *collaboration*) are more concerned with big enterprises, whereby the high degree of formalisation of the divisionary or matrix structure of the firm require reinforced coordination means.

The majority of studies on value creation and the development modes of companies within this sector address "visible" firms, listed on the stock market. Such companies having achieved a relative critical size, only constitute a minority amongst the numerous, small firms in a phase of emergence shaping the specificity of this high-tech sector within Quebec (Desmarteau and Saives, 2004).

The growth of these firms depends on their ability to enter into a virtuous cycle of growth (Niosi, 2003) brought about by determining factors of success such as: access to intellectual property (patents and licenses), venture-capital financing and strategic alliances and collaborations for the commercialisation of their knowledge.

Given that bio-industries are entering a "consolidation" phase (Nesta and Mangematin, 2004; Niosi, 2003), the literature on business models of biotechnology firms (Catherine *et al.*, 2003; Mangematin *et al.*, 2003; Fisken and Rutherford, 2002) tends to highlight how the majority of young firms, given their relative inexperience or risk associated in the

---

[2] These criteria may consist of: management style, organisational structure, managerial focus, control systems, characterization of the direction (capital propriety, strategic objectives, etc.) (Greiner, 1998; Churchill et Lewis, 1983).

development of products and platforms, focus on the management of invention in the first phases of their development cycle.

This cycle is based on a succession of stages (pre-start-up, start-up, growth). Typically, an established researcher within the academic field "comes out" with a university research project (pre-start-up) which constitutes an incubator "start-up" and subsequently evolves towards a semi-private mode of financing technology project oriented towards satisfying a potential precursor market need. These modalities of financings go through a succession of institutional start-up funds that precede a series of risk capital financing rounds and subsequent stock market IPO (initial public offering). This project may be more or less condemned to the degree of success obtained in the quest for these financing activities as well as to the ability of the organization to build a genuine innovation firm: i.e. an organization capable to judiciously direct R&D projects aimed at potential markets for products or processes ("platforms") (Fisken and Rutherford, 2002), as well as making the proper choice of intellectual property to be protected (typically patents).

It should be noted that the succession of phases overlap one another within the first of five growth phases as described by Greiner (1998), whereby the initial phase of "creativity" in which the firm is pre-occupied with, involves creation of both the "product" and its market.

Hamdouch and Depret (2003a) recently attempted to characterize this progressing organisation of the firm under "regimes of governance" differentiated at each stage of evolution (start-up, growth and routine types of governance). The description of the three phases (pre-start-up, start-up, growth) in terms of age indicators, critical size, modes of financing, modes of protecting intellectual property, etc. nevertheless reflects a very *financial* (i.e. investor) vision of company growth having little in common with the company founder (typically a researcher).

It is precisely within this understanding of the conciliation of these different points of views of governance (of the researcher, the manager and the investor) that our questions consist of.

If we return to the teachings of historians and economists on innovation (Baumol, 2002) who worked on the characterization of the invention/innovation/industrialisation cycle (Saives *et al.*, 2006, etc.), how does one characterize the type of governance practiced by a researcher and a researcher-entrepreneur? Does the passage from one type to the other automatically imply a "revolution" as defined by (Greiner, 1998)? How does the subsequent innovation phase occur (essentially governed by risk-capital investors today and by the routinisation within so-called innovating or "surviving" firms governed by financial markets (Churchill and Lewis, 1983))?

It is from the managerial stakes of the re-organisation of pharmaceutical innovation that our research questions derive from:

1) What are the different development modes of bio-techno-scientific knowledge-creating organizations?
2) What are the transformation logics within these high-tech organisations as they pass from one mode of development to another in the case of bio-technology firms?
3) What are the managerial implications from these transformations?

3. Methodology

The bio-industrial firms of Quebec make up one of the three principal bio-technology industrial clusters within Canada, along with Toronto and Vancouver (Niosi *et al.*, 2002). 73% of the total estimated population of industrial companies from the bio-industries are from Quebec. 60% of these firms' activities are in the human health sector. To analyse the stages of development of bio-technology companies within this system, we utilized some of our data (Niosi *et al.*, 2002) taken from a vast investigation conducted across semi-directive interviews within the numerous firms from the bio-industries of Quebec (that is, more than 52% of the estimated population of the province). This data is not longitudinal as the death rate of Quebec firms at the time was relatively high. 43% of those firms are very small firms (less than 10 employees) while 52% are SME's (10 to 250 employees).

During this work, over 110 semi-directed utilizable interviews were conducted with the managers of these firms of all sizes. Specific data were collected with regards to the stages of evolution and modes of governance within the firms according to the known cycles articulating the three dimensions of: 1) *organization* (culture (origin of spin-off), embeddedness (location and anchorage factors), size (age, number of employees)), governance type (financing modalities (public research funds, "seed-money", venture-capital, IPO)), core competencies (mechanisms of revenue generation (patents, licences, exportation)); 2) *strategy* (strategic vision (5 year perspectives)); and 3) *perception of the project pilot* (manager) of the difficulties encountered by these companies vis-à-vis the process of innovation (obstacles to growth) and financing of innovation (difficulties in the obtaining of capital). Details of our multifactorial analysis are described in Saives *et al.* (2005); while the following sections describe our constructivist observations.

4. Discussion

4.1. *The different "modes of development" of high-tech companies*

An in-depth dyadic examination of the qualitative data collected on our companies allows us to propose within the same spirit as Hamdouch and Depret (2003a) a series of three successive development modes within the growth "cycle" of knowledge intensive bio-technology firms. These three "modes" qualified as "pre-entrepreneurial", "entrepreneurial" and "managerial" are described in Table 1. Having yet to complete a longitudinal study of these firms, we can not affirm if these stages succeed each other in a sequential way.

These stages seem to differ depending on the nature of the pilot project of the organisation in itself. In the first "*pre-entrepreneurial*" case, a researcher-research service provider (from industry or public institutional research organisms) is responsible for a scientific project (to discover universal scientific laws), destined to perpetuate the exploration (as per March, 1991) and the creation of scientific knowledge with a simultaneous individual (that of "credit" in the sense of Latour (1988)) and collective (the maintenance of common goods/knowledge) goal. He acts, with a degree of relative freedom to pilot, in a scientific community of leading edge knowledge, within a more or less academic bureaucracy where key scientific competencies are organized and incarnated by research teams. Here, the evaluation of knowledge creation passes by the

Table 1. The development modes of bio-technology firms.

	Mode	PRE-ENTREPRENEURIAL	ENTREPRENEURIAL	MANAGERIAL
PROJECT	Project	Scientific project	Technological innovation project	Financial project
	Main Actor	Service provider-researcher	Entrepreneur-researcher	Manager
	Fonder	Heavy involvement	Shared implication	Targeted implication
	Objective	Perpetuity (of scientific knowledge creation)	Independence	Profit
	Aim	Credit (intellectual legitimacy and recognition (glory, social power))	Economic recognition (economic power, material recognition)	Economic power and recognition
ORGANISATION	Individual – Collective Link	Individual aim with a collective vocation	Individual aim with an individual vocation	Collective end-result with an individual vocation
	Culture	«Neo-academic»	Business	Capitalization
	Work Division	By knowledge	By inputs	By inputs and outputs
	Structure	Simple professional bureaucracy	Simple and/or organic	Mechanist bureaucracy
	Governance	Scientific community (peers)	Promoters (venture-capital, angels)	Institutional and private investors
	Embeddedness	Community of scientific knowledge	Innovation and expertal network +/– glocal	Value creating system
	Core Competence	Capacity of making unexpected links between ideas	Identification and exploitation of technological opportunities, integrated within a specific platform	Management of tangible and intangible assets (intellectual property, expertise relationships)
STRATEGY	Mission	Knowledge creation, knowledge exploration	Knowledge integration and exploitation, renewal of economy	Commercialization and intensive diffusion of strategic knowledge
	Position in the Innovation Cycle	Invention	Innovation	Industrialization / Merchandising

TELEOLOGICAL RUPTURE: commercial orientation of discovery process into a process of invention

CREATIVITY CRISIS: bureaucratization of the innovation process measure of known knowledge diffusion instruments by peers such as publications and patents.

In the second *entrepreneurial* stage, the researcher-entrepreneur is responsible for a technology project with an individual vocation wanting to accede to a form of independence (from the preceding bureaucracy for example). The latter is nevertheless relative to the degree of autonomy attributed by the interested parties of the deciding firm. The authority of governance is in effect shared between the founder, financiers (venture-capital) and advisory groups (board of directors, scientific board, innovation/expertise networks) committed to the commercial exploitation of scientific knowledge within a logic of complex, multi-disciplinary/multi-directional innovation. The intangible creation of value by these firms is often hard to evaluate across inherited instruments from the industrial era (measuring human and social capital, etc.)[3].

At the *"managerial"* stage, the organisational project, directed by managers but essentially piloted by the shareholders, becomes a financial project. The scientist, strongly involved in the founding of the company, has no more than a specific role, if not kept out of strategic decisions. The objective of the organisation consists of valorizing techno-scientific knowledge at no matter what stage of it being rendered operational, across a speculative exchange of parceled strategic knowledge[4] (patents, licenses on therapeutic molecules having attained phases I, II or III clinical trials, etc.). In this managerial scheme of speculative economic "corporate governance", the key competency of the organisation resides in the managing of its intangible assets (pipeline of patented products, expertise) within a merchant network of value creation. Innovation is routinized and/or fragmented to the benefit of a short term oriented commercial race committed to the merchandisation of knowledge and know-how. Indeed, the evaluation of the measure of contribution of the created knowledge towards the creation of value that is essentially financial runs up against the limits of the available measuring tools.

---

[3] "The investors under-value our company!" is what we often heard from the leaders of Quebec bio-technology firms in search of venture-capital.

[4] "Research outputs" for Chesborough, 2003.

The passage of a firm from one stage to the other is neither systematic nor strictly demonstrated (without any longitudinal data). Nevertheless, from the qualitative data collected, we have noticed that the transformation of firms from one mode of development to another might generate two states of "challenge" (Hite, 2001) or "rupture" in the sense of organisational crisis as defined by Greiner (1998): 1) a *teleological rupture* for the scientific researcher becoming an entrepreneur, often evoked (Tambourin *et al.*, 2003), and linked to the commercial orientation of the invention process; and 2) a *creativity crisis* linked to the bureaucratization of the innovation process of the company that is becoming managerial (Pignarre, 2004).

5. What is the Possible Conciliation between the Development Modes of High-Tech Knowledge Creating Firms?

5.1. *The "teleological rupture"*

A teleological rupture consists of the apparent distance between the two distinctive missions and objectives of the researcher, committed to the development and perpetuation of a common cognitive patrimony and search of credit on the one hand, and the entrepreneur who is in search of independence or perennial autonomy, as well as profit that is able to ensure economic and material power on the other. To conciliate these two caricaturised representations of research activity, it would seem useful to return to the work conducted by Latour during the 1970's, clearly identifying the motivations of the researchers and the transformation of the different managed capitals within a laboratory (the schematic central circle in Figure 1).

Hence, a *given* capital (proto-knowledge) is susceptible to lead to the proposal of scientific *arguments*, which, once published in the form of scientific articles enrich the *"recognition"* capital of the researcher; indispensable for the securing of research *subsidies* to finance the technical capital of the laboratory, permitting the collection of new data and information, thus initiating a new cycle generating recognition-credit etc.

The transformation of the organisation from the *pre-entrepreneurial* to *entrepreneurial* mode resides in the appearance of another source of "credit" outside of publications, that is the patent, and of the progressive denaturation of the notion of "credit". The search for "credit" on the part of researchers was found, according to Latour (1988), in their double quest of *recognition* and *credibility*. The notion of "credit" takes a new turn with the progressive denaturation of the patent vocation, which as a knowledge diffusion tool, becomes a tool of reassurance for investors and therefore an exclusive source of financial revenues once exploited under license(s).

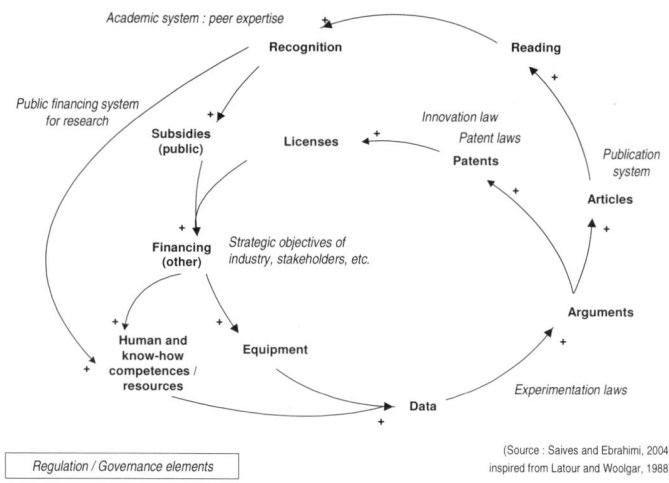

Figure 1. The transformation of capitals generated from academic research activities.

Commercialising the potential fruits of the patent at the disposal of the researcher-become-entrepreneur[5] therefore adds to the possibilities of intellectual and material recognitions, which are the drivers of entrepreneurship.

---

[5] (in a renewed regulatory framework brought about by recent laws on innovation, a context of a shortage of public and para-public research means and of a permeable, societal, ideological context)

5.2. The creativity crisis

Another source of paradox appears where innovations are turned into routines ("routinization") as a result of a short term vision adopted by the financial governance of firms at the "managerial" stage. The long bio-pharmaceutical research and development times (approximately ten years to complete pre-clinical and clinical phase I, II and III research) has nothing in common with the short investment cycle times (one to three years for venture-capital) and shareholder expectation. The race for generating revenues from intangible assets as encountered in current times while patents are expiring on *blockbusters* reinforces the system of finance management in choices that have paradoxically little risk.

The choices of knowledge creation to create new markets are oriented in the short term towards improved reproduction of existing products ("successor" medications) or the short term commercialisation of profitable products (e.g. diagnostic or laboratory products) rather than towards the creation of truly new or "novel" medications (Pignarre, 2004). With the standardisation of pharmaceutical research methods (e.g. HTS and combinatorial chemistry) resulting from the search for productivity coming from traditional industrial thinking, conjugated with the cogni-productive division of work between pharmas and biotechs (parcelling and specializing the knowledge rather than encouraging its circulation), and with the procedurizing of rules and regulations for the marketing of products (clinical trials), bio-pharmaceutical innovation suffers from a "bureaucratization" of its innovation processes (Pignarre, 2004) which blemishes its creativity.

6. Conclusion

The exploratory analysis of knowledge intensive bio-technology firms within Quebec permitted us to identify three development "modes" possibly involved in their growth cycle (*pre-entrepreneurial, entrepreneurial* and *managerial modes*). The development cycle perspective remains to be validated across a longitudinal study. Most of the companies sampled seem to be situated in transition between the last two modes of development. Two ruptures may mark the three development phases: a teleological rupture and a crisis of creativity. Both are inscribed within a dominant current of "corporate governance".

With information, expertise and knowledge being the engines of "value creation", how does one grasp and submit them to the mechanisms of finance?

Epingard (1999) pertinently reminds us that an immaterial investment must consider knowledge as an "economically useful detour within an act of production, but whose result, the *output*, is impalpable and non-directly observable, in the image of knowledge and aptitudes which make up human capital. It results in an incertitude that is partially irreducible, both on the reality of productive assets that are constituted as such and on its value". It is precisely this incertitude which poses a problem on financial thinking with respect to evaluating immaterial assets during the economic activities of a technological company. Currently, a research agenda may be proposed on the basis of the recent theories in the knowledge management field, on the appropriate methods for the evaluation of the effects of the knowledge creation process, whereby such an evaluation integrates non-measurable elements, such as motivation of the experts, the concordance of individual objectives and ambitions with those of the company, the articulation of individuals and of knowledge, and the potential for a cogni-productive network of companies.

Acknowledgments

The authors acknowledge financial support from the Social Sciences and Humanities Research Council of Canada (GEIRSO – Major Collaborative Research Initiatives) and FQRSC – Professor-Scholar start-up program (Fond Québécois de recherche sur la société et la culture). The authors also acknowledge David Holford, Ph.D candidate (HEC Montreal), for his translation of this work.

References

Baumol, W.J. (2002). The free-market innovation machine. *Princeton University Press.*

Catherine, D., Corolleur, F. and Coronini, R. (coll.) (2003). Les Fondateurs des nouvelles entreprises de biotechnologies et leurs modèles d'entreprise. *RIPME*, 12 (2), pp. 63-92.

Chesbrough, H. (2003). Open Innovation. Harv. Bus. Sch. Pub., Boston, MA.

Churchill, N.C. and V.L. Lewis (1983). The Five Stages of Small Business Growth. Harv. Bus. Rev., May-June.

Cohen, W. M. and D.A. Levinthal (1990). Absorptive capacity: A new perspective on learning and innovation. *Adm. Sc. Quart.*, 35, pp. 128-152.

Desmarteau, R. and Saives, A-L. (2004). Les très petites entreprises de biotechnologies sont-elles contre-nature? Proc. 7ième CIFEPME, Montpellier, 27-29 October.

Epingard, P. (1999). L'investissement immatériel. CNRS Éditions, Paris.

Fisken, J. and J. Rutherford (2002). Business models and investments trends in the biotechnology industry in Europe. *J. of Comm. Biotechnology*, 8 (3), Winter, pp. 191-199.

Godener, A. (2002). PME en croissance: peut-on prévoir les seuils organisationnels ? *RIPME*, 15 (1), pp. 39-63.

Grant R.M. (1996). Toward a Knowledge-Based Theory of the Firm. *Str. Mgt J.*, 17 (Winter 1996), pp. 109-122.

Greiner, L.E. (1998). Evolution and Revolution as Organizations Grow. Harv. Bus. Rev., May – June.

Hamdouch, A., and M-H. Depret (2003a). La gouvernance des jeunes entreprises innovantes : un éclairage analytique à partir du cas des sociétés de biotechnologies, to be published in Finance – Contrôle – Stratégie.

Hamdouch, A., and M-H. Depret (2003b). Innovation, coopération préemptive et concurrence réticulaire : les nouvelles dynamiques des relations inter-firmes, to be published in Économies et Sociétés, Série « Dynamique Technologique et Organisation », W, July.

Hamdouch, A., and D. Perrochon (2000). Formes d'engagement en R&D, processus d'innovation et modalités d'interaction entre firmes dans l'industrie pharmaceutique. Revue d'Économie Industrielle, 93, 4ème trimestre.

Hite, J.M. and Hesterly, W.S. (2001). The evolution of firm networks: from emergence to early growth of the firm. *St. Mgt J.*, 22 (3), pp. 275-286.

Lacetera, N. (2001). Corporate Governance and the Governance of Innovation: The Case of Pharmaceutical Industry. *J. of Management and Governance*, 5, pp. 29-59

Latour, B., and S. Woolgar (1988) (French Transl.). La vie de laboratoire. La Découverte, Paris.

March, J. (1991). Exploration and exploitation in organizational learning. *Organization Science*, 2 (1), February, pp. 71-87.

Nesta, L. and Mangematin, V. (2004). The Dynamics of Innovation Networks. SPRU Electronic Working Paper Series, no. 114, Freeman Centre, Univ. of Sussex, April.

Niosi, J. (2003). Alliances are not enough. Explaining rapid growth in biotechnology firms. *Research Policy*, 32 (5), pp. 737-750.

Niosi, J.E., M.L. Cloutier and A. Lejeune (Coord.) (2002). Biotechnologie et Industrie au Québec. Éditions Transcontinental, Montreal.

Nonaka, I. and H. Takeuchi (1995). The Knowledge-Creating Company: How Japanese Companies Create the Dynamics of Innovation. Oxford University Press.

Parolini, C. (1999). The Value Net. John Wiley & Sons, Chichester, England.

Pignarre, P. (2004). Le grand secret de l'industrie pharmaceutique. La découverte/Poche, Paris.

Pisano, G. P. (2002). The Life Sciences Revolution: A Technical Primer. Harv. Bus. Sch. Publishing, August 6th.

Powell, W. W. (1998). Learning From Collaboration: knowledge and Networks in the Biotechnology and Pharmaceutical Industries. *Cal. Mgt Review*, 40 (3), pp. 228-240.

Saives A-L., R.H. Desmarteau and D.A Seni (2006). Vers un nouveau concept de « bio-industries » ?. Économies et Sociétés, Série "Systèmes agroalimentaires", *A.G.*, 27 (5), pp. 957-968.

Saives A-L., Ebrahimi M., Desmarteau R.H., Garnier C., (2005). Les logiques d'évolution des entreprises de biotechnologie. RFG, 31 (155), mars-avril, pp. 153-171. A detailed and enriched version of this paper is to be published in english in a special issue on biotech in History and Technoogy.

Salman, N. and Saives, A-L. (2005) Indirect networks: an intangible resource for biotechnology innovation. *R&D Management,* 35 (2), pp. 203-215.

Tambourin, P., P.B. Joly, J-P. Dupuy et M. Berry (animateur) (2003). Les traditions françaises à l'épreuve des biotechnologies. Les Amis de l'École de Paris, compte-rendu de la séance du 6 octobre 2003 (Rédaction E. Bourguinat), École de Management de Paris.

Venkatraman, N., and M. Subramaniam (2002). In: Handbook of Strategy and Management (Pettigrew A., H. Thomas, R. Whittington, ed.), pp. 461-473, Sage, London.

CHAPTER 3

PARTNERSHIPS BETWEEN TECHNOLOGY-BASED START-UPS AND ESTABLISHED FIRMS: CASE STUDIES FROM THE CAMBRIDGE (U.K.) HIGH-TECH BUSINESS CLUSTER

Tim Minshall, Rob Valli, Pete Fraser, and David Probert

University of Cambridge, Centre for Technology Management, Institute for Manufacturing, Mill Lane, Cambridge, CB2 1RX, UK

> This paper summarises on-going research that seeks to improve understanding and practice in the use of partnerships between technology-based start-ups and established firms, drawing upon evidence from technology-based start-ups operating in Cambridge, U.K. The paper presents the rationale for this project, and seeks to place this topic within the wider literature on the use of partnerships. From this review of existing research, the resource-based view on partnerships is identified as a useful tool to help us understand the motives and operations of the partnerships between technology start-ups and established firms. The initial case studies allow us to structure the emerging issues around the five themes. Discussion of these themes focuses thinking for the next stages of the research which will map the different approaches taken by established firms to working with technology-based start-ups, and the views of investors into technology start-ups.

1. Introduction

Start-ups[1] drive economic growth by playing a significant role in the process of innovation, i.e., the successful exploitation of new ideas

---

[1] The term "start-ups" is taken here to mean firms that are less than 10 years old and which also conform to the definition of a "Small and Medium sized Enterprise" (or SME) provided by the European Commission (i.e., less than 250 employees, turnover less than €40 million and balance sheet total of less than €27 million).

(NCOE, 2002: 625). Research shows that the majority of radical innovations reaching the market since 1945 have been driven by start-ups rather than established businesses (Timmons, 1998).

Viewing start-up formation and growth from a resource-based perspective focuses attention onto a critical challenge faced by all start-ups – the ability to identify opportunities, and to access and exploit resources and competences to create new value from this opportunity. Established firms are generally able to access the core and complementary resources they need either internally, or through leveraging their existing resources to draw in additional resources. Start-ups typically are not usually able to operate in the same way and have either to bring the complementary resources they need in-house through spending their usually comparatively low levels of capital, or seek out a way of accessing indirectly via external sources – and this leads to the logic of using partnerships (Varis *et al.*, 2004). (For an overview of the non-partnership options available to early stage firms see Brush *et al.*, (2001) or Aldrich and Fiol, (1994)).

However, the setup and management of partnerships can be a significant drain on management time, and various data show that the majority of partnerships under-perform (George and Farris, 1999) and that partnerships between start-ups and established businesses present a particular set of management issues (Alvarez and Barney, 2001).

The remainder of this paper is structured as follows. We first present the background to this research by briefly reviewing the literature on start-up and growth of technology-based ventures and the role of partnerships, and the challenges of partnerships between new and established ventures. We then describe our research approach before moving to the discussion of some of the strategic and operational issues facing partnerships between start-ups and established businesses drawn from the case studies and background research. We conclude the paper by highlighting the further work that is currently being undertaken to drill deeper into some of the issues surfaced by this discussion.

2. Research Background

The creation of a new firm based around the commercial application of a new technology, or the new application of an existing technology, represents one specific form of innovative activity. This process can be viewed as comprising the pursuit of opportunity, the mobilisation of resources and the creation of a resource base for business activity to deliver value and capture returns (Druilhe and Garnsey, 2003). There are many models that seek to classify what are perceived to be the various stages of a new firm's evolution as it moves from opportunity recognition to sustainable commercial success. For example, Garnsey's model is based around a series of phases that are: "conceived as manifestations of critical problems that unfold as firms grow. They reflect the need to build the competence to address these key problems if the firm is to survive and succeed." (Garnsey, 1998: 530). These phases are resource access; resource mobilisation; resource generation; growth reinforcement; and growth reversal. During the early phases, start-ups have either to bring the resources or competences they need in-house while conserving scarce resources such as available funds, or seek out a way of accessing them from external sources (Varis *et al.*, 2004).

For the purposes of this research, we are taking the term "partnership[2]" to specify a range of inter-organisational relationships: *"[..] in which the parties [..] maintain autonomy but are bilaterally dependent to a non-trivial degree."* (Williams, 1991: 271). Partnerships can be observed in terms of a number of issues that include its function (Eisenhardt and Schoonhoven, 1996), form (Lorange and Roos, 1992), evolution (Jokela, 2004; Callahan and MacKenzie, 1999; Hoffmann and Schlosser, 2001), management and performance (Fraser *et al.*, 2001; Callahan and MacKenzie, 1999).

There are a number of conceptual views can be taken when seeking to explain the motives for firms in establishing partnerships. These differing conceptual views are discussed in depth in de Rond (2003) and summarised in Table 1. From these diverse conceptual views, we have

---

[2] In much of the literature the terms "alliance", "partnership" and "collaboration" are often used interchangeably. For the sake of clarity, we are using the single term "partnership" throughout this paper.

Table 1. Conceptual views on partnerships from the literature (de Rond (2003), summarised in Valli (2004)).

Theory / View	Key Concepts	Authors Include
Market power theory	A strategy of cooperation that might enable alliance partners to achieve a stronger position together rather than "going it alone"	Mason (1939); Child (1972); Porter (1980, 1985)
Transaction cost theory	Posits that a strategy of cooperation can be a cost reducing methodology	Coase (1937); Williamson (1991); Faulkner and de Rond (2000)
Resource-based view	Firm value is maximized through gaining access to the others firms' valuable resources. Task is that of adjusting and renewing resources as time, competition and change erode the resource value	Penrose (1959); Barney (1991); Eisenhardt and Schoonhoven (1996)
Agency theory	Concerned with the ability of "principals' to monitor and control agents. Agents seek to exploit or access resources	Child and Faulkner (1998); Das and Teng (2000)
Game theory	Social behaviour as a game. Players are interconnected and interdependent. Optimal outcome is gained through cooperation, not competition	von Neumann and Morgenstern (1944); Dixit and Nalebuff (1999)
Real options theory	Derived from finance modelling: spread risk, portfolio of low-risk options, incremental "wagers"	Amram and Kulatilaka (1999); Kogut and Kulatilaka (2001)
Resource dependence theory	Organisational behaviour is demand-centric, focusing on its environment to provide resources necessary for its survival	Donaldson (1995); Das and Teng (2000)
Relational contract theory	Exchange is not discreet; relationships need on-going interactions. Trust is critical to smooth exchanges	Child and Faulkner (1998); Das and Teng (2000)
Organizational learning theory	Experience and sense of community are preconditions to learning. Organisations partner to acquire, disseminate and retain knowledge by these methods of learning	Polanyi (1966); Simonin (1999)
Social network theory	Defined as "structured sets" of autonomous players who cooperate both to adapt to the environment and to coordinate and safeguard exchanges	Nohria (1992); Jones *et al.* (1997); Faulkner and de Rond (2000)

selected the resource based view (RBV) to provide us with a useful tool for analysing partnerships (as has been previously demonstrated by Eisenhardt and Schoonhoven (1996)). For the purposes of this paper, we are taking our definition of the RBV from that described in Barney (1991) and re-visited in Barney *et al.* (2001) where a firm's sustained competitive advantage is given to be derived from: "[..] the resources and capabilities a firm controls that are valuable, rare, imperfectly imitable and not substitutable. These resources and capabilities can be viewed as bundles of tangible and intangible assets, including a firm's management skills, its organizational processes and routines, and the information and knowledge it controls" (Barney *et al.*, 2001: 625).

This is based upon the assumptions that (a) firms within an industry are heterogeneous in terms of the strategic resources they control and (b) resources may not be perfectly mobile across firms and hence heterogeneity can be long lasting (Barney, 1991)

3. Research Approach

The first stage of this research was to review existing research in this area, to extract a useful conceptual approach (i.e., the resource-based view), and to conduct preliminary case studies to draw out the key issues that underpin the use of such partnerships from the perspective of the technology-based start-ups. These initial case studies are drawn from secondary (news archives, company reports, published case studies) but mainly primary sources (interviews with chief executive officers) in early-stage technology-based ventures operating within the Cambridge, U.K. high-technology business cluster. From the population of 1,500 high tech firms operating within the Greater Cambridge sub-region, a database of around 900 innovation based ventures has been acquired. This has then been filtered to provide us with a core of 50 companies that are (a) less than 10 years old; (b) are product based (including software, but excluding life-science); and (c) have formed partnerships with established companies. The initial 10 cases studies from this sample of 50 are summarised in Table 2.

Table 2. Summaries of 10 case study partnerships.

	Start-Up	Established Firm	Summary of Motive for Partnership(s)
Case 1	Developer of advanced audio technologies	Multinational producer of consumer electronics goods	Established firm worked with start-up to help convert technology to manufacturable product and then took licence to embed within own products
Case 2	Developer of wireless information access technology	Provider of products and services for visitor attractions	Established firm could provide access to market and could add start-up's technology to own portfolio
Case 3	Developer of novel electronic display technologies	Suppliers of materials, process technologies, and display device manufacturers	Start-up had core IP on potentially disruptive technology and production processes. Many established companies wished to "hedge bets" through forming partnership with this start-up
Case 4	Developer of web server software	Global provider of IT products and services	Partnership based around provision of start-up's software to be sold with established firm's hardware
Case 5	Developer of mobile location technologies and services	Mobile phone network operator	Start-up's technology enabled additional services that would be USP for mobile network operator, plus hedge bets on technology to enable phones to comply with new laws
Case 6	Developer of mobile device power technology	Mobile phone original equipment manufacturers (OEMs)	Start-up able to provide mobile phone OEMs with novel technology to help address demand for increasing power consumption caused by larger screens, and on and off network usage of phones
Case 7	Developer of novel cryogenic technology	Oil extracting and processing company	Start-up approached to provide licence to one specific technology, but led to consulting-based relationship
Case 8	Knowledge management technologies for clinical markets	Systems integrators operating in clinical markets	Start-up's specific knowledge management approach and dataset could form attractive additional element of broader IT system targeted at clinical markets
Case 9	Communications hardware developer and fabless manufacturer	Various IT hardware manufacturers' R&D divisions	Established manufacturers are end-users of start-up's technologies, but will only receive them via module manufactured by third-party. Start-up needs to have close working relationship end-user R&D labs
Case 10	Mobile content payment systems	Mobile phone network operator	Start-up delivered specific payment technology related service to network operator

4. Discussion

The issues raised in these initial case studies can be structured as in Figure 1.

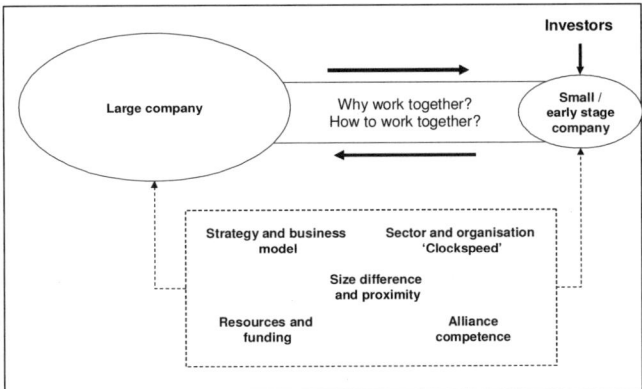

Figure 1. Structuring the issues emerging from the initial case studies.

4.1. *Strategies and business models*

For the start-up the partnership may be a critical element of their strategy for growth, but for the established firm it may be of relatively minor importance. This is not always the situation: in Case 3, the start-up's technology (and the fact that it owns the core IP that underpins one possible successor to LCD) may be of significant importance to the strategy of the established firms. In Case 4, the start-up's location technology for mobile devices may become mandated for use on all mobile phones and therefore address a critical need for their partners.

It was also noted in a number of the cases that each firms' strategic reason for forming the partnership may not remain static. Internal and external factors will influence both parties, and there will be "direct" and "indirect" motives for each of the partners. For example, in Case 1, while there was a clearly communicated strategy by the start-up for forming the alliance, there was also an indirect motive:

> "We viewed this as a "Trojan Horse" model; we were using our relationship with [large company] to better understand the industry" (Interview with Business Development Director).

The evolving nature of many of the start-ups' business models added an interesting dimension to some of the partnerships. There were examples of start-ups that had sought one partner with the aim of accessing complementary resources to execute their "develop only" business model (i.e., they wanted the partner to licence from them, and the start-up had no wish to be involved in actual production of devices based on the technology). However, the only way that the start-up could find an established firm willing to licence their technology was when the start-up agreed to take on responsibility for the early production runs. This required them rapidly to develop competence in the management of outsourced manufacturing; something that they had not intended to include as part of their business model.

From the start-up's perspective, there was sometimes confusion of what they were actually providing to the established firm, as revealed in Case 4:

> "We [start-up] delivered a great technology, but what [large company] wanted was a complete, documented, globally supported solution. With just six people in our company, that was almost impossible for us to do" (Interview with company founder).

The interplay between the partnership process and evolution of the start-up's business model was revealed in a number of other cases. In Case 10, the start-up intended at the outset only to seek to sell a package of their core services based on mobile content payment to the established firm. However, the implementation of this partnership revealed to both parties that there were a number of other, potentially much larger, areas in which the two companies could work together:

> "That first contract generated hardly any revenue for [start-up]. What it did though was provide a "green card" to get access to the inside of [established firm] and see where the real opportunities for working together lay" (Interview with investor in Case 10 start-up).

4.2. Organisational and industry "clockspeed"

From the case studies, it was noted that issues of "clockspeed" were relevant at two levels. Firstly, there was the clockspeed of the industry. If the start-up and the established firm were used to operating in commercial environments with markedly different paces of activity, then this could lead to problems. Secondly, there is the potentially differing clockspeeds of the two partner organisations. This is best summed up a quote from the founder of the start-up in Case 4 – where both they are the partners were IT-based:

> *"Our deal cycle is 2 months: For our partner is was 18 months. This caused us no end of headaches as we didn't appreciate this at the outset" (Interview with start-up founder).*

Clockspeed was just one of the aspects of each company's operations that influenced the process of partnering. The broader issue of company culture was clearly noted in some cases as influencing their working relationship.

4.3. Size difference and proximity

The management challenges posed by the difference in size between the start-up and established firm was raised in a number of the case studies. Phrases such those given below highlight this issue:

> *"We were like a fishing boat trying to dock with a super-tanker. [..]"We wanted to talk to someone from [established firm] – we looked in our address books and found 500 names" (Interview with start-up founder).*

There were also issues to do with the segmentation of activities within the larger organisations that posed challenges for some of the partnerships. Case 3 highlights how this issue can impact upon implementation:

"The deal was set up by technology people, but the deal was implemented by marketing people. The marketing people didn't buy-in to what we were doing at all" (Interview with start-up founder).

4.4. Resources and funding

The case studies reveal some interesting interplay between funding, resources and selection of partners. Firstly, the amount of time and effort to make a partnership work was revealed in the cases to have often been substantially underestimated:

"From when we started talking to [established company] to when we actually signed the deal was 18 months, 6 of those requiring the services of our lawyers. There was then a further 12 months of back-and-forth visits to the Far East before we actually started doing something that generated revenue for us" (Interview with start-up business development director in Case 1).

Secondly, there is the positive aspect of resource generation through the partnership. For example, the formation of the partnership may enhance the start-up's credibility in the eyes of potential investors:

"Working with [established company] nearly closed us down, and we ended the agreement after 12 months. However, the credibility that working with [established company] gave us in the eyes of our future investors was high. On the basis of this, we secured our next funding round" (Interview with start-up founder).

Thirdly, there are issues of changing resource appropriateness. The selection of partner for a start-up at its early phase of evolution may be driven by necessity to get "something rather than nothing". If investment then flows into the start-up, then they may be in a position to re-think partnerships more in terms of optimisation rather than adequacy. This can be illustrated by the following quote from Case 9:

"At the start, we had nothing but really needed to work with bigger companies to access design expertise and get knowledge of end-users.

After closing our first major funding round we could be much pickier about who we worked with. We made a clear decision to shift our partnerships to be with the best in the industry. Having backing from [major US venture capital fund] made this easy" (Interview with start-up CFO).

4.5. Partnering competence

The prior experience of each company in working on partnerships was highlighted in a number of the interviews. During the discussions of the challenges presented by the partnership with a larger firm, the following quotes were revealing:

"One of the first questions we should have asked them [established firm] is "Have you ever worked with a start-up before?" We both made some really basic mistakes based on our understanding of how each other operates" (Founder of start-up in Case 2).

Examples were also noted in the interviews of established firms with substantial experience of working with start-ups, and how they manage this process:

"Working with [established firm] has been great. They are aware that it is really hard for a start-up to understand their complex operations, so they found a way to insulate us from most of it. They have a team which is setup to run almost as a small business to interface with start-ups" (Interview with start-up founder).

5. Conclusions

Many of the issues identified through this initial case study interviews reflect themes already identified in the literature on partnerships as described earlier in this paper. However, a number of themes can be identified from these cases that provide us with a structure for the next stages of this research.

The resource-based view provides us with a useful language for examining the interplay between partnering and the start-up's development through the early phases of resource access; resource mobilisation; and resource generation. This allows us draw in analysis from the perspective of investors in start-ups to add to the research consideration of the relative value of different types of partnerships at different stages of a start-up's growth. This will provide us with the opportunity to focus the project onto one relatively weak area in existing knowledge: the interplay between the ability of start-ups to access the resources they need to grow; the role of early-stage technology investors in supporting this growth; and the evolution of a start-ups business model in the light of resource access, mobilisation and generation.

The case studies revealed examples of perceived good and poor practice by the established firms from the perspective of the start-ups. One strand of current research is to investigate the role that partnering with early-stage technology-based firms can play in the strategies of established firms. This will encompass areas such as the established firms' activities in technology management (in particular, technology scanning and technology acquisition) and the diverse forms of corporate venturing.

References

Aldrich, H. E. and C. M. Fiol (1994). "Fools rush in? The institutional context of industry creation." *Academy of Management Review* 19(4): 645-670.

Alvarez, S. A. and J. B. Barney (2001). "How entrepreneurial firms can benefit from alliances with large partners." Academy of Management Executive 15(1): 139-148.

Barney, J. B. (1991). "Firm resources and sustainable competitive advantage." *Journal of Management* 17(1): 99-120.

Barney, J. B., M. Wright and D. J. Ketchen (2001). "The resource-based view of the firm: Ten years after 1991." *Journal of Management* 27: 625-641.

Brush, C. G., P. G. Greene and M. M. Hart (2001). "From initial idea to unique advantage: The entrepreneurial challenge of constructing a resource base." *Academy of Management Executive* 15(1): 64-78.

Callahan, J. and S. MacKenzie (1999). "Metrics for strategic alliance control." *R&D Management* 29(4): 365-377.

de Rond, M. (2003). Strategic alliances as social facts: Business, biotechnology and intellectual history. Cambridge, Cambridge University Press.

Druilhe, C. and E. W. Garnsey (2003). "Do academic spin-outs differ and does it matter?" University of Cambridge Centre for Technology Management Working Paper Series 2003/02.

Eisenhardt, K. M. and C. B. Schoonhoven (1996). "Resource-based view of strategic alliance formation: Strategic and social effects in entrepreneurial firms." *Organization Science* 7(2): 136-150.

Fraser, P., T. Horsfall and M. J. Gregory (2001). "Taken on trust: The role of contracts in product development collaborations involving small firms". EIASM: Trust within and between organisations, Vrije Universiteit Amsterdam, The Netherlands.

Garnsey, E. W. (1998). "A theory of the early growth of the firm." *Industrial and Corporate Change* 7(3): 523-556.

George, V. and G. Farris (1999). "Performance of alliances: Formative stages and changing organisational and environmental influences." *R&D Management* 29(4).

Hoffmann, W. H. and R. Schlosser (2001). "Success factors of strategic alliances in small and medium sized enterprises: An empirical survey." *Long Range Planning* 34: 357-381.

Jokela, P. (2004). "Challenging cooperation: Why do small high technology firms fail to form technology partnerships?" 12th High Tech Small Firms Conference, University of Twente, The Netherlands.

Lorange, P. and J. Roos (1992). Strategic Alliances, Oxford/Blackwell.

NCOE (2002). American formula for growth: Federal policy & the entrepreneurial economy, 1958-1998, National Commission on Entrepreneurship.

Timmons, J. A. (1998). America's Entrepreneurial Revolution: The demise of Brontosaurus Capitalism, Babson College, F.W. Olin Graduate School of Business.

Valli, R. (2004). "Building investment readiness for university spin-out companies through partnerships." PhD First Year Report, University of Cambridge Centre for Technology Management.

Varis, J., V.-M. Virolainen and K. Puumalainen (2004). "In search for complementarities: Partnering of technology-intensive small firms." *International Journal of Production Economics* 90: 117-125.

Williams, O. E. (1991). "Comparative economic organisation: The analysis of discrete structural alternatives." *Administrative Science Quarterly* 36: 269-296.

CHAPTER 4

IMPACTS OF RFID ON WAREHOUSE MANAGEMENT IN THE RETAIL INDUSTRY

Louis-A. Lefebvre, Elisabeth Lefebvre,
Samuel Fosso Wamba, and Harold Boeck
École Polytechnique de Montréal, Canada

Based on a field study conducted in the retail industry, this paper examines the impacts and the potential benefits generated by an RFID application in a warehousing environment. Through a detailed investigation of the underlying business processes, we will demonstrate how process optimization can be achieved when integrating RFID technology.

1. Introduction

Even though RFID (Radio Frequency Identification) technology has been around for decades, it has only recently gained a strong academic and industry interest (Sheffi, 2004; Collins, 2003). Industrial applications of RFID are spreading as very large organizations such as Hewlett-Packard, Wal-Mart and the American Department of Defense are now pressuring their most important suppliers to adopt this technology (Sliwa, 2004; Dignan, 2004; Barlas, 2003). Although RFID represents a simple technology, its numerous applications can become quite complex. A basic RFID system is composed of three layers: a chip/tag, a reader and a computer. The tag is attached to or embedded in a physical object and communicates wirelessly (e.g. without line of sight) with the reader. A network of readers can therefore follow the object throughout the physical world. The readers send the location and the identification of the object to a computer which adjusts or initiates business processes automatically (Kärkkäinen *et al.*, 2003).

This paper focuses on the impact of RFID technology on warehousing activities in the retail industry. These activities represent one of the critical elements of the supply chain: opportunities to optimize warehouse business processes are numerous (Mills, 2000; Moore, 2004) and can have an impact on the whole supply chain performance.

2. Background

2.1. *Current context of the retail industry*

The retail industry represents one of the largest industries worldwide. In the United States, it is the second-largest industry (in terms of the number of establishments and the number of employees), with $3.8 trillion in sales annually and 11.7 percent of U.S. employment (Vargas, 2004). This industry is characterized by globalization, aggressive competition, shorter product life cycles, increasing cost pressures and the rise of customized demand with high product variants. The short shelf-life of grocery goods presents some of the biggest challenges for the retail supply chain management due to the strict traceability requirements and the need for temperature control in the supply chain (Kärkkäinen, 2003). The short shelf life of grocery goods makes warehousing management a time-critical operation (Eleni and Vlachos, 2004).

Despite the high number of products handled in the supply-chain, a large number of American businesses are still using manual and thus error-prone methods to collect data (Quinn, 2004), causing inventory inaccuracies (Fleisch and Tellkamp, 2004; Raman *et al.*, 2001). However, in order to stay competitive, companies must optimize their internal (intra-organizational) and external (inter-organizational) processes. This later phenomenon is particularly evident in the consumer product good (CPG) branch, where multiple players are involved in the delivery of the products, contributing to the complexity of the supply chain. In addition, data inaccuracies in the supply chain are costly. Incorrect or outdated data used in invoices, bills of lading (a document from the carrier indicating the description of the goods being shipped) or purchase orders can result in product delivery errors and lost sales estimated to more than $50 billion annually (UCCnet, 2004).

With the intent of streamlining their supply chain processes and of controlling costs, leading CPG distributors around the world are relying more heavily on the use of information technologies. Applications such as enterprise resource planning (ERP), materials requirement planning (MRP), manufacturing resources planning (MRPII), warehouse management system (WMS), advanced planning and scheduling (APS), Electronic Data Interchange (EDI), automatic identification and data collection (AIDC), etc. are currently used to support the intra- and inter-organizational business processes, decision-making, workflow management and automatic information exchange with their supply chain partners. In addition, new customer-focused concepts are progressively introduced into the management of retail supply chains in order to improve performance (Sparks and Wagner, 2003). These concepts cover for example quick response (QR), efficient consumer response (ECR), vendor-managed inventory (VMI), point of sale (POS) and collaborative planning, forecasting and replenishment (CPFR) (Seifert, 2003; Sparks and Wagner, 2003).

2.2. RFID early adopters in the retail industry

RFID technology does indeed have the potential to be quite beneficial for the retail industry. Wal-Mart, for example, has been analysing its potential for more than a decade (Roberti, 2003). In fact, Wal-Mart has now become a major reference in the adoption of RFID in the retail industry since it officially disclosed its RFID initiative in 2003 (Sullivan, 2004). Coined as the "Wal-Mart effect", the company detains sufficient power to influence its suppliers (Goldman and Cleeland, 2003) and thus determines to a large extent the technology adoption paths of these suppliers: it had set January 2005 as the deadline for its 100 top suppliers to incorporate RFID technology for all their deliveries at the pallet level (Wired News, 2003). While a few have even successfully incorporated RFID at the product level (Wal-Mart, 2005), the latest information suggests that not all suppliers have completely met the 2005 deadline (Kevan, 2004). Yet the vast majority are trying to comply although it may be with little enthusiasm (O'Connor, 2005). Wal-Mart stands to save substantial amounts from supply chain optimization, just-in-time deliveries and disappearance of stock-outs. The retail giant will know exactly where a product is throughout its entire supply chain and could theoretically track a product from its suppliers' supplier location.

Wal-Mart states that the business goal derived from this RFID initiative is to "increase customer satisfaction in the near-term and ultimately play an important role in helping us control costs and continue offering low prices" (Wal-Mart, 2004). Additionally, benefits could be realized such as informing customers that the products they are looking for are in stock, monitoring product expiration dates or acting quickly during recalls, improving supply chain processes so that the right products are in the right places at the right time, and, cutting inventories by 5% and labor costs in the warehouse by 7.5% (Pruitt, 2004; A.T. Kearney, 2003).

Other companies in the retail space are also interested by the potential benefits that the RFID technology offers. In 2003, Metro Group in Germany had already gone beyond the pilot phase as it had opened its first "Extra Future Store" where RFID is used in a live supermarket environment (Collins, 2004). This phenomenon has created much excitement in the retail industry among technology suppliers and people wanting to optimize their processes. In fact, so much excitement has been created in a relatively short timeframe that tag suppliers have not been able to completely respond to the demand, resulting in a shortage of RFID tags in 2004 (Trebilcock, 2004).

2.3. *Warehousing and the potential of RFID*

Basically, there are three types of warehouses (Van Den Berg and Zijm, 1999): (i) a distribution warehouse where products from different suppliers are collected and sometimes assembled for delivery to a number of customers, (ii) a contract warehouse which performs the same activities as a distribution warehouse for one or many customers, (iii) a production warehouse which is located in a production facility and is dedicated to the storage of raw materials, semi-finished and finished products. Warehouse management can be analyzed from three perspectives (Rouwenhorst *et al.*, 2000) namely the process perspective (steps undertaken in a warehouse), the resource perspective (equipment and personnel needed) and the organization perspective (planning and control procedures).

In this study, we have investigated one distribution warehouse and have chosen the process perspective. This deliberate choice arises from our intention to understand HOW the work is carried out within one type of a distribution warehouse in order to fully grasp the impacts of implementing RFID. Four distinct warehousing activities are usually identified (Rouwenhorst, *et al.*, 2000; Van den Berg and Zijm, 1999) and can benefit from RFID technology:

(i) The receiving process is handled when products or items arrive at the warehouse. At this step, received items or products are checked with documents such as purchase order (PO) or an advanced shipping notice (ASN). When there is a match, the receiver can apply a label to the pallet/item for tracking throughout the warehouse. When there is no match, the receiver has to confirm the discrepancy which often involves another person (a manager for example). Receiving is time consuming and subject to human error (Keith *et al.*, 2002), making it a good candidate for RFID technology.

(ii) The put-away process consists of moving and placing products or items to specific storage locations. This process can be greatly improved by automation. For example, the use of a mobile terminal allows to scan different bar codes on pallets or products and automatically display their storage locations. Cost savings related to the put-away process in a warehouse derived from the adoption of RFID technology could reach 50% (Capone *et al.*, 2004), thus positioning this process as another candidate for RFID technology.

(iii) The picking process refers to the retrieval of products or items from their storage locations for the consolidation of customer orders. This process is labor intensive and prone to human error. For example, case picking can occupy up to half the staff in a distribution center and requires many verifications (Keith *et al.*, 2002). Globally, order picking represents 50–75% of the total operating costs in a warehouse (Petersen and Aase, 2004).

(iv) The shipping process is performed before the products reach the end customers. Customer orders are checked, packed and loaded in trailers, trucks, trains or any other transportation unit and have to match with outbound orders all the time. Again, the verification attached to this process highly depends on the level of automation.

3. Methodology

Our methodology builds on previous work (Strassner and Schoch, 2004; Subirana *et al.*, 2003) and focuses on one specific "open-loop" supply chain RFID initiative in the retail industry.

Table 1. Steps undertaken in the field study with emphasis on scenario demonstration and analysis.

	Detailed Activities
Phase 1: Opportunity Seeking	
Step 1	**Determination of the primary motivation** to consider the use of RFID technologies (**WHY?**)
Step 2	**Analysis of the Product Value Chain (PVC)** specific to a given product (**WHAT?**)
Step 3	**Identification of the critical activities in the PVC:** Identification of critical PVC activities (**WHICH** activities to select and **WHY?**)
Step 4	**Mapping of the network of firms supporting the PVC;** to understand the links within the network of firms supporting the product (**WHO and WITH WHOM?**)
Step 5-6	**Mapping of intra- and inter organizational processes** for the identified opportunities as they are carried out now ("As is") (**HOW within and between organization?**)
Phase 2: Scenario Building and Validation	
Step 7	**Evaluation of RFID opportunities** in the PVC with respect to the product (level of granularity), to the firms involved in the SC and to the specific activities in the PVC
Step 8	**Evaluation of potential RFID applications including scenario building** and process optimization ("As could be") (**HOW within and between organizations?**)
Step 9	**Mapping of intra- and inter-organizational processes integrating RFID technology**
Step 10	**Validating business and technological processes integrating RFID technology with key respondents:** Feasibility analysis including ERP and middleware integration and business process redesign
Phase 3: Scenario Demonstration and Analysis	
Step 11	**Proof of concept (POC) in laboratory simulating RFID physical environment and interface between supply chain players:** Feasibility demonstration and evaluation including ERP and middleware integration and process redesign at all the supply chain members' level **Proof of concept post-analysis and decision to go for the** pilot replicating POC scenarios in a real-life setting
Step 12	**Pilot** project and evaluation of anticipated vs. realized benefits and impacts of RFID. **Appropriation by the different organizations involved**

As the main objective of this study is to examine the impacts and the potential benefits generated by an RFID application in a warehousing environment, the research design corresponds to an exploratory research initiative. Field research was conducted in 12 consecutive steps (see Table 1, adapted from (Lefebvre *et al.*, 2005)).

4. Results

Within the scope of this paper, we will present and discuss only one of the scenarios proposed for the "receiving" and "put-away" processes.

Figure 1 summarizes the results obtained from the field research. Processes displayed in Figure 1 are drilled down, i.e. from the more general to the more detailed: for instance, the overall process corresponds to "receiving process", first level process to "1. receive bill of lading (BOL)", and second level process to "1.1. create a BOL in the ERP", and so on.

On the left hand side of Figure 1, the actual or existing processes are presented. The following observations can be made:

(i) the two overall "receiving" and "put-away" processes consist respectively of 17 and 5 second level processes, for a total of 22;
(ii) most existing processes involve numerous interventions from the employees such as pallet scans, visual count of boxes in each pallet or data input;
(iii) the "put-away" process starts with the end of the "receiving" process. Received products are in a staging area while a message has been manually sent from the warehouse management system to a dedicated forklift terminal via radio frequency in order to initiate the put-away.

On the right hand side of Figure 1, the potential impacts of RFID on the same two overall processes are investigated. This corresponds to the following scenario: all products (boxes and pallets) have an RFID tag, the warehouse is equipped with RFID readers and the existing ERP has a middleware to integrate data read from RFID tags. Interesting observations emerge from this scenario:

(i) the number of processes drops from 22 second level processes in the existing context to 8 second level and 5 third level processes with RFID technology;
(ii) the two overall processes are merged together;
(iii) all information-based processes (1.1.1., 1.1.2., 1.1.3., 2.5.1., and 2.5.2.) are now automatically performed. In fact, the use of RFID automates verification procedures during the receiving process and provides accurate information at a very high level of granularity (pallet, box) allowing the possibility to generate efficiency measures in real-time and making transparent the flow of products. RFID can also eliminate most paper-based documents generated from traditional receiving and put-away processes. Finally, operational improvements such as removing manual checks and eliminating human errors can also derive from the RFID technology.

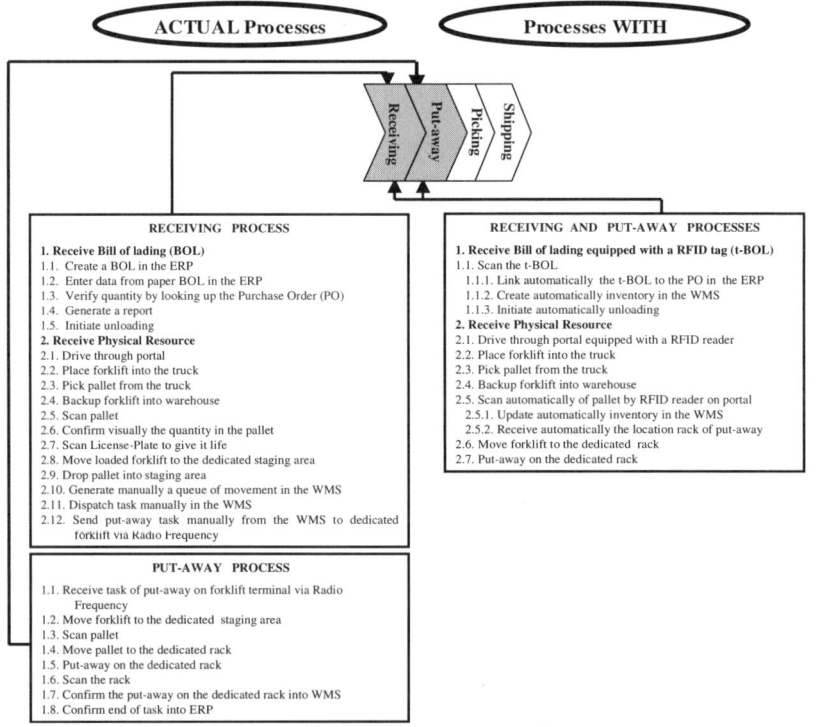

Figure 1. The impact of RFID on two warehouse processes.

5. Conclusion and Future Research Avenues

The results from the three-step field research demonstrate the potential of RFID technology in warehousing activities. The next steps will be to test the technology in a Beta site and then in a real-life environment.

This paper makes several contributions. First, in order to capture the real potential of RFID, the business process approach seems quite appropriate. Second, RFID can improve the "receiving" and "put-away" processes in the warehouse. By extension, RFID technology could optimize the entire warehouse processes as well as the entire supply-chain processes. In fact, RFID can be seen as a support for information sharing among the players in the supply-chain. In our scenario, the same RFID tag used for the shipping by the suppliers could be used during the receiving in the warehouse. The same tag could also be used during the shipping process to the final customer and ultimately respond to different interests of the business partners in one supply-chain. Third, the preliminary results from our study show that RFID technology triggers automatically some business processes. As products attached with RFID tags become "intelligent or smart", a world of possibilities unfastens. Despite considerable technological and organizational challenges, the future of RFID seems promising.

References

Barlas, D. (2003) "DoD's RFID Commitment". Line 56 [online]. 4 December. Available from: http://www.line56.com [Accessed 31 January 2005].

Capone, G., D. Costlow, W. L Grenoble and R. A. Novack. (2004). The RFID-Enabled Warehouse. Center for Supply-Chain Research, Penn State University.

Collins, J. (2003). "Estimating RFID's Pace of Adoption". RFID Journal [online] http://www.rfidjournal.com [Accessed 31 January 2005].

Collins, J. (2004). Metro Launches RFID Test Center. RFID Journal [online] 7 July. Available from: http://www.rfidjournal.com [Accessed 31 January 2005].

Dignan, L. (2004). "Wal-Mart RFID Suppliers to Top 100". Baseline: The Project Management Center [online]. http://www.baselinemag.com [Accessed 31 January 2005].

Eleni, M. and I. P. Vlachos. (2004). The changing role of information technology in food and beverage logistics management: beverage network optimization using intelligent agent technology. Journal of Food Engineering, In Press.

Fleisch, E. and C. Tellkamp. (2004). Inventory inaccuracy and supply chain performance: a simulation study of a retail supply chain. International Journal of Production Economics, accepted February 2004.

Goldman, A., and N. Cleeland. (2003). An Empire Built on Bargains Remakes the Working World. LA TIMES [online] 23 November. Available from: http://www.latimes.com [Accessed 28 January 2005].

Kärkkäinen, M. (2003). Increasing efficiency in the supply chain for short shelf life goods using RFID tagging. International *Journal of Retail & Distribution Management.* Vol. 31, No. 10, pp. 529-536.

Kärkkäinen, M., J. Holmstrom, K. Framling and K. Artto. (2003). "Intelligent products - a step towards a more effective project delivery chain". *Computers in Industry.* Vol. 50, No. 2, pp. 141-151.

Kearney, A.T. (2003). Meeting the Retail RFID Mandate: A discussion of the issues facing the CPG companies. Chicago: A.T. Kearney, Inc.

Keith, A., G. Tig, K. Gramling, M. Kindy, D. Moogimane, M. Schultz and M. Woods. (2002). Focus on the Supply Chain: Applying Auto-ID within the Distribution Center [online] 1 June. Available from: http://www.autoid.org [Accessed 28 January 2005].

Kevan, T. (2004). Dispelling the Myths of Wal-Mart's RFID Initiative. Frontline Solutions [online] 29 November. Available from: http://www.frontlinetoday.com [Accessed 31 January 2005].

Lefebvre, L.A., Lefebvre, E., Bendavid, Y., Fosso Wamba, S., Boeck, H. (2005). The potential of RFID in warehousing activities in a retail industry supply chain. *Journal of Chain and Network Science* 5, 2, 101-111.

Mills, A. (2000). Where's the warehouse? *Manufacturing Engineer.* Vol. 79, No. 5, pp. 214-215.

Moore, B. (2004). AIDC in the warehouse: Hardware's Easy; Software's Hard. Vol. 59, No. 8, pp. 25-31.

O'Connor, M.C. (2005). Suppliers Meet Mandate Frugally. *RFID Journal* [online] 3 January. Available from: http://www.rfidjournal.com [Accessed 31 January 2005].

Petersen, C.G. and G. Aase. (2004). A comparison of picking, storage, and routing policies in manual order picking. *International Journal of Production Economics.* No. 92, pp. 11–19.

Pruitt, S. (2004). Wal-Mart begins RFID trial in Texas. Computerworld [online] 30 April. Available from: http://www.computerworld.com [Accessed 31 January 2005].

Quinn, P. (2004). Bar code: Stronger than ever. *Supply Chain Systems Magazine,* Vol. 24, No. 10, pp. 16-20.

Raman, A., N., DeHoratius and Z. Ton. (2001). Execution: the missing link in retail operations. *California Management Review,* No. 43, pp. 136–152.

Roberti, M. (2003). Analysis: RFID - Wal-Mart's Network Effect. CIO Insight [online] 15 September. Available from: http://www.cioinsight.com [Accessed 28 January 2005].

Rouwenhorst, B. Reuter, V. Stockrahm, G.J. Van Houtum, R.J. Mantel, W.H.M. Zijm. (2000). Warehouse design and control: Framework and literature review. *European Journal of Operational Research,* No. 122, pp. 515-533.

Seifert, D. (2003). Collaborative Planning, Forecasting and Replenishment. How to create a Supply-Chain Advantage. AMACOM, New-York.

Sheffi, Y. (2004). "RFID and the Innovation Cycle". *International Journal of Logistics Management.* Vol. 15, No.1, pp. 1-10.

Sliwa, J. (2004). HP, Sun Launch RFID Test Centers. Computerworld [online] 10 May. Available from: http://www.computerworld.com [Accessed 30 may 2006].

Sparks, L. and B. A. Wagner. (2003). Retail exchanges a research agenda. *Supply Chain Management,* Vol. 8, No. 1, pp. 17-25.

Staff. (2004). WMS reaches out beyond the four walls. Modern Materials Handling (Warehousing Management Edition). Vol. 59, No. 12; pp. 9.

Strassner, M., Schoch, T. Today's Impact of Ubiquitous Computing on Business Processes. Institute of Information Management of University of St. Gallen, 2004.

Subirana, B., Eckes, C., Herman, G., Sarma, S., Barrett, M. Measuring the impact of information technology on value and productivity using a process-based approach: The case for RFID technologies. MIT Sloan Working Paper, December, 2003.

Sullivan, L. (2004). Wal-Mart Takes RFID To Sam's Club. InformationWeek [online] 28 October. Available from: http://www.informationweek.com [Accessed 28 January 2005].

Trebilcock, B. (2004). Will there be enough RFID tags in 2005? Modern Materials Handling [online] 17 November. Available from: http://www.mmh.com [Accessed 28 January 2005].

UCCnet. (2004). UCCnet History and Background [online]. Available from: http://www.uccnet.org [Accessed 28 January 2005].

Van Den Berg, J.P. and W.H.M. Zijm. (1999). Models for warehouse management: Classification and examples. *International Journal of Production Economics.* No. 59, pp. 519-528.

Vargas, M. (2004). Retail Industry Profile Overview of the Retail Sector. [online]. Available from: http://www.retailindustry.about.com [Accessed 28 January 2005].

Wal-Mart. (2004). Wal-Mart Begins Roll-Out Of Electronic Product Codes in Dallas/Fort Worth Area. Press release. [online]. Available from: http://www.walmartstores.com [Accessed 31 January 2005].

Wal-Mart. (2005). Supplier Information: Your guide to becoming a Wal-Mart Supplier. [online]. Available from: http://www.walmartstores.com [Accessed 28 January 2005].

Wired News. (2003). "Wal-Mart, DOD Forcing RFID" [online] 4 November. Available from: http://www.wired.com [Accessed 28 January 2005]

CHAPTER 5

FACTORS DRIVING THE BROADBAND INTERNET GROWTH IN THE OECD COUNTRIES

Petri Kero

Lappeenranta University of Technology, Finland

Broadband Internet diffusion varies widely between the OECD countries. This paper examines the influence of several factors on the diffusion of broadband Internet. The crosssectional regression models are estimated using the data on broadband Internet subscribers in the OECD countries for the years 2001 and 2004. The basic findings are that preparedness for the broadband Internet, monthly price and the local loop unbundling are the most robust factors when explaining the broadband diffusion. Income and population density do not seem to have influence on broadband Internet penetration rates. Neither is the dummy variable for government ownership of the main telecommunication operators a statistically significant predictor of broadband diffusion.

1. Introduction

The focus in this paper is on the factors that determine the broadband Internet diffusion in the OECD countries. These factors can be divided into three different categories. Firstly there are economic factors like income and access costs. Secondly there are political factors. These political factors can include e.g. competition policies and subsidy programs. The competition policies may be aimed to promote competition between technologies (inter-platform competition)[1] or between service providers who offer broadband Internet connections

---

[1] E.g. the competition between cable access and Digital Subscriber Line (DSL).

using the same technology (intra-platform competition). Thirdly, the country's infrastructure level can set up barriers to broadband diffusion. The cable television network can be used as an example. The number of cable modem subscribers is limited by cable television network coverage.

The economic factors used in this study are country's gross domestic product and broadband access costs. The effect of local loop unbundling and the government ownership of major telecommunication operators are also examined. The broadband readiness index was calculated to measure the infrastructure level. Subsidy programs were not included in the analysis.

Local loop unbundling[2] has been one of the main means in hastening the diffusion of broadband Internet. A number of OECD countries have taken a major regulatory initiative by requiring the incumbent operators to offer local loop unbundling to new Internet service providers. EU regulation requiring 15 member states to unbundle access to the local loop came into force January 2001. Local loop unbundling is anticipated to increase competition between Internet service providers offering DSL accesses and thus cut down access prices and therefore increase DSL penetration. However, some previous studies have come to the conclusion that local loop unbundling has not had a positive effect on the diffusion of the broadband Internet.

2. Previous Studies

Howell (2002) argues that local loop unbundling hasn't been successful in promoting broadband rollout. García-Murillo and Gabel (2003) studied International broadband deployment and the impact of unbundling. Also in their study unbundling wasn't a statistically significant factor. The income level of a country, the price, privatization of the incumbent carrier, and the presence of competition were all factors that can facilitate and promote the deployment of broadband. In contrast

---

[2] *Local loop unbundling* means that incumbents have to offer access to their competitors on the last segment of telephone wire linking the network with the subscriber.

to these two studies Ikeda (2003) found out that the number of subscribers of DSL in Japan has grown phenomenally due to the unbundling regulation.

Ismail and Wu (2003) found out that in many countries the cable modem services provided strong competition for the DSL services and the resulting competition has been effective in generating the rapid take-up of broadband. The research by Distaso and Lupi (2004) confirmed this result. They suggest that while this kind of inter-platform competition drives platform growth, competition in the market for DSL services does not play significant role. Their results also confirm that lower unbundling prices stimulate broadband uptake.

Kim et al. (2003) studied the influence of economic and policy variables on the diffusion patterns of broadband. Preparedness of a nation and the cost conditions of deploying networks were the most consistent factors explaining broadband uptake. However, the price of broadband, competition and the relative income position were either less or not significant factors at all.

These previous studies concerning broadband diffusion and deployment offer quite diverse and even contradictory conclusions. In these six studies under discussion not one factor could get support in all studies.

3. Data Analysis and the Models

The *broadband subscribers per capita* is a widely used measure of the broadband diffusion. South Korea has been the global leader in broadband adoption and South Koreas broadband Internet adoption rates have been far ahead of other OECD countries (Figures 1a and 1b). However, in the year 2004 these differences became smaller. In the year 2004 Denmark was second and United States 11th when ranked by broadband adoption rates. In Iceland, Denmark, Belgium and Sweden the diffusion process has been fast and they have overtaken the U.S. In the OECD countries the growth rate in 2000-2001 was 118%, in 2001-2002 30.1% and in 2003-2004 the growth rate was about 50%.

The two leading technologies used in the 2003 were digital subscriber lines (DSL) and cable modems. In 1999, the respective shares of DSL and cable modems were 16% and 84%. By the end of 2001, the balance for the two was 49% for DSL and 51% for cable modems[3]. However, in the June 2003 the share of DSL subscribers was 62% and the share of cable modem subscribers only 34%. The share of other technologies is still small (4%). Figures 1a and 1b illustrates broadband Internet penetration rates for years 2001, 2003 and 2004.

Multivariate OLS regression was used to estimate model parameters for following models:

1) $Q_{BB} = \alpha + \beta_1 P + \beta_2 GDP + \beta_3 LLU_1 + \beta_4 LLU_2 + \beta_5 Gov + \beta_6 read + \beta_7 Pop + \varepsilon$

2) $Q_{BB} = \alpha + \beta_1 P + \beta_2 Rural + \beta_3 LLU_1 + \beta_4 LLU_2 + \beta_5 Gov + \beta_6 read + \varepsilon$

3) $LnQ_{BB} = \alpha + \beta_1 P + \beta_2 Rural + \beta_3 LLU_1 + \beta_4 LLU_2 + \beta_5 Gov + \beta_6 read + \varepsilon$

The model 1) was estimated using data on December 2001. The models 2) and 3) were estimated using only data from the year 2004. The dependent variable Q_{BB} was defined as the number of broadband subscribers per 100 inhabitants. In the model 2) LnQ_{BB} is the logarithmic transformation of the dependent variable Q_{BB}.

The price of broadband P[4] was calculated as weighted average of the monthly access price of either DSL or cable modem service. Gross domestic product per capita was used as an estimate for income. In order to test whether public policy has had any influence on broadband diffusion, variables LLU_1, LLU_2 and Gov were included in the models. The local loop unbundling variable LLU_1 gets value 1, if local loop unbundling was introduced before the year 2000 and LLU_2 gets value 1, if unbundling was introduced between the years 2000 and 2005. The government ownership dummy variable Gov was used to estimate if government's ownership of telecommunication operators has any influence on broadband diffusion. The variable Pop measures the country's population density and the variable $Rural$ is the share of population living in the rural areas[5]. ε is a random disturbance.

---

[3] OECD (2001).
[4] Statistics for broadband prices in a year 2003 are from OECD (2004a).
[5] Statistics for the rural population are from OECD (2004b).

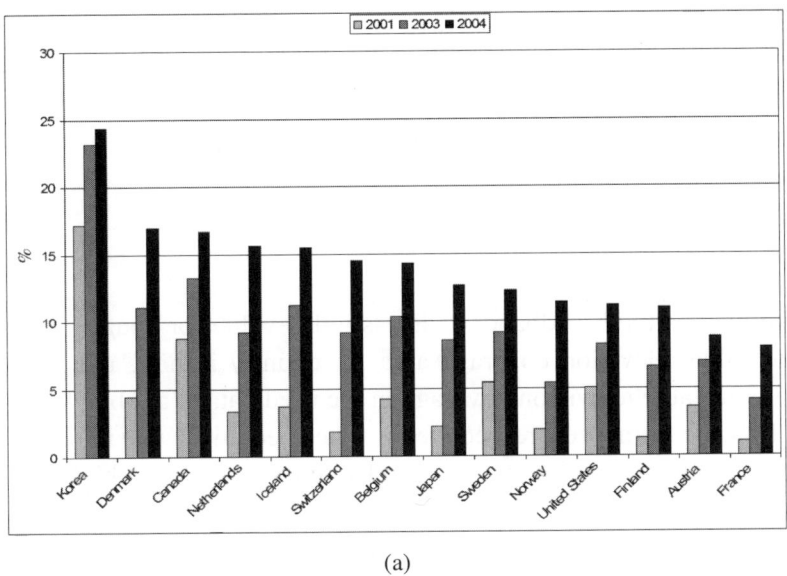

(a)

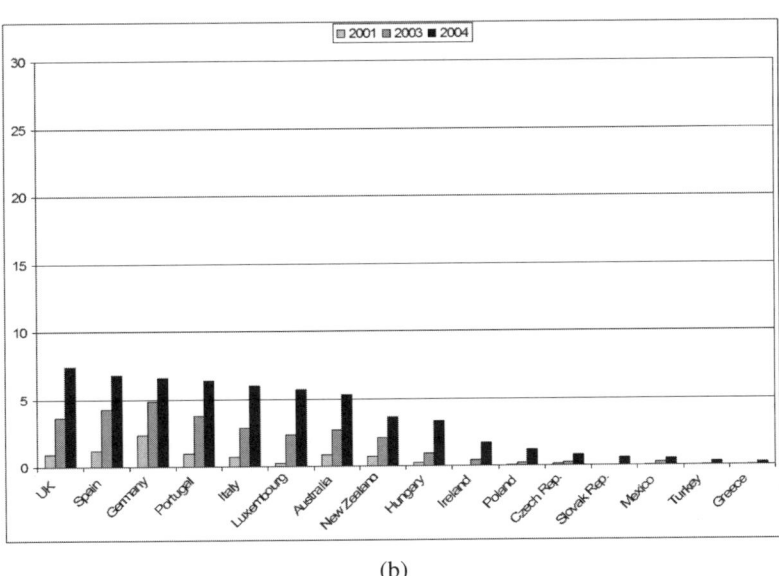

(b)

Figure 1. The number of broadband subscribers per 100 inhabitants in December 2001, in June 2003 and in June 2004.

The readiness index *read* was composed of three separate variables. The first one is the cable television coverage area, the second one is the number of access channels per household and the third one is number of personal computers per household. The index was calculated as follows:

4) $$read_i = \frac{Cable_i / Cable_{max} + Channels_i / Channels_{max} + PCs_i / PCs_{max}}{3},$$

where $read_i$ is the broadband readiness index value for country i, $Cable_i$ is the cable television coverage area for country i and $Cable_{max}$ is the maximum cable television coverage in the used data. Variables Channels and PCs are composed respectively. The theoretical maximum of the index is 1 and the minimum is 0. The United States got the highest value and Turkey the lowest. Korea was ranked 9th although it is far ahead of the rest of the OECD countries in the broadband penetration. Figure 2 presents the correlation between calculated values for the broadband preparedness index and the number of broadband Internet subscribers. Figure 3 illustrates the correlation between the monthly price and the broadband penetration in 2004.

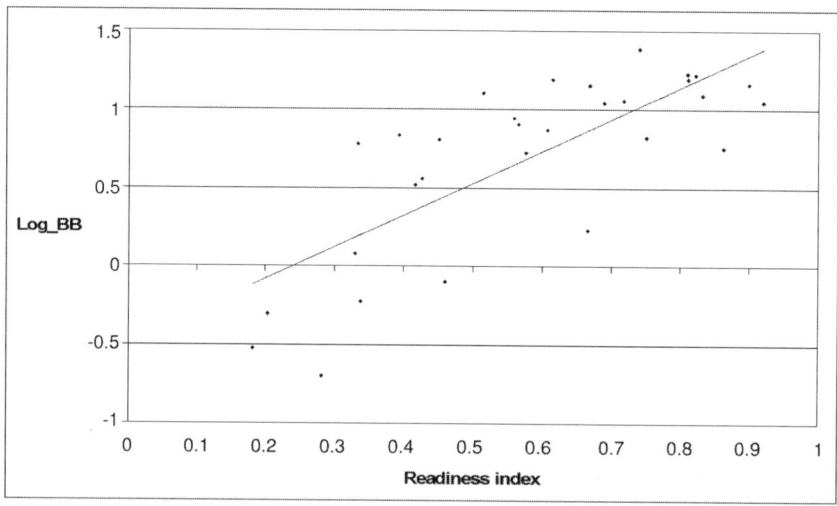

Figure 2. Broadband preparedness index.

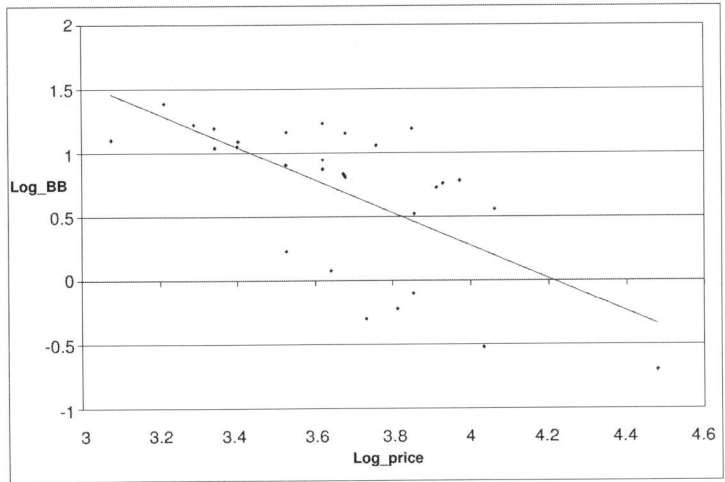

Figure 3. The correlation between the broadband penetration and the price (logarithm of monthly fee in US dollars).

4. The Determinants of the Broadband Internet Diffusion in the OECD Countries

Multivariate OLS regression was used to estimate three empirical models. The results of these estimations are shown in Table 1. For the year 2001 (model 1) four variables were statistically significant. The local loop unbundling seems to have positive effect on the diffusion of the broadband Internet. The coefficient for the dummy variable LLU_1 was statistically significant at the level 0.01, but the coefficient for the variable LLU_2 was significant only at the level 0.10. The broadband readiness index is also statistically significant and it has a positive sign as expected. The variable GDP was statistically significant only at the level 0.10 and it has a negative sign, opposite to what was expected.

In the model 2 (year 2004), the coefficients for four variables were statistically significant. The coefficient for the LLU_1 is significant at the level 0.05, but the coefficient for the variable LLU_2 is not significant any more. Unexpectedly also the coefficient for the variable price is now statistically highly significant. The coefficient for the share of the rural population is statistically significant at the level 0.10. For the logarithmic model 3 the results are in a line with the results from the model 2.

Table 1. Regression results.

	Model 1	Model 2		Model 3	
Dependent Var	BB01	BB04		LnBB04	
	Coefficient	Coefficient	t	Coefficient	t
Constant	-0.629	15.04**	2.12	1.206**	2.09
LUL1	3.089***	6.56**	2.69	0.316**	1.79
LUL2	1.683*	1.79	0.955	0.364*	2.07
Gov	-0.141	0.783	0.50	-0.11	-0.89
Pop	-0.0019	-	-	-	-
Read	8.07***	6.41	1.24	0.868**	2.07
Price	-0.0017	-0.17***	-1.952	-0.17***	-2.88
GDP	-0.13*	-	-	-	-
Rural	-	-2.43*	-1.91	-0.182*	-1.76

*Significant at 0.1 level, **Significant at 0.05 level, ***Significant at 0.01 level.

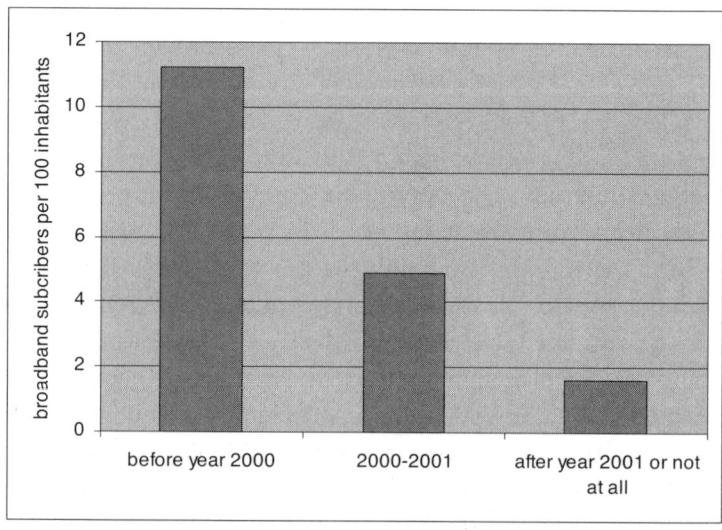

Figure 4. The effect of unbundling on the broadband penetration.

In Figure 4 is shown the effect of the local loop unbundling to the broadband penetration. To the group 1 (7 countries) belong all countries where local loop unbundling has been made mandatory before the year 2000 and to the group 2 (15 countries) those countries where it have been made mandatory in the year 2000 or 2001. In the group 3 (8 countries), are those countries where unbundling is not mandatory or it have been mandatory only after the year 2002. The bars represent the mean values of the broadband penetration inside the group. The penetration rate in the group 1 is over 11% and in the group it is only 5%. The lowest penetration rates are in the group 3, only 1.6%.

5. Conclusions

The main goal of this paper was to examine what are the main factors that determine the diffusion of the broadband Internet. The multivariate OLS regression was used to estimate three broadband Internet models. Model 1 was estimated for the year 2001 and models 2 and 3 for the year 2004. The data consisted of information from 30 OECD countries.

The main findings were that the most consistent factors explaining the level of broadband Internet diffusion are the readiness of the country for the broadband, broadband Internet access costs and the early implementation of local loop unbundling. Other variables that were used in the analysis were statistically less or not at all significant. These variables were the gross domestic product, the population density, share of rural population and the dummy variable for the government ownership of the main telecommunications operators.

In the future when data for a longer period is available also the panel data regression could be used for better estimation of broadband Internet diffusion.

References

Distaso, Walter and Lupi, Paolo (2004): Platform Competition and Broadband Evidence from the European Union. Economics Working Paper Archive at WUSTL, number 0403005.

García-Murillo, M. and Gabel, D. (2003): International broadband deployment: The impact of unbundling. Paper presented at the 31st Telecommunications Policy Research Conference, September 2003.

Howell, B. (2002): Broadband Uptake and Infrastructure Regulation: Evidence from the OECD Countries. ISCR Working Paper BH02/01.

Ikeda, N. (2003): The Unbundling of Network Elements. RIETI Discussion paper series, 23.

Ismail, S. and Wu, I. (2003): Broadband Internet Access in OECD Countries: A Comparative Analysis. A staff report of the Strategic Planning and Policy Analysis and International Bureau.

Kim, J. H., Bauer, J. M. and Wildman S. S. (2003): Policy lessons from comparative statistical analysis. WWW-document, http://intel.si.umich.edu/tprc/papers/2003/203/Kim-Bauer-Wildman.pdf.

OECD (2001): The Development of Broadband Access in OECD Countries. Working Party on Telecommunication and Information Services Policies, report 2001-2.

OECD (2003): Developments in local loop unbundling. Working Party on Telecommunication and Information Services Policies, report 2002-5.

OECD (2004a): Benchmarking Broadband Prices in the OECD. Working Party on Telecommunications and Information Services Policies Report DSTI(2003)/8.

OECD (2004b): The Development of Broadband Access in Rural and Remote Areas. Working Party on Telecommunications and Information Services Policies Report DSTI(2003)/7.

CHAPTER 6

WHAT COMES AFTER "NEW-TO-THE-WORLD" PRODUCT SUCCESS FOR A SMALL FIRM? UTILIZE MOT ANALYSIS AND IMPLEMENTATION FOR INNOVATIVE PRODUCTS AND COMPETITIVE LEAD

Myra Urness

Department of Radiology, University of Minnesota, Minnesota, USA

A Management of Technology (MOT) strategy analysis should be used to position this firm to remain competitive in the medical device industry and to maximize the probability for sustaining a competitive advantage. This analysis was built from MOT knowledge and experience in R&D where 'new to the world' medical device products were consistently being developed. These medical devices are now the preferred treatment by patients and the medical specialties of Interventional – Radiology, Cardiology and Neurology. This MOT strategy analysis includes: continuous scanning of landscapes, SWOT analysis, scenario analysis, understanding patient and physician needs, and balance of the firm's capabilities (current and future) for the best possible outcome. This analysis is not a one time effort but a continuous or quarterly review so that the firm remains competitive and positioned to maximize opportunities. The outcome for this firm is: 1. Understand methods of MOT strategy analysis to build innovative products and competitive lead. 2. Four key points to implement immediately.

1. Introduction

The firm can benefit from the specialty of Management of Technology (MOT) for corporate road mapping and sustained and planned growth where a company is driven by new product development. This is an effort to help guide the firm in resuming development of 'new to the world' products and devices for interventional (minimally invasive), medical treatment for patients.

There is no one tool or path for the firm to sustain a competitive advantage. A MOT strategy analysis framework can be used for corporate road mapping to maximize the probability for a sustainable competitive advantage. The firm must practice due diligence to:

- Continuously <u>scan landscapes</u> and their relationships globally and locally (external/internal) looking at the past/present/future (Van Wyk, 1997)
- Continuous update both internal and external <u>SWOT analysis</u> (Kotelnikov)
- Analyze and discuss <u>scenario planning</u>
- <u>Listen, analyze, respond to physicians/patients for best outcome</u>
- <u>Balance the above with the firm's current and future capabilities</u>.

Recommendations are the understanding of strategy tools used for positioning the firm to remain a competitive leader and four key points for the Firm to implement immediately.

Proposed outcomes of MOT analysis process enables:

- Strategic planning process – sheds light on new opportunities or problem areas within and outside an organization.
- Corporate decisions and direction – for partnership/alliances, acquisitions, additions to core competencies, and intellectual property/knowledge for business spin offs
- Project and/or monitor business strategy – timing and positioning for targeting new product platforms measured with company's or industry's capabilities and discernment for acquiring new capabilities and reinforce/integrate strengths and opportunities and improve weaknesses
- Insight for R&D strategy – to discern prioritization and product or treatment mix in interventional treatment – new market niches, portfolio balance and novel approaches for improving patient care
- Networking/knowledge sharing – of technical/medical and other business communities

- Discern impact of decisions – from scenario planning and reasonable outcomes/outgrowths
- Rational to reach agreement – for team work mission and vision and to empower/re-energize individuals to use their gifts and talent

2. Background – The Firm and Interventional Treatment

The firm's foundation has been built on the innovative preclinical R&D. Our university's laboratory has been a key service provider for the firm's R&D product development.

The firm is at a pivotal point in its strategy and growth. It is a small to medium sized medical device company. The firm is number one in the "new to the world" market niche of implantable devices for treating heart conditions in children and adults. This medical treatment is performed by trained Interventional Cardiologists. Examples are: heart catherization and placement of implanted devices like stents or occluders placed by using radiology equipment and catheter delivery systems.

The firm's current product line is fast replacing traditional surgical methods. These devices used interventionally have less surgical risk and quicker recovery and healing time (one to three days for interventional versus eight weeks to two years for surgery), thereby, reducing length of illness and reducing the cost of medical treatment.

What makes the firm's device preferred over others? Early to mid market entrance; device material is considered bio-compatible; catheter delivery systems can easily be navigated to the target treatment site; devices can be retrieved and repositioned before final placement; easy to deploy for a good fit (reduces operator error).

A Research & Development division within the firm needs to be more formally established and expanded so that the firm does not loose core competencies and their current competitive advantage.

3. MOT Strategy and Analysis Framework – Examples

The firm must decide where to spend their effort/energy and finances for sustaining a strategic competitive advantage (SCA) involving Strategy,

Technology and Innovation (STI). This involves: <u>Scanning landscapes</u>, <u>SWOT analysis</u>, <u>Scenario planning</u>, <u>Listening and responding for a patient's best outcome</u> and <u>Balancing and growing current and future capabilities of the firm</u> (Van Wyk, 1997; Kotelnikov).

R. J. Van Wyk's seven landscapes are used to see past, current and future trends that affect a firm or industry. The technology landscape alone is growing exponentially, often in matrix combinations. MOT analysis of scanning these landscapes can help a firm to navigate and drive a medical device company. Technology is in any field or effort for understanding and using inanimate objects and discovery to improve the human condition. The Technology landscape impacts our lives through paradigm shifts or convergence of other landscapes influences or combining fields within the technology landscape such as chemistry, material sciences, basic sciences, engineering, medicine, and medical specialties to name a few. Technology examples of everyday use can be transformed to medical use, for example: bridges or tunnels for improving flow (stents for blood vessels); Rotor-Rooter for opening up drainage pipes (opening clogged veins); Blenders/waterpicks/lasers/drugs for breaking up or dissolving blood clots; electrical networking and hubs for communications inside and outside a body. The evolving technology landscape and interventional field of medicine provides opportunities and competition or collaboration for optimizing patient care.

As a sampling of the technology landscape in and outside this industry one can see the growth and forecast paradigm shifts in medical treatments. For example, one can compare the size of companies; their products and comparable products; and the competitions capabilities. One must also look at the interventional industry as a whole and the medical community for diagnosis and treating these heart maladies. Another area to combine in this analysis is the timing and possible impact of other technologies that could replace the firm's current products.

3.1. Example of one area in the technology landscape: The interventional field – Firm's current competitive analysis. Size of current competitors in interventional field (private and public)

The billion dollar companies all have the capability of entering the firm's market. They all have products for interventional treatment throughout the body. They may choose to enter this market niche once the treatment for these heart maladies become more widely used. Two of the billion dollar companies have a somewhat similar product but they have limited use because of their limited technology specifications.

The million dollar companies have products in this market and are currently the firm's direct competition. The firm's success has been built on their proprietary design and the devices have been tested and modified for attributes that put the firm's products at the forefront.

The competition has improved their products over the last several years to become more competitive but they are usually not the preferred (by physician) devices. The competition is also working on design modifications for operator easy to deploy a device and clinical testing internationally for governmental approval.

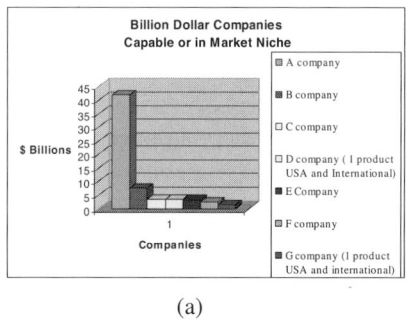
(a)

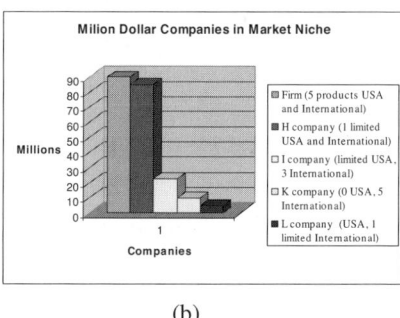
(b)

Figure 1. Size of current competitors in interventional field (private and public).

3.1.1. *Example of product matrix of firm and current competition*

Using the Boston Consulting Groups growth share matrix (between competitions) this graph gives a quick snap shot of the firm's product offerings compared to the competition. The positions are determined by physician preference seen in the medical literature and medical scientific meetings and the positive and negative aspects of each product. The position of the competitions products can change depending on new materials and technologies and development design breakthroughs. <u>From this view the firm looks very good but competitions improvements and the firm's current lack of preclinical R&D could change the future outlook.</u>

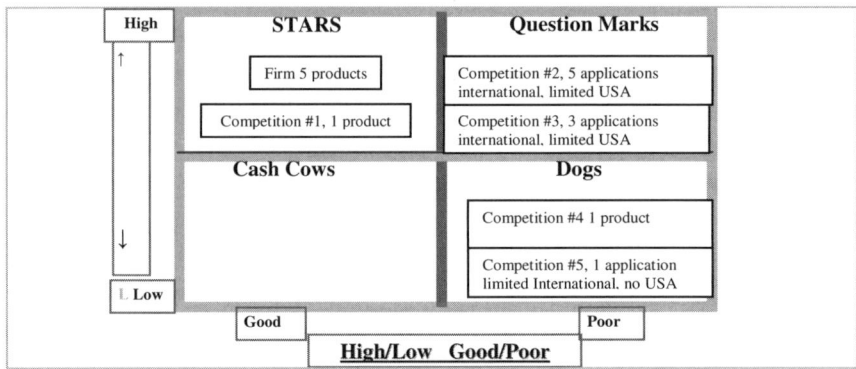

Figure 2. Example of product matrix of firm and current competition.

3.1.2. *Forecast of diagnosis and treatment (gained from scanning/reviewing multiple landscapes)*

Each year more and more children and adults are diagnosed and treated interventionally (as seen in the medical literature, device companies growth, and increased patient diagnosis) with fewer deaths from these medical maladies occurring. This trend is increasing as the Interventional Radiology/Cardiology/Neurology specialties as a whole grows (locally/nationally/and internationally). There is still a need for qualified physicians for treating these cases. Patients may have to travel to medical centers nationally or internationally for treatment. General physician awareness and public awareness from the news media or web based information for patients and their families increases the opportunity for interventional treatment. The firm will have to continue to partner in physician training, public awareness and legislation.

There are still reimbursement difficulties with some of the Health Maintenance Organizations (HMOs) this can be dealt with by legislative awareness and cost comparisons between surgery and interventional procedures and physician education. There are approximately 22 articles in the medical journals that pertain to cost reimbursement comparisons between surgery and this type of interventional treatment. There is not a large enough body of knowledge as yet for an overwhelming change or paradigm shift for this interventional treatment as there are multiple types of this heart condition to analyze the cost benefit analysis.

HMOs are increasingly accepting this new trend on a case by case or by medical center volume of treatment. Diagnosis is increasing where as interventional treatment has a slower upstart curve. Increasingly the trend shows that patients and physicians are switching from surgical treatment to interventional treatment and within the next five years interventional treatment will most likely dramatically outpace surgical treatment in some types of patient cases.

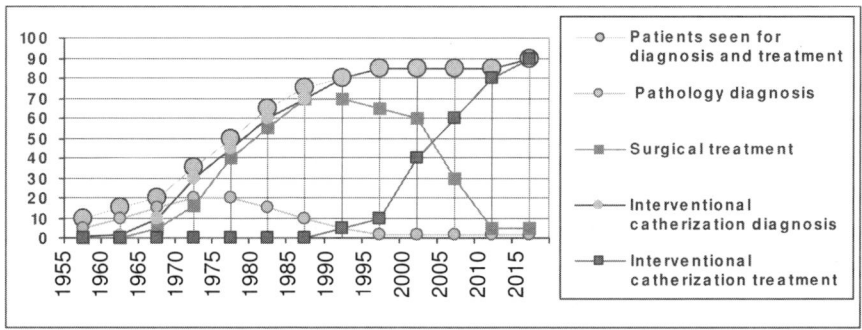

Figure 3. Trends/Projections for diagnosis and treatment.

3.1.3. Forecast of device life cycles for treatment (gained from current industry and university technology capabilities; the following graph is estimated)

The firm's current material designs should be profitable for the next ten years unless there is a severe paradigm shift. This shift could be brought on by competitor's improvement or lower costs; new matrix lattice of absorbable material or devices coated with stem cells or cardiac tissue. These materials are still in the infancy of development (estimating 10-15 years). Gene alteration may be another angle for treating this condition. There are also other advantages that the firm should consider from other landscapes and technologies that could impact their products or services.

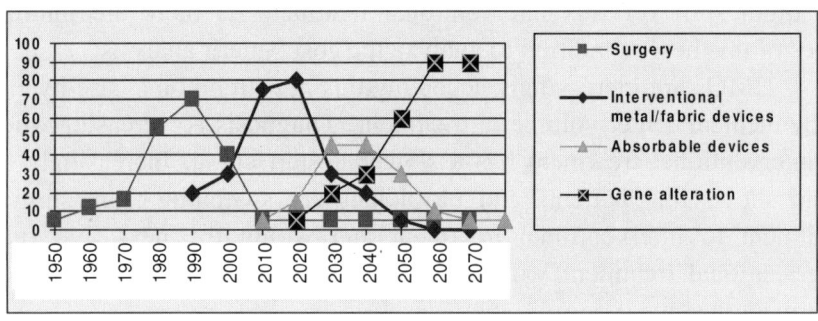

Figure 4. Trends/Forecast of device life cycles for treatment.

3.2. SWOT analysis – Strengths, weaknesses, opportunities, threats

After scanning landscapes one should look at SWOT analysis. This opens one's eyes to the relationships and possibilities. The basic SWOT analysis looks at strengths and weaknesses within a firm and looks at opportunities and threats from outside a firm (Kotelnikov). As a general example the firm's SWOT is as follows:

Strengths
- Device design and platform – preferred
- Material: unique varying flexibility of proprietary properties to the material
- Production ramp up: current focus

- Access to knowledge of "tailor fit" and technique to capitalize on new emerging technologies.

Weaknesses
- Operational: one area – proprietary
- Leadership: pending resolution current mission focus change
- Management of R&D is not formalized in-house
- Example of areas to check out for weaknesses and strengths: current management skills, principles, value chain, capabilities, range of competencies.

Opportunities
- Expansion of product line, new to the world product lines: to balance portfolio mix – short and long term prospects
- Core competencies; expanding customer base; distribution methods; partnerships; leverage knowledge management and database sharing across the Firm.

Threats
- The key threat is the pending shareholder dispute
- Current corporate direction reducing innovation and new to world applications
- Length of time R&D is on hold – current and new competition
- Weaknesses not improved.

To expand this SWOT analysis with the technology landscape, the firm can make a stronger strategic plan that includes and builds on short to long term goals. This format can also be used for scenario analysis.

Table 1. Scheme for SWOT analysis.

Local, National, International	Firm	Competition	Interventional-Landscape/New Technologies on the Horizon
Strengths			
Weaknesses			
Opportunities			
Threats			

3.3. Scenarios

With knowledge of landscapes and SWOT analysis, scenarios can be used for decision analysis based on possible conditions. Scenario analysis is used for: corporate, global and business strategy – mixture of timing and positioning; product strategy for corporate growth; and tools to add value to the R&D process.

Table 2. Scheme for scenario analysis.

Scenarios	Products	Firm	Competition	Firm Focus	Needs	Possible outcomes
Current direction	Ramp up production	Build capital, stable platform extension	Improve products for larger share of market	Implement lean manufacturing and grow bottom line	Establish in-house R&D and strategy analysis	
Remain privately held						
Merge/Acquisition						

Currently the firm is at a legal crossroad and a pivotal position:–the firm is becoming recognized in the field. Other companies are looking for acquisition/merger or possibly public offering.

Along with scanning landscapes, SWOT analysis and scenario planning, the firm needs to continually listen and responding for a patient's best outcome. This is accomplished through physician contact and training, outcome based research analysis, regulatory feedback, medical scientific meetings, and a patient's option for a feedback loop to the physician and the firm. One also needs to balance all of this with the firm's current and future capabilities.

4. Firm's R&D

For the firm's current and future capabilities, one of the key moves is to further develop their R&D division in-house. The firm's foundation has been built on the innovative preclinical R&D. Our university's laboratory has been a key service provider for testing the firm's R&D product development. A Research & Development division within the firm needs to be formally established and expanded so that the firm does not loose core competencies and their current innovative and competitive advantage.

To transfer the R&D portfolio knowledge, a review of the R&D portfolio includes (Cooper, 2001): using a modified Stage-Gate process for discussion of projects, stages of development and communicating the derivative intellectual knowledge gained. A matrix or cluster of projects is used to discuss how key proprietary material properties are similar and different across product platforms or for other product extensions.

4.1. *Summary of R&D portfolio*

From 1999 to the beginning of 2004 there were 20 independent projects. The following is a general summary of product development.

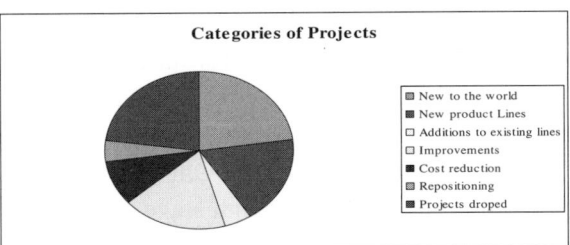

Figure 5. Summary of product development.

Before 1999 preclinical R&D projects were mostly new to the world projects of devices and new treatments that had never been used or commercially sold. Those projects were part of the ongoing race internationally for materials and design changes and testing to produce the first commercial products for treating these maladies interventionally.

4.1.1. *Modified stage-gate*

The R&D portfolio can be summarized using Robert G Cooper's Stage-Gate Process (Cooper, 2001). Cooper's Stage-Gate process is used by many companies for manufacturing and product ramp up. With some modifications, this Stage-Gate process can be used to summarize and monitor the firm's research. The Modified Stage-Gate Process for research will include: Stages of Discovery, feasibility development and a new ¾ cycle for optimization, GLP Studies (Good Laboratory Practices) preclinical and clinical for documentation to the FDA. The gates between stages are used for prioritization and decision points (they generally deal with quality issues and action plans); a key leader from regulatory, manufacturing and marketing are involved as needed for efficient and quality transfer from R&D to final sales. Product post launch review is brought back to R&D from medical scientific meetings, physicians, patients, and the firm's internal needs or changes.

Modifications are defined in stages and gates that fit for a specific firm but are understood throughout the firm. One modification is the ¾ cycle between stages three and four which is the development-testing-validation-optimizing cycle.

A modified Stage-Gate process (appropriately used) can streamline and improve the overall R&D effort by reducing bottlenecks and improve transitions and communications between R&D, regulatory, finance, marketing and manufacturing. Once a product/procedure is invented and proven preclinically (through stage ¾ development cycle) a product lifecycle cost or preliminary return on investment (ROI) could be estimated.

The R&D results can be presented and discussed by project tables; stages and gates – for more detail on device process; by portfolio matrix mix to see quickly the makeup and spread of commercial products and R&D projects; and project/platform clusters to discuss how some projects are related.

4.1.2. Example of project table: 20 projects

Table 3.

Project Name	Stage-Gate	Outcome/Status of Devices	Priority Recommendation 0-5+
A	¾ cycle	Good/fine tune	Continue 5+
B	3 drop	Fair-poor/needs modifications	Discontinue at this time 0
C	1,2 gate 3	Poor/modify design and plan	Continue 5+
D	¾ cycle	Good/begin test for final design	Continue 5+
E	4,5 discuss	Ok/not as good as in project T	Discuss …
F	3 drop	Fair-poor/major modify	Discontinue at this time 0
G	3 drop	Fair-poor/modify	Discontinue at this time 0
H	Post launch	Excellent/clinical proof	5+ nugget of knowledge
I	¾ cycle	Fair/modify, has possibilities	Continue 3+
J	Post launch	Ok/not as good as in project A	Discuss...
K	¾ cycle	Fair/modify, looks promising	Continue 3+
L	Post launch	Ok/not as good as in project M	Modify for next gen. 5+
M	¾ cycle	On-going/design modifications	Continue 5+
N	1	Needs some modification	Continue 5+
O	3 drop	Fair-poor/modify	Discontinue at this time 0
P	3 drop	Fair/sizing difficult	Discontinue at this time 0
Q	1	Proposed/check prototype	Continue 3+
R	3	Feasibility good/modify device	Continue 3+
S	3 drop	Fair-poor/modify	Discontinue at this time 0
T	¾ cycle	Ok, technique needs work	Continue 5+

4.1.3. Example of stages and gates

- Stage 3 – Projects: B, F, G, O, P, S; development dropped at this time – not a high priority
- Stage ¾ cycle – Projects: A, I, J, K, R, T; continue, looks promising, some modifications needed
- Stage ¾ cycle – Projects: D, L, M; development close to final, prototype needs slight modifications, <u>ROI can be estimated here</u>
- Stage 4 gate 5 – Project E; Close to final, discuss trade offs (ease of manufacturing versus optimization) before launch this also effects product profiles of projects J, L, M, N, P, and T
- Post Launch Review: Projects: J, H, L; Products J and L could be optimized if too similar to competition, project H is important nugget of knowledge.

4.1.4. *Example of portfolio matrix products and projects*

Using the Boston Consulting Groups growth share matrix (between competitions) this graph can be modified to give a quick snap shot of the firm's product offerings and current status of R&D portfolio. The position of products and projects can change depending on the market, competition, new materials and technologies and development design breakthroughs. From this snap shot view the firm looks very good but the competitions improvements and the firm's current lack of preclinical R&D could change this present outlook. (Notice that P and p are the same project. P was a failure in testing conducted but it closely relates to E's success or manufacturing ease compromise.)

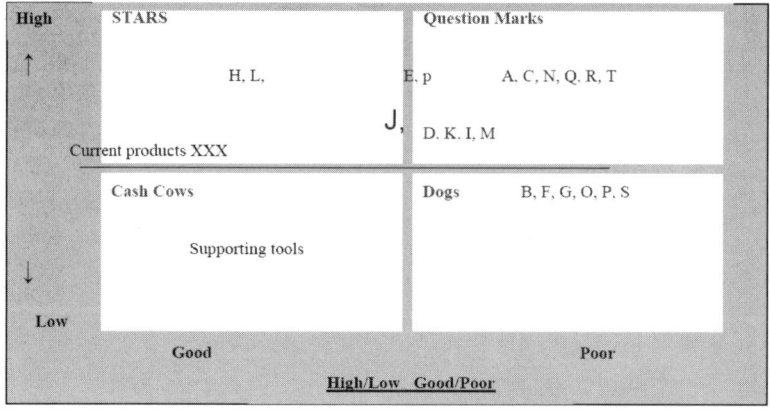

Figure 6. Matrix of products and projects.

4.1.5. *Example of projects in related clusters*

Overall device design strengths and weaknesses can be discussed across different project applications. Projects can also be discussed by material similarities or their application to specific areas of the body or organ.

Derivative intellectual property that is gained from discovery and learning in one project can be applied and communicated to another project or across multiple projects. Also new discoveries or applications

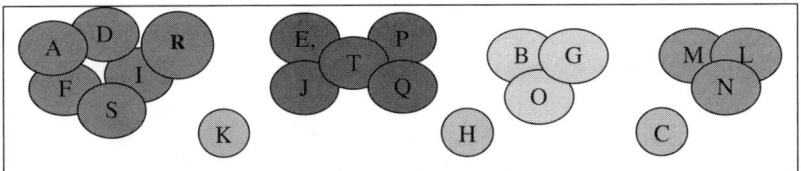

Figure 7. Projects grouped in related clusters.

in techniques and materials and changes in other business fields can inspire product improvements and new treatments.

- Project clusters: A, D, I, and R look promising and should be a prioritized in this order. Projects F and S should be discontinued at this time. If the engineering of device F and S can be improved it could be used neurologically in relation to project A.
- Project clusters: E, P, J, T, and Q are related devices. Project Q should be discussed for rational and prioritization.
- Project clusters: B, G, and O are related to other organ treatments. These projects should be discontinued, not a priority at this time.
- Project clusters: L, M, and N are related to a type of treatment for applications in humans and animals. Continue to fine tune design.

4.1.6. *Recommendations and estimated timeline of projects to continue*

These preclinical R&D portfolio recommendations are based on technical knowledge and the MOT Strategic Analysis Framework for innovative and sustainable competitive advantage for the Firm.

Four key moves to implement right now would be to:

1. Review with the firm their preclinical portfolio and make recommendations for projects short term - long term.
2. Implement a modified Stage-Gate process for R&D and bring in representatives from manufacturing and marketing and regulatory at defined points as needed so that ramp up for the "S" curve of manufacturing to life of product is smooth and timely.
3. Implement software for relational database to capture data and intrinsic knowledge gained.

Table 4. Timeline of projects to continue.

Projects	1999	2000	2001	2002	2003*	2005	2006	2007	2008
A	3/1/99								
B	3/1/99			7/31/2002					
C									
D	3/1/1999				12/30/2003				
E				4/22/2002					
F	11/1/99				12/30/03				
G	3/1/1999				12/31/2003				
H	3/1/99								
I				4/22/02					
J									
K		12/1/2000							
L	7/1/99								
M									
N					11/1/03				
O	3/1/1999				12/30/2003				
P			4/1/2001						
Q					11/1/2003				
R	2/1/1999				9/30/2003				
S	11/1/1999			11/30/2002					
T	3/1/1999								

* R&D on hold 2003-2005.

4. Assess knowledge and ownership of R&D division for in-house core competency and the use of the MOT strategic analysis.

5. Conclusions and Outlook

A MOT analysis is valuable to the firm's innovation and completive lead. A hidden bottleneck for many companies is in management of innovation where technology leadership is not leveraged for net future profit. Knowledge gained from MOT needs to be integrated throughout the corporate structure and shared across departments or divisions so that intrinsic knowledge can be integrated in departments with a focus on best possible product and service for customer needs and satisfaction (Edge, 1995; Davis, 2002).

The firm must practice due diligence to scan landscapes; continually update SWOT analysis and scenario planning; listen, analyze, respond for patient's best outcome; and balance the above with the firm's current and future capabilities. If a firm continuously engages in this Management of Technology (MOT) analysis, it will be better able to: deal with inflection points; redeploy assets; and reinvent when necessary. Essential to the firm's success is to make tough decisions, empower employees, and clearly communicate strategy to the tactical level. The firm can benefit from the specialty of Management of Technology (MOT) for corporate road mapping where a company is driven by new product development. This road mapping will help the firm leverage and improve strengths/opportunities and weaknesses while minimizing threats.

References

Cooper, R. G. *Winning at New Products– Accelerating the process from idea to launch*, Third Edition Perseus Books Group, New York. (2001).

Davis, Knowledge management: the four pillars of success. Univ. of MN General Business File ASAP, Vol. 15, No. 7, pp. 44-46. (July 2002).

Edge, G. *Thinking about the technology future*, R&D Management, Vol. 25, No. 2, pp. 117-128. (1995).

Kotelnikov, *Strategy Formulation SWOT analysis*, www.1000ventures.com/business_guide/crosscuttings/swot_analysis.html

University of MN Ovid Medline database search 1996 to 2004, http://www.biomed.lib.umn.edu/articles

Van Wyk, R. J. *Strategic technology scanning*, Technological Forecasting and Social Change, Vol. 55, pp. 21-38 (1997).

http://www.ama-assn.org/
www.quickmba.com/strategy/matrix/bcg/

CHAPTER 7

DEMAND FORECAST AND ADOPTION FACTORS FOR PORTABLE INTERNET SERVICE

Moon-Koo Kim and Jong-Hyun Park

Electronics and Telecommunications Research Institute, Korea

Amid the momentum of digital convergence taking place in Korea, both between fixed and mobile networks and at the industry level, the portable Internet is quickly emerging as a leading next-generation communications service. To increase its odds for success, the portable Internet service must meet potential users' needs and deliver the kind of benefits they expect, and correspond to related preference factors and patterns of usage. In this paper, we forecast the demand for the portable Internet service, based on the results of a consumer survey, and conduct an in-depth investigation of adoption factors, to provide strategic suggestions for improving market prospects for this service.

1. Introduction

The key paradigms leading the industry evolution in Korea's communications service market are 'broadband' and 'multimedia'. The convergence between wired and wireless services and between different industry segments is giving a whole new impetus to this trend. Our perception of the communications service industry has also been transformed. No longer considered simply as one among other sectors making up the national industry, the communications service market is viewed as a segment which can beef up the competitiveness of the value chain in the overall industry, and is able to generate a tremendous economic ripple effect. It is touted as one of the mainstay industries which will help Korea advance toward its goal of US$30,000 in GNI per capita. Portable Internet service, baptized in Korea as WiBro, is included among the 8 new core service categories designated by a project named "u-IT 8-3-9 Strategy," carried out by the Korean Ministry of Information and Communication (2005). In addition to the sizeable ripple effect to be expected, the portable Internet service is highly valued for its business potential, technological impact and global relevance.

Innovative new communications services have been hitting the market, in fact, since the mid-1990's, thanks to a series of big breakthroughs in communications technology. Not all of these services proved to be viable. Only a handful of corresponded to actual user needs, and had a profitability structure able to guarantee survival amid existing services. Market failures were ascribable, on the one hand, to the way these services were conceived: heavily technology-oriented, virtually heedless of all concerns relating to users, in terms of benefits or technology acceptance. On the other hand, to untimely market entry that compromised or doomed their chance for competitiveness (Ahn *et al.*, 2002). The lesson from these past failures is unambiguous: for the successful market launch of the portable Internet, service must closely correspond to potential needs among users and benefits expected by them.

2. Portable Internet Service and Demand Forecast

2.1. *Service deployment*

Portable Internet service (WiBro: Wireless + Broadband) is the Korean brand of next-generation Internet, known outside South Korea as mobile WiMAX. It is a mobile broadband Internet that provides a virtually ubiquitous high-speed access to the Internet and optimal support for multimedia content. The portable Internet service offers an unprecedented level of mobility. Giving its users an upload of up to 1Mbps and download of up to 3Mbps, the service guarantees a stable connection in urban areas, even during movement at the speed of 120kmh (Korean Ministry of Information and Communication, 2004). After the final decision on its technical standard was reached by the Telecommunications Technology Association (TTA) in 2004, preparations for the commercial rollout of the service officially kicked off. Related technologies, equipment and devices were developed by a R&D team made up of researchers from Electronics and Telecommunications Research Institute (ETRI) and Samsung Electronics. The portable Internet service was officially approved, in December of 2005, as mobile WiMAX and compliant with IEEE 802.16e. The service was commercially rolled out in June of 2006, by KT and SK Telecom in Korea.

2.1.1. *Survey*

The survey carried out by the Network Economy Research Team of ETRI, to collect data needed for forecasting demand for the portable Internet service and analyze user intention to adopt this service was set up as follows:

The sample group of respondents for the survey was chosen among people aged 15 to 60, residing in Seoul and other major cities of Korea, including the six metropolitan cities and Jeju Island. The selection was made through a quota sampling by gender, age and region. A total of 1,200 respondents were surveyed by trained staff from a leading Korean research firm over a period of approximately one month between March and April 2004. Personal interviews were conducted face-to-face with respondents using a structured questionnaire.

2.1.2. *Demand forecast*

To predict the demand on the portable Internet service, the survey measured the intention of subscription (I) using a five-point scale to rate the purchase intention of respondents:

                    0: No, I will not subscribe.
                    1: I probably will not subscribe.
                    2: I feel divided about it.
                    3: I will subscribe.
                    4: I will definitely subscribe.

The results of the survey broke down to 17.1% positive answers (I=3, 4), 15.0% neutral answers (I=2), and 68.0% negative responses (I=0, 1). These results were then entered into the Morrison model (1979). We obtained a subscription intention for the portable Internet service of 31.1%, which corresponds to a maximum number of subscribers of 9,244,000 at the saturation time, accounting for 19.2% of the national population in Korea.

To estimate yearly subscription numbers of the portable Internet service, we used the Bass model of new product/service diffusion, as shown below:

$$Y_t - Y_{t-1} = (p + q\frac{Y_{t-1}}{N})(N - Y_{t-1}) \qquad (1)$$

In Equation (1), Y_t corresponds to the number of subscribers at the point in time t, and N is the maximum number of subscribers, which may be understood as the potential market size. 'p' is the coefficient corresponding to the diffusion accounted for by the effect of innovation, and 'q', to that by the effect of imitation. In our analysis, we set the maximum number of subscribers at 9,244,000, the estimate obtained above. For the innovation coefficient (p) and imitation coefficient (q), we took into account diffusion patterns among similar services and the results of existing Delphi surveys. The resulting estimates of demand by year are provided in Figure 1.

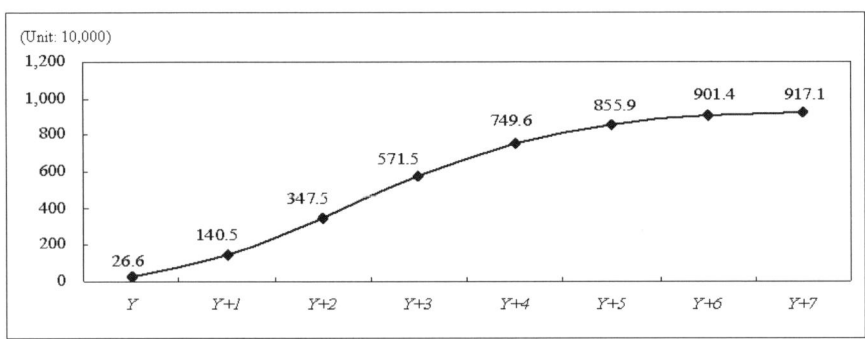

Figure 1. Yearly forecast on the portable internet demand.

3. Adoption Factors

3.1. *Reasons for subscribing and not doing*

The single-most important reason (53.5%) why potential users would subscribe to the portable Internet service was, as shown in Figure 2, was outdoor access to the Internet, followed by mobility (14.5%) and faster transmission speed (13.5%). These three attributes outweighed other advantages of the portable Internet service, such as expandability of the device, wealth of information and multimedia contents. As for reasons for not subscribing to the portable Internet service, as provided in Figure 3, 'insufficient need' topped the list, selected by as many as 68.2% of respondents. This was followed by high service charges (14.0%) and financial burdens from having to purchase a new device (8.8%). These results suggest that an increase of the diffusion rate will require providers to develop service products that address existing needs among users. Further, this has to be done in such a manner that users perceive portable Internet as a necessity, similar to the way how consumers perceive broadband Internet and mobile internet. Furthermore, they must reduce the barriers of adoption having to do with costs by bundling portable Internet-based services with existing communications services, by introducing diversified rate schedules and discount programs, and perhaps also by participating toward the purchase cost of devices.

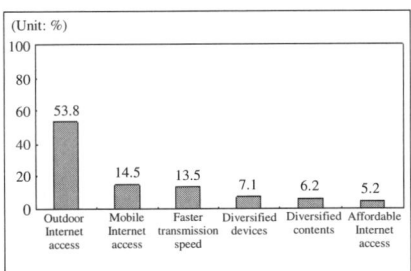
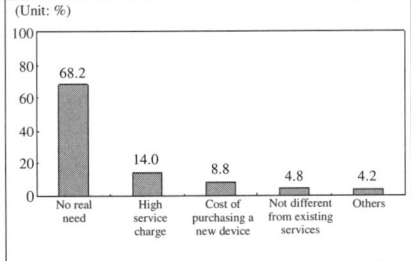

Figure 2. Reasons for subscribing to the portable internet service.

Figure 3. Reasons for not subscribing to the portable internet service.

3.2. Primary uses and user preferences

As illustrated in Figure 4, the list of main uses of the portable Internet service was topped by 'work, study and business' (29.4%), and nearly as many respondents chose the answer 'improving daily convenience' (29.3%). 22.2% of respondents chose 'communication', and 19.1%, 'entertainment'. All in all, no one use of the portable Internet service prevailed over the rest, and responses were evenly distributed. This means that the portable Internet must not offer services that are overly communication-centered, and should try to maintain a balance between the Internet and multimedia content offerings so as to meet the broad range of uses and benefits expected by users.

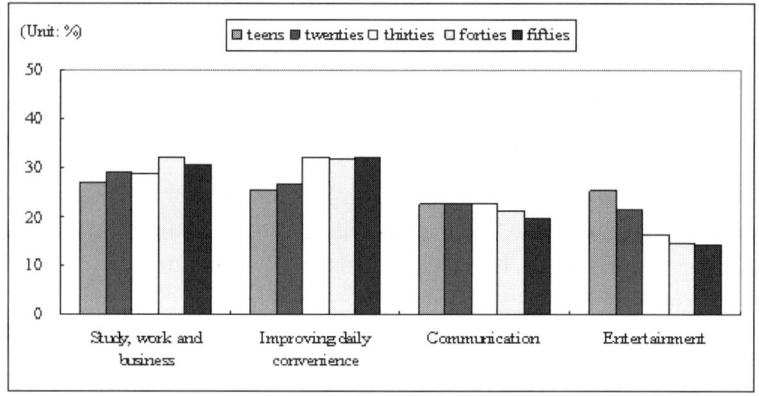

Figure 4. Primary uses of the portable internet service.

The most-preferred reception device, as indicates Figure 5, was PDA (Personal Digital Assistant), selected by 36.4% of respondents. Smart phones and notebook computers accounted for 23.9% and 22.6%, respectively. Finally, 16.2% chose handheld PCs as their preferred device. PDA was the favorite device across all ages, enjoying roughly the same level of popularity. Men exhibited stronger preference for handheld PCs and notebook computers than women, whereas smart phones were more favored by teens than by other ages. Finally, notebook computers were most popular among respondents aged 20 to 40. This result speaks positively of the potential for PDAs, a device of proven competitiveness combining wireless LAN and mobile telephony, to emerge as the killer device for portable Internet service, provided a price decline is accompanied by functional improvements. Concerning the pricing of the portable Internet service, provided in Figure 6, 66.3% favored flat rate systems. Usage-based pricing plans based on hours of usage or frequency of access drew relatively low percentages of positive responses, amounting to 12.5% and 10.6% respectively.

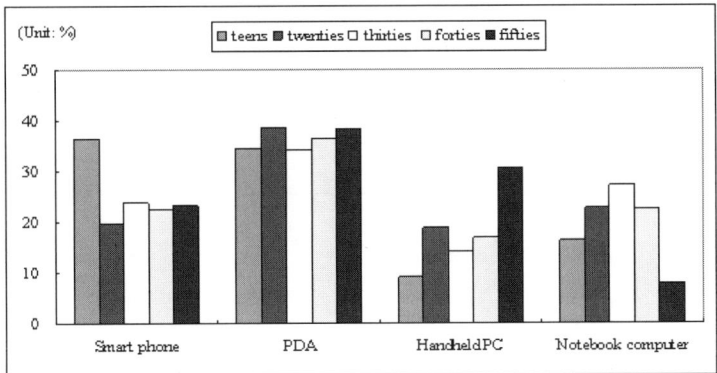

Figure 5. Preferred devices among the potential portable internet users.

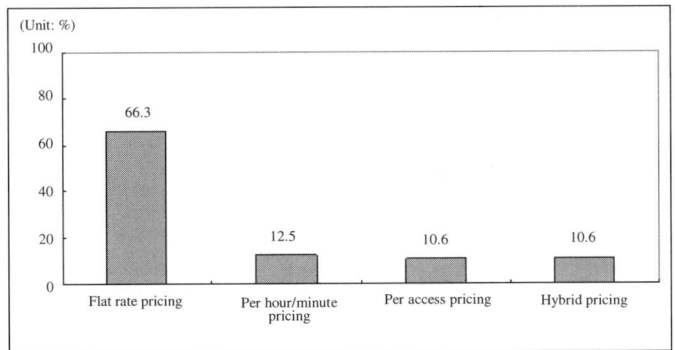

Figure 6. User preference in the portable internet pricing system.

The distribution of maximum monthly prices, which users were willing to pay (WTP) for the portable Internet service, is shown in Figure 7. The monthly mean stood in the range of 29,000 KRW (1 USD=900

KRW). The maximum WTP tended to be higher among early adopters, and late adopters appeared more sensitive to pricing. Although a substantial number of subjects in this survey responded positively to flat rate pricing and the monthly WTP range of 30,000 KRW, this pricing structure and WTP range, in actual situations, may nevertheless turn out to be burdensome for some users. Accordingly, service providers must also examine pricing structures such as those consisting of a basic fee and surcharges for additional use, or hybrid ones combining a flat rate system with usage-based pricing. They should consider offering different pricing structures adapted to different ages, professions and usage patterns. Judicious pricing policies at initial stages of service launch may be critical for early diffusion of the service.

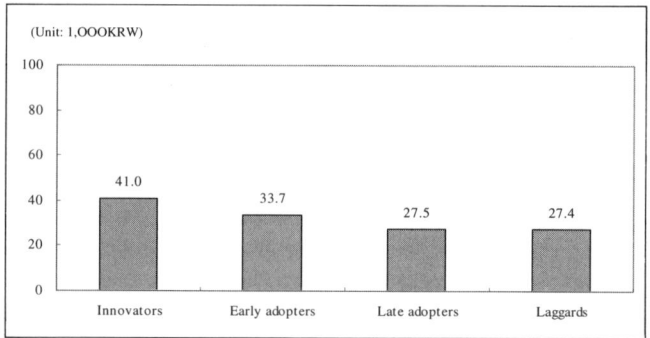

Figure 7. Monthly willingness to pay for the portable internet service.

3.3. *Likely adopters and promising services*

3.3.1. *Likely adopters*

Based on the results of the survey, the portable Internet users were divided into different user groups. There are two most likely adopter groups of the portable Internet service, as given in Table 1. Firstly, those in their late 20s and early 30s; a group characterized by interest in conducting business online. Secondly, students in their late teens and early 20s; a group characterized by interest in multimedia contents.

This demographic segment with interest in conducting business online, generally accesses the Internet using PDAs and notebook computers. They search for information, shop online, use e-mail, conduct banking, or trade stock, and are likely to constitute a core user pool for the portable Internet service during the introductory stage. The other group characterized by their interest in multimedia contents, who connects to the Internet through PDAs and smart phones to search information, use messenger, MMS (Multimedia Messaging Service) and VOD (Video On Demand), and play online games. This group, noted for extensive and consistent use of basic features of the Internet, including e-mail and information search, is expected to account for a large segment of the portable Internet user pool during the growth and post-growth stages.

Table 1. Characteristics of likely portable internet adopters.

Category	Group with Interest in Conducting Business Online	Group with Interest in Multimedia Contents
Demographic characteristics	Office/clerical workers, professionals in their late 20s to early 30s	High school or college students in their late teens to early 20s
Life-style	Practical benefits	Self-centered and trend-responsive
Main usage stages	Introductory and growth stages	Growth and maturity stages
Preferred devices	PDA, notebook computer (functionality-oriented)	PDA, smart phone (design-oriented)
Preferred types of services	Internet access, information search, e-commerce, banking and financial transactions	Internet access, e-mail, information search, messenger, MMS, VOD and games
Preferred pricing plan	Flat rate pricing	hybrid pricing (flat+usage based)
Preferred environments	Outdoors, indoors	Outdoors and in moving situations
Determinants of adoption	Service rate, device purchase price, selection of contents	Device purchase price, service rate and adoption by peers

3.3.2. Killer services and desired services for bundling

Figure 8 illustrates the user usage intention corresponding to different the portable Internet-based services. The results suggest that killer services are likely to be data search, e-mail, messenger, news and information access and games. The portable Internet-based services (wireless) that the largest numbers of respondents expressed an intention to use were virtually identical to those services most used within the high-speed wired Internet environment. This tendency was more marked among the respondents in their late 20s. Subjects in their teens and early 20s showed a stronger intention than other ages to use multimedia services such as MMS, VOD, P2P (Peer to Peer), LBS (Location Based Service), and broadcasting.

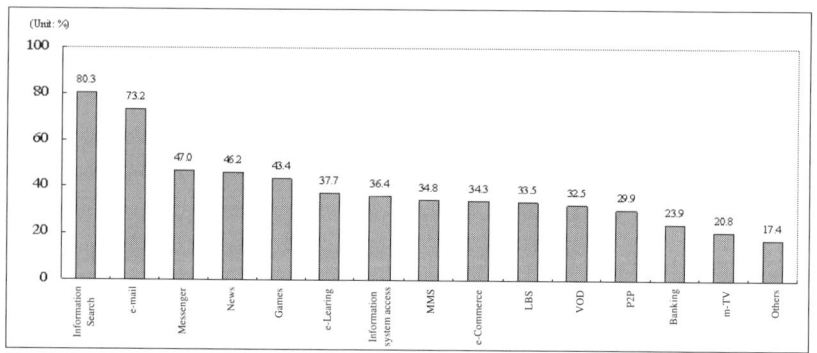

Figure 8. User usage intentions among the portable internet-based services.

Figure 9 gives user choices for services to be bundled with the portable Internet service. Telematics appeared as the prevailing choice, drawing 40.3% of responses, followed by terrestrial DMB (34.3%) and satellite DMB (26.5%). Telematics was a choice particularly popular among respondents in their 30s and older, and by occupation, among professionals, self-employed, office workers and sales professionals. DMB (Digital Multimedia Broadcasting) services were favored among those younger than 30, and by occupation, among students and professionals. These results pointed toward the potential of the portable

Internet as a convergence service platform, offering service bundles combining services such as broadcasting and telematics.

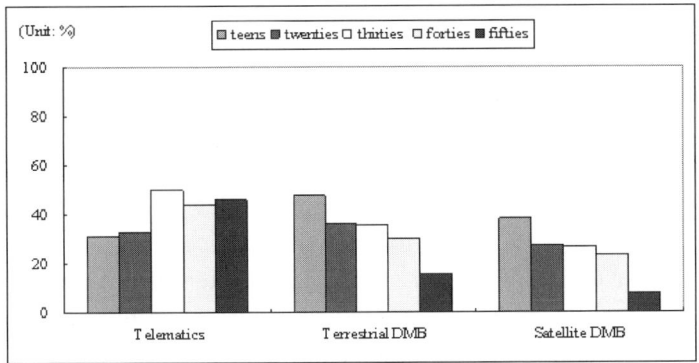

Figure 9. Desired services for bundling with the portable internet service.

4. Conclusions and Implications

The following is a list of observations produced based on our survey of the market prospects of the portable Internet service and adoption factors among potential users, as well as analysis of the related results. Various points relating to market activation strategies for the deployment of the portable Internet service are considered:

First, our estimation of the subscriber pool and market size in Korea following the commercial rollout of the portable internet service, and analysis of needs among potential users, indicate that this service possesses sufficient market potential as an independent service segment. Rather than a service supporting other communications services, the portable Internet service - as a high-speed mobile Internet service accessed in outdoor environments and from public transportation systems - appeared capable of forming an independent service sector and growing as a stand alone service segment.

Second, the portable Internet service appeared to be able to position itself within the telecom service market as a distinctive service segment. Spatial freedom, mobility, and high transmission speed - which are the primary attributes of the portable Internet service according to the perception of potential users - are a set of features that none of high-speed wired Internet, wireless LAN-based Internet, or mobile Internet can fully offer. Accordingly, the service will be viewed as an upgrade to high-speed wired Internet with the added advantages of freedom from spatial restrictions and mobility, and thus emerge as a wireless high-speed Internet for heavy traffic users.

Third, the killer services in the portable Internet will be the wireless versions of the traditional services from the fixed-line broadband Internet environment, and multimedia contents. These services, furthermore, are expected to reflect characteristics of different devices. The results of our survey revealed that the great majority of services that users expressed their intention to adopt within the portable Internet environment were almost identical to those most frequently used within the fixed-line broadband Internet. Multimedia services also appeared as a promising service category. And as for devices, while no single device prevailed over others, multifunctional devices such as PDAs, and information devices such as handheld PCs and notebook computers, were among the popular choices. Compact devices such as smart phones were also favored by some respondents.

Finally, concerning killer applications, there is a strong possibility that, in addition to Internet access and multimedia services mentioned earlier, the portable Internet service may come as service bundles in the forms of communications/broadcasting convergence services and services combining telematics. A competitive business model for the portable Internet service providers would be one that operates through an IP-based open system, making the most of the high-speed Internet platform and contents, and which combines Internet broadcasting, on-demand services and multimedia services. Moreover, by integrating terrestrial and satellite DMB, service providers must seek to appreciate synergy effect produced from the marriage of the portable Internet and broadcasting. They must also look to expand connectivity, for instance, with home information appliances and compatibility with fixed and mobile communications systems. Accordingly, they must develop a networking model, supporting telematics, fast emerging as a new growth engine sector, to expand the capability for two-way multimedia transmission.

References

Ahn, J. H., Gweon, J.W., Kim, M. S., and Lee, D. J. (2002). Learning from the Failure: Experiences in the Korean Telecommunications Markets, Korean Journal of Management Science, 27(3),115-133.

Korean Ministry of Information and Communication (2004). Proceeding of Workshop in Policy Directions for WiBro (Portable Internet Service). Korea.

Korean Ministry of Information and Communication (2005). What is the u-IT 8-3-9 Strategy? Korea.

Morrison. D. (1979). Purchase Intentions and Purchase Behavior, Journal of Marketing, 43, 65-74.

CHAPTER 8

THE USE OF SCENARIO ANALYSIS IN STUDYING EMERGING TECHNOLOGIES — CASE GRID COMPUTING

Mika Lankila*, Liisa-Maija Sainio**, and Jukka-Pekka Bergman***
* Deloitte, Finland; ** Department of Business Administration, Lappeenranta University of Technology, Finland; *** Department of Electrical Engineering, Lappeenranta University of Technology, Finland

> The goal of this study was to explore how scenario analysis may be used in evaluating the future business environment of an emerging technology, namely Grid computing, that also contains features of a disruptive technology. Grid computing may change the traditional product-based computing to service-based activities which would have a major impact on the industry. The scenarios were created in an expert workshop in Geneva in December 2003. Four environmental, explorative scenarios in the 2010 timeframe were designed, by firstly identifying the main driving forces and evaluating their uncertainty and impact. The most important driving forces that emerged in voting were the issues of security and value creation, and they formed the dimensions for the scenarios to be created: strong added value vs. publicly supported development, and mistrust concerning the security issues vs. blind trust. The scenarios were named Brave New World, Commercial Dodo, Trust Me, and Dead Duck. After refining the scenarios, they were approved by the expert workshop participants. The scenarios brought up various issues that will have to be solved before the commercialization of Grid services. Because of its flexibility, scenario analysis was regarded as a suitable method for examining the complexity related to the future of Grid computing.

1. Introduction

The importance of technological futures research in today's fast changing global environment is perhaps greater than even before. This is reality especially in the ICT industry, where a great deal of uncertainty and risk surrounds every new promising technology. Although retrospective statistical prediction methods give some information for decision-making, tools breaking today's business logic are also required to examine the possible futures.

The fundamental interest in futures research is instrumental. In this context, futures research may have either a direct effect as a tool in decision-making and planning or it can have an indirect effect and work as catalyst for general discussion (Mannermaa, 1999). Given the uncertainty inherent in the future, expectations placed upon these visions should be realistic. This means that futures research should indicate ranges of future possibilities, not point values (Courtney et al., 1997).

The futures research method studied and used in this paper is called scenario analysis. Scenarios are focused descriptions of fundamentally different futures presented in coherent script-like or narrative fashion (e.g., van der Heijden et al., 2002; Schoemaker, 1995; Schwartz, 1996). Their purpose is to illustrate possible pictures of the emerging future based on the analysis of the key factors and driving forces surrounding the examined issue. With the scenario approach companies can benefit from seeing the possible discontinuities ahead and better understand their operating environment. This should lead to better decisions, to enhanced conversation in the organization, and eventually to better results.

2. The Scenario Process – A Method for Exploring the Dynamic Environment

Creating new knowledge presumes that individuals recognize useful data and information, and are then able to transform it, through some process, into knowledge that brings future value for the organization (Senge, 1990). The fundamental idea behind scenario planning is to provide a structured way to create a dynamic interaction between the environment and the organization to cover a broad range of future possibilities to confront the future uncertainties and expand people's thinking (Ellis and Shpielberg, 2003; Schoemaker, 1993; Wack, 1985a,b; Weick and Quinn, 1999). According to Godet and Roubelat (1996), Schwartz (1996), and Wilson (2000), scenario planning makes it possible to share and reassemble personal knowledge to build a holistic understanding between the internal and external environment of the organization. Scenarios explore the simultaneous impact of various uncertainties by changing multiple variables at a time, and describe very complex models that cannot be formally modelled (Coates, 2000; Schoemaker, 1997). Scenario planning makes it possible to assess the competitive landscape and strategic segments of an organization in a new light and renew organizational capabilities towards the future needs under a created future strategic vision (Godet, 2000; Schoemaker, 1997; Teece *et al.*, 1997).

A typical scenario development process involves organizational leaders and managers, R&D centers, associations, different groups of experts and many other stakeholders in the form of a structured and facilitated network to share knowledge and to create alternative presentations of the future (Roubelat, 2000; Schoemaker, 1995). In terms of Nonaka *et al.* (2000; 2003), knowledge creation is a dynamic synthesizing process through interaction between individuals, and individuals and the environment in a shared context. Scenarios reflect beliefs, expertise, and intuition of individuals and organizations concerning the future (Aligica, 2003; Bood and Postma, 1997; Johannessen *et al.*, 1999; Kulkki and Kosonen, 2001; Scharmer, 2001).

According to Day and Schoemaker (2000), technologies where (1) the knowledge base is expanding, (2) the application to existing markets is undergoing innovation, or (3) new markets are being tapped or created are emerging technologies. In the case of these emerging technologies there are three inherent and distinct challenges that scenarios can confront better than other planning techniques:

Uncertainty. Scenario planning embraces it as the central element in its process.

Complexity. Scenarios explore how a diverse set of forces (from social to economic) dynamically influence each other over time as a complex system.

Paradigm shift. Scenario analysis challenges the prevailing mindset and core assumptions by amplifying weak signals that would otherwise be unnoticed (Schoemaker and Mavaddat, 2000).

2.1. Scenario creation

According to Masini and Vasquez (2003), creating scenarios means carrying out an ongoing cumulative process, establishing a project periodically or when needed. The scenario process has a certain and common structure involving a varying number of steps (e.g., Godet, 2000; Masini and Vasquez, 2003; Schoemaker, 1993; Schwartz, 1996; van der Heijden et al., 2002). Phelps et al. (2001) argue that the scenario process can be conceptualized into four stages (Figure 1):

1. Preparation of a scenario process: background analysis and delimitation of the focus. The analysis of the industry explains the dynamics of the external business environment. In this phase, the shared collective knowledge base, a common language and practicalities for the basis of the scenario process and knowledge creation are provided. During this phase, the scope and the goals around a shared loose vision are created (Godet, 2000; von Krogh et al., 2001).

2. Knowledge base construction. When the backgrounds have been examined, the stakeholders, driving forces, and the required capabilities of future business are recognized, and the working group explores their significance and logical implications (Schwartz, 1996; van der Heijden et al., 2002). In the end of this phase, the disseminated and articulated tacit knowledge within the working group is transformed into explicit knowledge and is presented in the form of initial scenarios.

3. Scenario creation. The scenario process is continued by integrating the existing explicit organizational and collected external knowledge into created new knowledge (Godet and Roubelat, 1996; Nonaka and Toyama, 2003; Schoemaker, 1992). The final scenarios are created through the evaluation and combination of knowledge gathered during the preceding phases on an intuitive, heuristic, or statistic basis (Masini and Vasquez, 2003; Schoemaker, 1997). The goal is to identify the logics of the business environment that are important and relevant to the issue, and then organize the possible outcomes of the scenarios around these logics (Schoemaker, 1993; van der Heijden et al., 2002). This phase provides alternative scenarios of the future related to the issue considered.

4. Implementation of scenarios. The scenario process stimulates strategic thinking and challenges present routines and mental models, and leads to continuous and cumulative knowledge creation and learning process (Bennett, 1998; Masini and Vasquez, 2003; van der Heijden *et al.*, 2002; Wack, 1985b). Finally, the scenarios are also used as a communication tool to diffuse explicit future-oriented knowledge throughout the organization and accelerate the process of organizational learning and innovation development (Bood and Postma, 1997).

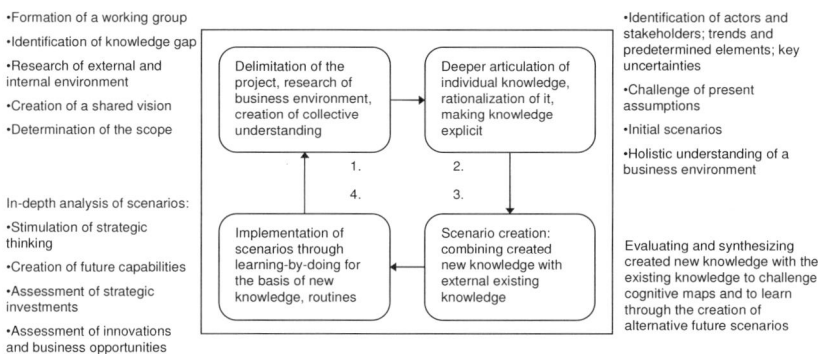

Figure 1. Ongoing scenario process and knowledge creation (Source: Bergman *et al.*, 2004).

O'Connor and Veryzer (2001) state that when exploring and creating new innovative ideas it is essential to have appropriate methods and tools for linking expertise and future opportunities. Innovations seek information about technological and market changes in the future (Watts *et al.*, 1997).

3. GRID Computing – Definition, Benefits and Challenges

A Grid computing system gathers geographically remote resources and makes them accessible to users and applications in order to reduce overhead and accelerate projects. Grid computing enables users to collaborate securely by sharing processing, applications, and data across the systems (Grimshaw *et al.*, 2003). This Grid infrastructure will provide "the ability to dynamically link together resources to support the execution of large-scale, resource-intensive and distributed applications" (Berman *et al.*, 2003). Grids integrate networking, communication, computation, and information to provide a virtual platform for computation and data management. The resource sharing is done via a set of additional protocols and services that are built on Internet protocols and services (Foster *et al.*, 2003).

The benefits of Grid computing range from enhanced collaboration and more rapid computational results and thus increased productivity to more flexible and scalable computing. Examples of today's and tomorrow's uses of Grid computing include applications in bioinformatics, data-set visualization, simulation-based test and analysis and distributed database coordination (Berman *et al.*, 2003). The greatest potential and challenge of Grid computing lies in the possible transformation of the orientation of the computing business from product-based to service-based: in the future, instead of buying more servers, the organization may be able to reserve computing power flexibly from the network. This approach is called utility computing; and if it gains ground in the market, it would be a marketing discontinuity at both the macro and micro level. That is why many large ICT companies, such as IBM, HP and Sun, are active in this development. However, there are still many technical and market obstacles in the way to security, standardization, customer trust and general management of extremely complex systems.

4. GRID Business 2010 - Scenarios

In this study the scenario analysis method was used to study Grid computing and its future implications. The main purpose was to create new knowledge about Grid computing by gathering existing knowledge and exploring the possible paths of development. The time frame of the scenarios was set in the year 2010, because a too long scenario perspective can make scenarios inapplicable (Mannermaa, 1999; Kleiner, 1999).

The workshop activities started with gathering the needed experts to form the scenario team. As a prerequisite, every participant had to have a strong basis in technical issues, because otherwise the limits of the technology and possible development paths would have been too difficult to understand (Levary and Han, 1995). The working group consisted of eight persons who all had very good technical knowledge and up-to-date information on Grid computing and its environment. Most of the participants were Grid computing researchers. In addition, two participants held executive positions in the leading ICT companies.

4.1. *The structure of the project*

The main event of this scenario project was a one-day workshop held in Geneva at CERN (Conseil Européen pour la Recherche Nucléaire = European Organization for Nuclear Research) in the 16th of December, 2003. The participants of the workshop were experts of the Grid technology representing international ICT companies and universities. The main focus of the workshop was to create sketches for the scenarios to be refined later. During the workshop, *the Instant scenario* approach (van der Heijden *et al.*, 2002) was applied to enhance the process of the workshop because of the time restriction. The research project included six main stages including the four phases of the scenario creation, as shown below.

Stage 1 – definition of the scope of the study, agreement on methods and tools.

Stage 2 – an environmental scan based on the literature and workshop brainstorming session and preliminary consultations with Grid and ICT experts: analysis of the major drivers likely to shape the commercial Grid computing over the next six years.

Stage 3 – a facilitated workshop involving the participation of Grid computing experts to develop four draft scenarios.

Stage 4 – identification of the key issues for Grid computing over the next years in each scenario.

Stage 5 – expert interviews to refine the scenarios and to identify the implications of Grid computing.

Stage 6 – preliminary action plan for further research and R&D projects.

It has been noticed that the applied approach is very practicable to get well-defined results within the expert group in a short period of time, especially in an extremely uncertain and rapidly changing business environment. In our study, the workshop was applied to conduct the third phase of the scenario process, namely scenario creations, as shown in Figure 1.

4.2. *The key drivers of the development*

The workshop started with a brainstorming session in order to capture the important elements that will shape the future operating environment of Grid computing (i.e. *drivers*). After this, the identified drivers were clustered into ten groups, which were named after their primary focus. These clusters included, e.g., value creation, security, global economic development, and regulation.

The drivers were voted on and organized by their uncertainty and impact on the future in order to find out which two of these clusters play the most important role in creating scenarios as *key uncertainties*. According to the scenario literature (e.g., Schwartz, 1996; van der Heijden *et al.,* 2002) those forces or factors are the ones with the *highest uncertainty and importance.*

After voting there was a short discussion about the results and the most interesting factors. It was agreed that the most important and the most uncertain issue in order to create commercial success was *security*. Here, security refers to not only technical issues which have to be solved but also trust issues which are more or less social and psychological. Trust is the key point when users (corporate and government IT departments) decide whether to use Grid computing enabled services and solutions. If the users do not have confidence and accept the technology, the commercial solutions are likely to be failures. The polar outcomes or dimensions for the security axis were *blind trust* and *mistrust*.

The other most important and uncertain issue according to the group voting was *value creation*. Services and solutions enabled by Grid computing have to have strong value added for the customers in order to be commercially successful. The polar outcomes for the value creation were *public support* in the other end of the axis and *strong added value* in the other. In the *public support* dimension the governments and public instances, e.g. universities, are driving the development, because the corporations have not achieved clearly more value or gains from the Grid computing to be provided directly to consumers. The public sector is willing to invest in these technologies and services in order to improve research capabilities and build better services to enhance the development of the information society. This creates opportunities as well as the business communities. In the *strong value added* dimension companies have succeeded in getting the real benefits out of these technologies. This strongly drives the diffusion of Grid solutions in the market.

After this phase the four following scenarios were sketched out. Each group were asked to come up with strikingly memorable names (see e.g. Coyle, 2004) which would capture the essence of each scenario. Some of the groups used animal names to describe the scenarios. Scenarios are illustrated below in Figure 2 with the chosen dimensions as the dividing axes.

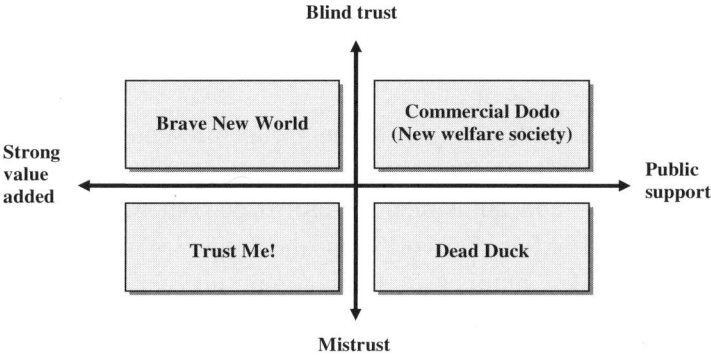

Figure 2. Scenarios for Grid computing.

4.3. *Creating the scenarios*

The written scenarios are *environmental*, concentrating on describing the possible future worlds of the Grid computing market or industry, so they do not take a firm-specific point of view. These scenarios are *explorative*, which means that they are directed towards the future on the basis of certain set of current assumptions. By their logic, the created scenarios are intuitive with some formality (i.e. *heuristic* approach). In this research, the main reasons to choose the intuitive logic approach as the scenario methodology were the time restriction, the ease of use, and the need to exploit expert knowledge.

In turning the list of descriptions and outcomes into scenarios, the teams were told to project themselves to the final year of our scenarios, 2010, and develop the story of how they came to be in the situation they find themselves in. Teams were also looking for some actions or events, *milestones,* which could show the path or direction of the development in the near future (e.g. Schwartz (1996) uses the concept of *leading indicators* or *signposts*).

The development of scenario outlines was undertaken in sub-groups, so the attendees were divided into three groups, with each group having about an hour to work on their storyline. Sub-groups had two or three members each. The scenario called "Dead Duck" was decided to be written later by the researcher based on the notes and ideas from the session because of our time restriction and number of participants (with eight members we got two groups of three and only one group of two instead of four groups of two).

After sketching the storylines, the groups presented their scenarios to each other, with the larger group commenting with a critical attitude. Some questions and adjustments were made, but major discontinuities were not found. After the workshop, the created scenarios were sent to participants and a number of Grid computing experts as an iteration cycle, which is an important element of the scenario analysis. On the basis of the received remarks, the created scenarios were rewritten and the implications were pointed out.

4.4. *Scenario descriptions*

Each original scenario is two pages long and illustrated by a figure. This chapter shortly presents the created scenarios and shows a summary table of each scenario: *preconditions* (how the environment develops in order to enable this scenario), *characteristics* (key features of each scenario), *motivation factors* (why actors want this scenario to happen) and *milestones* (events that show this scenario is emerging) of every scenario. The scenario themes can also be found in Table 5, where the key themes of all the scenarios are summarized.

Scenario 1: Brave New World. The dimensions of this scenario are strong added value and blind trust. Grid computing has definitely got the momentum, and the Inter-grid development is on the horizon. Grid computing is widely considered as one of the most important developments in the history of computing. This scenario is clearly the most positive scenario and describes the world where many of the great visions have realized and Grid computing business is flourishing (Table 1).

Table 1. The key issues in the "Brave New World" scenario.

Preconditions	Motivation
- Actions against cyber terrorism work - Positive economic development - Open source dominates - Strong diffusion of broadband networks - Increasing interest in outsourcing	- Customers want to enhance their productivity and competitiveness - Companies need to concentrate on their core competencies and outsource IT activities - Vendors want to create markets, push new business models and make profits before other vendors (fear of losing the business)
Characteristics	**Milestones**
- The emergence of open, virtual organizations - Benefits and concrete applications of Grid computing can be seen in the early phase of emergence - Effective cooperation both in research and business and between them - Successful standardization process - Profitable business with variety of vendors in the market	- First Extra-grids appear in enterprises - Small Grid computing vendors' stock prices outperform the markets - Building of nationwide high speed networks - The service business of large vendors becomes more important for them (more turnover comes from services)

Scenario 2: Commercial Dodo (New welfare society). The limits of possibility are public support and blind trust. The business environment has not accepted Grid solutions, but governments are pushing the development further. Trust issues have largely been overcome. In this scenario the public sector is a major actor as they support the emergence of the Grid computing environment. Business is created around the modern information society services, which are pursued by the governments and the public sector. Governments invest greatly to create an infrastructure which could deliver digital civic services to citizens. This creates business opportunities to hardware providers and service developers. Authorities provide network identities to users and collect charges for it (Table 2).

Table 2. The key issues in the "Commercial Dodo (New welfare society)" scenario.

Preconditions	Motivation
- Failure of private security - Governments want secure computing - Clear decrease in cyber terrorism - Open source dominates - Digitalization of civic services	- Need to cut costs in healthcare and care of the elderly - To enhance equality between citizens by providing Grid computing to everyone - To get better research results - Need to provide secure digital services for citizens in building the new kind of welfare state
Characteristics	**Milestones**
- Governments are active in the Grid development and invest in these technologies - National Grid strategies with e.g. partly publicly supported broadband networks - The public sector has an important role in delivering data security - Tight price competition between the vendors - Emphasis on basic research (computing and storage-intensive research, current development continues) and civic services - Users have strong trust in Grid-based services	- Governments announce clear information society and/or Grid computing strategies - Launch of first civic Grid services

Scenario 3: Dead Duck. Key drivers in this scenario are public support and mistrust. Grid computing is still relatively small business, and companies offering "traditional" computing services are winning the game. Many USP (Utility Service Providers) outsourcing contracts and deals have been called off (Table 3). "Dead Duck" clearly shows the future where the Grid business takeoff has failed.

Table 3. The key issues in the "Dead Duck" scenario.

Preconditions	Motivation
- Major problems with cyber terrorism - Poor global economic development - Public sector interested in exploiting the advances of Grid computing	- To get better research results - To enhance national competitiveness and advancement (e.g. China) - High computing-intensive users can experience major benefits
Characteristics	**Milestones**
- Customers cannot afford to invest in Grid computing and keep using more robust technologies - Collaboration between vendors breaking up and large Grid projects launched by governments fail - Strong mistrust in Grid computing services leads to many different security solutions - Added value is not realized and major vendors deliver managed or private utility computing solutions that are more readily accepted by the enterprises	- Utility computing vendors do not push and invest in Grid computing technologies - Cooperation organs fail to build robust technology - Bankruptcies or mergers of the small "pure" utility computing vendors

Scenario 4: Trust Me! The dimensions of this scenario are strong added value and mistrust; i.e., Grid computing creates a great deal of added value to the companies but the trust in the technology is missing. The emergence of the *trust boundaries* is characteristic of this scenario. It means that companies lacking the trust make deals and contracts only with their trusted suppliers and partners (Table 4), which results in customer lock-in and vendor-specific security solutions.

Table 4. The key issues in the "Trust Me!" scenario.

Preconditions	Motivation
- Major increase in cyber terrorism - Low trust in IT solutions and services - Positive economic development - Improvements for the security requested	- Vendors want to lock in customers - Customers want to exploit the possibilities of Grid computing inside their 'trust boundaries' - Companies outsource their IT departments (to trusted vendors)
Characteristics	**Milestones**
- Big players invest in own secure Grid techniques - Cooperation in the standardization process does not work - Many different security solutions with interoperability problems - Grid technology itself develops strongly and creates strong added value compared to 'traditional' technologies - Companies purchase IT inside their 'trust boundaries'	- Failure of basic Grid computing security or short-term collapse of a major data system (e.g. the Internet) - Launch of vendor-specific security solutions - Applications with clear benefits of Grid computing

Table 5. Scenario themes.

	"Brave New World"	"Commercial Dodo"	"Dead Duck"	"Trust Me!"
End user markets	Consumers have multiple options to access Grids	Public sector buys Grid services	Some success in small computing-intensive industries	Users buy only from trusted partners
	Business users pay for new services	Users have strong trust on Grid services	Users have strong mistrust towards Grid computing services	Consumers are willing to pay for trustworthy services
Technology	Most technological challenges resolved	Public sector funds the development of Grid technologies	Interoperability problems, large Grid computing projects fail	Major increase of cyber terrorism
	Strong diffusion of broadband networks Inter-grids	Open source development dominates	Grid technologies are not deployed widely	Grid computing develops rapidly
Industry players	Emergence of many new players	Niche strategy is an opportunity	First-movers have financial troubles	Vendors position themselves as secure outsourcing partners
	Large vendors invest strongly in Grid computing	Tight price competition	Large vendors restructure or disinvest their Grid-focused SBUs	Company-specific security solutions
Business models	New business models, such as resource brokering, emerge	Launch of civic services based on Grid technologies	Offering the highly specialized services	Deals and contracts are made inside the trust boundaries
	Business users in certain vertical markets enhance their competitiveness and efficiency	Differentiated service to public sector	Private utility computing services	Customer lock-in
Legal issues	Privacy and security concerns have been overcome	Authorities distribute and control Grid identities	Fierce fighting over patents and technologies	Few patent wars and lawsuits
		Governments have Grid computing strategies	Cyber terrorists are being sued with no real effect	

5. Conclusion

Although Grid computing has got analogies with extensively used technologies, such as the Internet or power distribution network, it is still long way from having the same extent of impact on people's lives.

In the research community, Grid computing has already showed its potential and it is currently used in many different science and industry-wide collaboration projects all over the world. At the same time a number of ICT vendors are pushing these technologies strongly forward and trying to "cannibalize" their own old business, before someone else does it for them. If Grid computing destroys the current competence base in the industry, the first-movers will have a clear advantage. That is why major ICT vendors are collaborating with their competitors.

Many open questions and a great amount of uncertainty remains surrounding the possible new paradigm of the computing technology and business. At this point, no one can really say if there is going to be any major Grid business or if it will remain "a great promise of the ICT industry" still in the year 2010. The written scenarios tried to expose those options and tell feasible stories about the future.

The scenario "Trust Me!" was created on the assumption that Grid computing has a great deal of added value to the companies but the trust in the technology is missing. The role of the governments can be quite surprisingly very important, and especially the scenario "Commercial Dodo (New welfare society)" explores the possibilities that countries actively utilize Grid computing in delivering welfare services and regards it as an important tool in building the modern information society. "Brave New World" is clearly the most positive scenario and describes the world where many of the great visions have realized and Grid computing business is flourishing. "Dead Duck" instead showed the future where the Grid business has not taken off almost at all. Among other things, the role of the government and public sector was found highly important and the possible emergence of trust boundaries was noticed. Milestones were identified for each scenario as well.

The scenario creation process enhances organizational learning, reveals blind spots in the decision-making challenging the present assumptions, and simplifies the complex reality. Scenario analysis is suitable and applicable in a variety of situations in any organization because it is a very flexible method. The final structure and method should be chosen with respect to the available resources. In technology evaluation, the applicability and the value of scenario analysis grows when the level of technological change is more radical and one has to exploit expert knowledge because codified knowledge is not useful and/or available. In those situations scenario analysis may give additional insights for strategic decision-making.

References

Aligica, P. D. (2003). Prediction, explanation and the epistemology of futures studies. Futures. 35, 1027-1040.

Bennett III, R. H. (1998). The importance of tacit knowledge in strategic deliberations and decisions. *Management Decision.* 36, 589-597.

Bergman, J.-P., Jantunen, A., and Saksa, J. M. (2004). Managing knowledge creation and sharing – Scenarios and dynamic capabilities in inter-industrial knowledge networks. *Journal of Knowledge Management.* 8, 63-76.

Berman, F., Hey A. and Fox, G. (2003). The Grid: past, present, future. In: Grid Computing - Making the Global Infrastructure a Reality, F. Berman, A. Hey and G. Fox (eds.). 1, 9-50, John Wiley & Sons, Chichester.

Bood, R. and Postma, T. (1997). Strategic learning with scenarios. *European Management Journal.* 15, 633-647.

Coates, J. F. (2000). Scenario planning. Technological Forecasting and Social Change. 65, 115-123.

Courtney, H., Kirkland J. and Viguerie P. (1997). Strategy Under Uncertainty. *Harvard Business Review.* 75, 66-80.

Coyle, G. (2004). Practical Strategy: Structured Tools and Techniques. Bell & Bain, Ltd., Glasgow.

Day, G.S. and Schoemaker P. J. H. (2000). Preface: Looking for the Edge. In: G. S. Day, P. J. H. Schoemaker & R. E. Gunther (eds.). Wharton on Managing Emerging Technologies. John Wiley & Sons, Inc., New York.

Ellis, S. and Shpielberg, N. (2003). Organizational learning mechanisms and managers' perceived uncertainty. *Human Relations.* 56, 1233-1254.

Foster, I. Kesselman, C. and Tuecke, S. (2003). The Anatomy of the Grid. In: Grid Computing - Making the Global Infrastructure a Reality, F. Berman, A. Hey and G. Fox (eds.), 171-197, John Wiley & Sons, Chichester.

Godet, M. (2000). The art of scenarios and strategic planning: Tools and Pitfalls. *Technological Forecasting and Social Change.* 65, 3-22.

Godet, M. and Roubelat, F. (1996). Creating the future: The use and misuse of scenarios. Long Range Planning. 29, 164-171.

Grimshaw, A. S., Natrajan A., Humprey M. A., Lewis M. J., Nguyen-Tuong A., Karpovich J. F., Morgan M. M. and Ferrari A. J. (2003). From Legion to Avaki: the persistence of vision. In: Grid Computing - Making the Global Infrastructure a Reality, F. Berman, A. Hey and G. Fox (eds.). 265-299, John Wiley & Sons, Chichester.

Johannessen, J.-A., Olsen, B., and Olaisen, J. (1999). Aspects of innovation theory based on knowledge-management. *International Journal of Information Management.* 19, 121-139.

Kleiner, A. (1999). Doing scenarios: A heartful look at thinking the unthinkable. Whole Earth. 96, 76-82.

Kulkki, S. and Kosonen, M. (2001). How tacit knowledge explains organizational renewal and growth: The case of Nokia. In I. Nonaka and D. J. Teece (eds.). Managing industrial knowledge: creation, transfer and utilization. 244-269. Sage Publications, New York.

Levary, R. R. and Han D. (1995). Choosing a technological forecasting method. *Industrial Management.* 37, 14-18.

Mannermaa, M. (1999). Tulevaisuuden hallinta – skenaariot strategiatyöskentelyssä. (in Finnish) (Managing the future – scenarios in strategy projects) Werner Söderström Osakeyhtiö, Porvoo.

Masini, E. and Vasquez, J. (2003). Scenarios as seen from a human and social perspective. Technological forecasting and social change. 65, 49-66.

Nonaka, I. and Toyama, R. (2003). The knowledge created theory revisited: knowledge creation as a synthesizing process. *Knowledge Management Research & Practice.* 1, 2-10.

Nonaka, I., Toyama, R., and Konno, N. (2000). SECI, Ba and leadership: a unified model of dynamic knowledge creation. Long Range Planning. 33, 5-34.

O'Connor, G. C. and Veryzer, R. W. (2001), "The nature of market visioning for technology-based radical innovation", *Product Innovation Management.* 18, 231-246.

Phelps, R., Chan, C., and Kapsalis, S. C. (2001). Does scenario planning affect performance? Two exploratory studies. Journal of Business Research. 51, 223-232.

Porter, A. L., Roper A. T., Mason T. W., Rossini F. A. & Banks J. (1991). Forecasting and management of technology. John Wiley & Sons, Inc, New York.

Roubelat, F. (2000). Scenario planning as a network process. Technological forecasting and social change. 65, 99-112.

Scharmer, O. C. (2001). Self-transcending knowledge: sensing and organizing around emerging opportunities. *Journal of Knowledge Management.* 5, 137-150.

Schoemaker, P. J. H. (1992). How to link strategic vision to core capabilities. Sloan Management Review. Fall, 67-81.

Schoemaker, P. J. H. (1993). Multiple scenario development: its conceptual and behavioral foundation. *Strategic Management Journal.* 14, 193-213.

Schoemaker, P. J. H. (1995). Scenario planning: A tool for strategic thinking. Sloan Management Review. Winter, 25-40.

Schoemaker, P. J. H. (1997). Disciplined imagination - From scenarios to strategic options. *International Studies of Management & Organization.* 27, 43-70.

Schoemaker, P. J. H. & Mavaddat V. M. 2000. Scenario Planning for Disruptive Technologies. In: G. S. Day, P. J. H. Schoemaker & R. E. Gunther (eds.). Wharton on Managing Emerging Technologies. John Wiley & Sons, Inc., New York.

Schwartz, P. (1996). The Art of the Long View - Planning for the Future in an Uncertain World. Doubleday Dell publishing Inc., New York.

Senge, P. (1990). The fifth discipline: the art and practice of the learning organization. Doubleday, New York.

Teece, D. J., Pisano, G. P., and Shuen, A. (1997). Dynamic capabilities and strategic management. *Strategic Management Journal.* 18, 509-533.

van der Heijden, K., Bradfield, R., George, B., Cairns, G. and Wright, G. (2002). The Sixth Sense - Accelerating Organizational Learning with Scenario. John Wiley & Sons, Ltd., Chichester.

von Krogh, G., Nonaka, I., and Aben, M. (2001). Making the most of your company's knowledge: a strategic framework. *Long Range Planning.* 34, 421-439.

Wack, P. (1985a). Scenarios uncharted waters ahead. Harvard Business Review. Sept-Oct, 73-89.

Wack, P. (1985b). Scenarios: shooting the rapids. Harvard Business Review. Nov-Dec, 139-150.

Watts, R. L. and Porter, A. L. (1997), "Innovation forecasting", Technological Forecasting and Social Change. 56, 25-47.

Weick, K. E. and Quinn, R. E. (1999). Organizational change and development. *Annual Review Psychology,* 50, 361-386.

Wilson, I. (2000). From scenario thinking to strategic action. Technological Forecasting and Social Change. 65, 23-29.

CHAPTER 9

COMPUTER SERVICES AS INNOVATION AGENTS FOR LOCAL PRODUCTION AREAS

Enrico Scarso and Ettore Bolisani

DTG, University of Padova, Vicenza, Italy

In order to achieve a better understanding of the computer services as a catalyst of innovation for local economies, some issues have to be addressed, as follows: a) the links between the processes of innovation, and the cognitive exchanges underpinning such processes; b) the peculiar operational methods and activities performed by computer services; and c) the constraints and problems that can hinder the use of such services in a specific local context. The paper assumes a perspective that considers the spread of an innovation as a process of knowledge transfer and dissemination, and applies it to a case-study investigation about the role of innovation agents played by some computer services companies in the North-east of Italy.

1. Introduction

The importance of business services to the performance of the economy is largely acknowledged. In 2002, this sector constituted a large part of the EU 25 non-financial business economy, with the 18.1% of total employment and the 20.8% of the total value added (Eurostat, 2006). A key feature of business services is that they contribute to - and are integrated into - every stage of the value chain: all companies, in fact, require a broad and mixed set of services to remain competitive. But the very reason for their relevance consists in being able to stimulate innovation among clients. Services, in fact, are increasingly recognised as a main focus and driver of technological and managerial advancements (Rodriguez and Camacho, 2004). Among the different business services, a particular attention deserves the computer services sector, since it directly relates with the new paradigm of the information economy. Computer services prove to be important in the case of small manufacturers, which generally lack knowledge and resources to adopt and use the new information technologies effectively.

Our analysis aims to verify whether and how "local" computer services can be a catalyst of innovation for SMEs. The assumption is that hardware and software solutions "invented elsewhere" cannot be easily transferred to small users. In principle, the geographical proximity, added to the attitude to service small clients, can be favorable factors that facilitate computer services in their role of innovation diffusion agents. Nevertheless, the experience shows that this is a complex and difficult task, whose analysis and understanding deserve further investigation.

2. The Role of KIBS in the Innovation Transfer Process

Recent studies have focused on the main role played by *intangible factors* when an innovation is transferred from primary sources (i.e. research centres, R&D laboratories, Universities, etc.) to the potential users. As a matter of fact a complex bundle of *knowledge* needs to be transferred in order to successfully implement and exploit *any* new solution (Bozeman, 2000). Hence, technology transfer can be seen as a special kind of *knowledge communication process* (Williams and Gibson, 1990; Amesse and Cohendet, 2001; Gorman, 2002), constituted by the following elements (Bolisani and Scarso, 2004):

- *The transfer object.* The traditional notion of technology as a pure physical device is inappropriate, since the crucial ingredient is the knowledge required to use it. This calls for an appropriate determination of what is the real transfer object. In case of non-standard and highly customised applications, knowledge assets may be much more significant than the physical device itself;
- *The direction of the transfer process.* Transfer cannot be regarded as a *one-way*, but a *two-way communication process*, where source, transfer agent, and recipient continuously exchange information to

correctly design and customise the technology, according to the requirements and capability of the end user;
- *The transfer agent*. The main task of the transfer agent is to fill the users' knowledge shortage that may hinder the actual exploitation of the transferred technology. For that reason, the agent operates not only on as a technical but also as a business assistant;
- *The proximity between sources and users*. When technology transfer is seen as a communication process, the proximity between the various actors is thus critical. Such proximity should not be intended just in geographical terms (although this can make the difference), but especially in cognitive or relational terms.

To sum up, if transferring a new technology means transferring knowledge, the action of a transfer agent, able to fill the cognitive gap between the technology source and user is crucial, as well as its physical location. Exchanging knowledge (especially the tacit component), in fact, requires frequent communication, direct contacts, mutual trust, shared understanding of problems, and similar ingredients.

More frequently the role of diffusion agents is played by knowledge-intensive business service (KIBS) firms, i.e. private companies or organisations which help other organizations to solve problems for which external sources of knowledge are required (Miles, 2005). In playing such a role, KIBS companies rely heavily on knowledge or expertise related to specific (technical) discipline or (technical) functional domain.

There are as many kinds of KIBS as the areas of knowledge (a broad classification can be found in Thomi and Böhn, 2003). A first distinction can be made between the so-called T-KIBS (technology-based KIBS, such as e.g. computer services) and P-KIBS (pure professional KIBS - Miles *et al.*, 1995). In general, KIBS sectors are populated by a few large transnational companies and a vast majority of small firms dealing with specific localities or niches (Miles, 2005).

KIBS can play multiple roles inside innovation systems (Den Hertog, 2000). They are commonly believed to serve as "bridges for innovation", since they connect knowledge producers and users (Muller and Zenker, 2001; Kuusisto and Meyer, 2003). In particular, KIBS are actively involved in adapting the more generic technical and commercial knowledge and experience to the specific needs of the clients (Wood, 2005). Such bridging function is so vital for the capacity and dynamism of local systems, that their lack of development may hinder the growth of entire economy. It must be reminded that the delivery of a service needs a knowledge interaction between service providers and clients, thus producing a mutual learning. Such co-production process makes service delivery distinct from other types of buyer-supplier relationships.

KIBS can also be complementary to public research organisations, by serving as a conduit to link public R&D centres with firms that are not able to collaborate with such institutions directly (Wong and Singh, 2004). All this explains the growing interest in analysing the contribution given by KIBS to the local development (Drejer and Vinding, 2003; Kuusisto and Meyer, 2003; Wood, 2006), and the need of further investigations to fully understand the role (with associated advantages and problems) they can play as innovation catalysts.

3. Computer Services as KIBS Companies

The term "computer services" generally includes the supply of computer software and all the related services (e.g. bureaux services, facility management, system analysis, consulting and training in IT, etc. - see Howells, 2000). Indeed, the boundaries of this industry are not unambiguous, though firms generally classified in other industries (e.g. original equipment manufacturers, management consultants, etc.) might also provide services. Here, we will explicitly focus on firms whose core activity is the supply of computer services.

There are some relevant issues of computer services seen as a T-KIBS sector (Howells, 2000). Firstly, a flourishing software industry is of great importance in contributing to the employment in national economies, and in developing other industries. Software business has remarkable effects in manufacturing and retail sectors, because IT systems are essential for their business processes and competitiveness.

Furthermore, computer services are a highly labour-intensive industry, based on the exploitation of knowledge workers and specialised competencies. Although efforts have been made to structure and formalise the design of new software, or to reduce the cost of labour by outsourcing such activity to developing countries having highly qualified professionals (e.g. India), this is still a very costly and time-spending activity. Consequently, while the price of hardware has progressively fallen for at least two decades, this has not occurred to software and services. Indeed, the major portion of the investments of a new project of computer system regards those elements, rather than hardware.

In addition, the process of providing software and computer services to business users raises special problems. The implementation of a new information system in a company implies issues such as an analysis of user business requirements, a selection of appropriate technologies and suppliers, a proper design and implementation of the system, training and maintenance activities, etc. Any implementation is the combination of standard technologies and modules designed or adapted to the specific application. For this reason, a subdivision of specialisations can be found in the industry. Beyond the conventional distinction between hardware, software, and service suppliers, there is also a pronounced subdivision of activities among different software producers and service providers.

As an example, computer services may include specialised firms as e.g.: large software companies and small software houses, outsourcing services and maintenance operators, dealers and consulting companies, etc. Although some vertically integrated companies can provide different services, there is often a distinct specialisation, also in relation to the specific market. As regards business software, for instance, there is an important distinction between:

- large companies providing standard software solutions;
- integrated companies providing tailored solutions for large users;
- small or medium sized software houses that develop (at least in part) new solutions for smaller or local customers;
- companies (generally, small software houses) that customise standard solutions;
- pure resellers of standard software.

In all those companies, the way software is developed and transferred to the customers is clearly different.

The *proximity* to the client can be an important element here, since interactions with actual and potential customers are essential for effective "localised" implementations. Some studies (e.g. Jones, 1994) calculated that the distance between a producer of standard software and its customer can reach thousands of kilometres. A developer of (e.g.) a particular MIS solution has to be "stuck to the clients", to better understand their needs and translate them into appropriate solutions. In some cases, even important producers of business solutions (e.g. the German company SAP) need a network of small firms that customise and resell their products.

With regard to this, some studies (Egan, 2001) highlight the importance of *application districts*, i.e. areas where computer service firms operate in direct contact with specific customers, and can thus specialise in software solutions for a particular industry or application field. As a matter of fact, the localised nature of innovation requires that in most cases the user is served through a direct interaction. In turn, the supplier can exploit this interaction to learn from the customer. Very often, a solution experimented or implemented for a client is then adapted and transferred to other firms in the same user industry.

Thus, various firms operating at different levels compose the computer service sector. At one extreme, we can place large multinational, proving standard technologies and platform; at the other extreme there are local software providers and "application districts" that elaborate and combine standard technologies to provide a solution that fits the specific user needs. This study focuses on the importance of the last category of firms and their potential as innovation agents. In particular, the customer-supplier interaction, seen as a knowledge-transfer process, is deemed to be a key aspect for the innovation process of specific areas of industries. In the next sections, this analysis is applied to the economic area of Veneto, in the North-East of Italy.

4. The Empirical Analysis: Context and Approach

This section describes the computer services sector, seen as a "local industry", in Veneto. In other words, the assumptions are that: a) there is a "local" demand for software products and computer services that has peculiar characters and dynamics, and b) there is a "local" supply of computer services, i.e. firms that specialise in servicing the local market. The analysis provided is based on recent reports combined with additional information collected in interviews with company managers.

The Italian ICT market still suffers from a difficult business climate (see Figure 1). After a decade of growth, the last years showed a decline that, according to the current forecasts, is expected to continue in the near future. One reason of this decline can be traced to the structural characteristics of the business demand. As a matter of fact, SMEs, that account for approximately 90% of the total number of Italian firms, represent a minor part of ICT investments and, what is worse, are the least dynamic segment, showing significant reduction in expenditure (see Figure 2). Also, it is especially the manufacturing companies, which are the backbone of the national economy and represent the largest ICT market (about 4,500 millions euro – 2003), that showed the most impressive decrease (-7% between 2002 and 2003). A critical point is that the decline now extends beyond hardware (whose market is subject to a continuous fall in prices but apparently not in terms of quantities). The decline also reaches software and services (i.e. the computer services sector) which, from our viewpoint, are the "very carriers" of innovation.

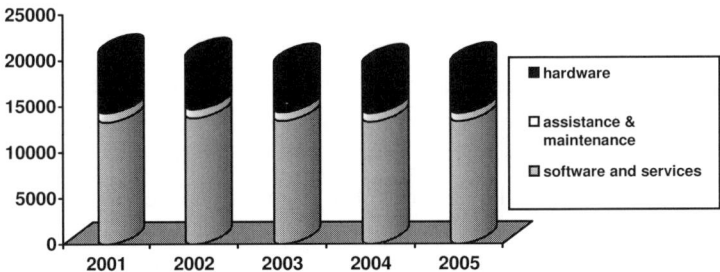

Figure 1. The Italian ICT market (millions of euro – excluding telecommunications) (Source: Assinform/Netconsulting).

This situation is due to a variety of factors which, combined together, have a severe impact on the overall trends. i.e., a) a general reduction in investments by SMEs; b) a competitive downpricing by vendors to retain their customer base; c) a reduction in the number of innovative projects, while expenditure is mainly directed to maintenance and IT management; and d) a focus on short-term returns.

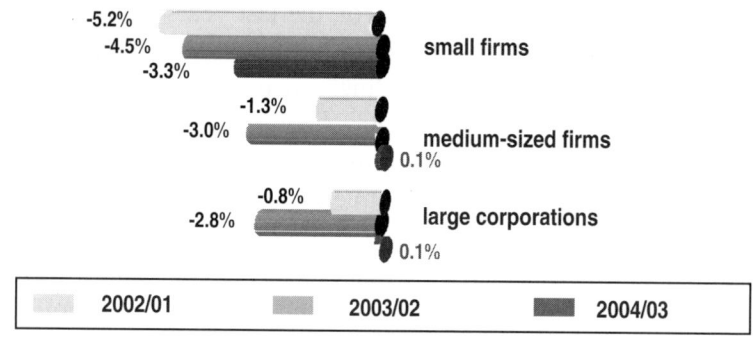

Figure 2. Variations in expenditure per segment (Source: Assinform/Netconsulting).

Geographically, Veneto is the fifth region for ICT spending (8.08% of the total market – 2003), although its contribution to the national GDP is more relevant (around 9% of national GDP). Indeed, ICT spending is low, and with a negative trend (1.8 of regional GDP in 2002, and 1.45 % in 2003), especially when compared to other industrialised areas of the country (i.e. Lombardy and Piedmont).

As regards the supply, the number of ICT firms directly based in Veneto amounts to some thousands (around 6,500 in 2002 – source: Assinform/Unioncamere). This may be intended as a point of weakness, because the percentage of ICT firms (7.7% of Italy) is less than the market share (8.6%). What's more, the firms' average size is very small. More than one half of firms have a turnover that does not reach millions of Euro, with not more than 10 employees. In Veneto, there is no large firm operating at an international level. Instead, the general picture of the computer services includes a majority of small vendors of standard products, small application developers specialising in the customisation, commercialisation and servicing of standard suites, and small software houses or service firms that work in collaboration with larger companies (source: Assinform). Such firms have a marginal and "passive" role in the transfer of innovation. There are also a restricted number of independent software developers that produce proprietary technologies directly for the customers. According to our analysis, this category of firms appears to be the most important as agent of innovation transfer.

4.1. *Supplier-client relationships*

The basic assumptions of our study are as follows:

- The "great" innovation streams of software and computer services are mainly produced outside the area (and, often, abroad);
- However, the majority of small users (and, especially, traditional manufacturers) do not possess the market power to deal with large software multinationals, nor the competence to develop ICT innovations internally;
- Thus, an effective transfer of ICT innovations, from the "international sources" to the small users, can be underpinned by intermediate assistants (i.e. KIBS firms), that can exploit both a technical competence, and a capability to understand the needs of the local users, and keep effective relationships with them;
- It is thus important to understand the mechanisms that enable the process of transfer of the particular kind of innovation represented by ICT applications.

At this level, the most proper method of investigation is probably the direct analysis of the ways computer service suppliers interact with the local customers, how they can perform an effective innovation transfer, and the related issues. A case-study analysis was conducted, by investigating a restricted number of providers, selecting out the most dynamic ones in the market (see Table 1). In other words, pure resellers were excluded, and the focus was on suppliers capable of (partly) producing fresh innovation, or interacting with customers effectively.

Table 1. An outline of the cases examined.

Company	Specialisation	Main Markets	Size
A	ERP; CRM	Machinery; food	220
B	ERP; prod. configurator	Furniture; marble	130
C	ERP	Manufacturers; distributors	70
D	ERP components	SMEs; professional services	40
E	CRM; MIS application	Various	20
F	System integration; KM; BI	Retailing	11
G	ERP	Service and manufacturing	10
H	ERP components; CRM	Service and manufacturing	5

General information was collected by various documental sources. After that, semi-structured interviews were conducted with key managers, with an emphasis on the following issues:

- category of products and their innovativeness;
- market specialisation and kind of customers;
- supplier-client relationships and related issues;
- exchange of knowledge and other issues in the innovation transfer;
- sources of innovations, technical knowledge, and existing ICT solutions.

5. Investigation: A Summary

Nature of innovation. Innovativeness is generally indicated as a critical point by all the managers interviewed. A critical point despite being intended as the introduction of incremental upgrading (e.g. new functions to existing software packages) or adjustments to new technological platforms (i.e. new operating systems, new programming languages, emerging standards, etc.). Typically, any implementation by a customer is "new", since it is specific. Innovation is, first, an essential change in the way of working. In relation to this, a major problem highlighted during the interviews is that the sale of a new computer application does not resolve the problem of innovation transfer, or it is better to say, a new software can be innovative but does not result automatically in innovation for the purchasers.

Sources of innovation. Although many providers that operate in Veneto are still strictly dependant on large multinational suppliers of technology, the companies interviewed stress the importance to develop "in-house" proprietary technologies and applications. Indeed, a large number of firms, specialising in reselling or customising standard software, now face a negative climate. The development of a (at least partly) proprietary technology is thus seen as a way to maintain a complete control over the systems sold, and other advantages are possible in terms of easy upgrading and better customer service. Clearly, the development of proprietary technologies is a costly and time-spending activity, that implies significant investments. In addition, since small companies do not have an established and widely known brand, proprietary applications should be "as interoperable as possible" with different standard technologies, as well as easy to customise, so that the client does not feel "locked" in a closed system. The relationships with producers of global software solutions are often difficult, but essential. Although there are examples of products that the interviewed firms just resell, such global providers are generally seen as sources of platforms for developing specific applications, new ideas, models, and concepts. In general, the firms of our sample prefer to maintain their independent business which, up to now, is considered an essential ingredient of their survival. In some cases there are close relationships with small local software houses or service firms, especially for secondary time-spending activities (e.g. the development of parts of software), and for a more punctual service to the single client. Controversial are also the relationships with local Universities. While Universities (there are several faculties in the Region and around) are seen as a primary source of educated people with competencies of informatics or management, research collaborations are much more difficult. In particular, the attitude of University researchers is considered "too academic" and not oriented to the practical problems of business.

Organisational issues in innovation management. Although one firm differs from another, a significant part of their personnel is assigned to organisational activities, customer relationships, sales, etc. As a consequence, while the technical skills still represent the "kernel" of a computer service firm, it is the organisational-managerial competence that underpins a proper business activity. The understanding of the business and information needs of the customers, their organisational structure, strategic visions, and so forth is considered as an essential starting point. Some firms also have different subsidiaries scattered in the Region, which ensure a constant presence "near to the client."

Knowledge transfer and customer-supplier relations. The transfer of knowledge to customers is essential for the success of a new project. Both technical and managerial knowledge has to be transferred, but especially the latter is sometimes difficult. Generally speaking, the real issue is not the lack of technical competence by customers, but the knowledge of how the technology can be used for business. This is the reason why it is the entrepreneur or the CEO that represents the necessary point of reference for any project, while technical personnel (e.g. the EDP manager) is involved in the specific implementation activities. The main channel of knowledge transfer is represented by direct contacts (i.e. joint teams, training, face-to-face contacts, etc.), while formal channels (e.g. technical literature) are provided but generally less critical. An important element of knowledge transfer is the *feedback from customers.* Indeed, the information collected from customers (with formal or informal mechanisms) is essential not only for developing the specific project, but also for transferring such experience to other situations or clients. All the firms reported that, to some extent, experience contributes to consolidate the specialisation in a specific market or sector. Indeed, the focalisation in an industry or class of applications is regarded as a key factor of competition. In general, any project requires a long-lasting interaction with customers. Surprisingly, the duration of the single project is not related to its complexity or to the dimension of the client firms, but rather to the capability to make their needs explicit, and to their propensity to innovate.

Peculiar issues and drawbacks in the innovation process. All the managers interviewed highlight the low propensity of local enterprises (and, especially, the smaller ones) to invest in ICT applications. In addition, there is no long-term vision, and the clients generally put an emphasis on very specific problems. ICT applications are rarely seen as a key element for re-engineering the business processes, or reformulating the strategies. The privacy of internal data, and the reservations by clients to permit the access to internal databases, is also seen as a barrier to the adoption. With regard to this, a trustworthy climate is a critical element, which, again, raises the issue of client-supplier relationship management.

6. Conclusion

The analysis conducted here highlights the main critical points in the development of a computer services sector capable of catalysing innovation in a traditional manufacturing economy. Indeed, the study shows that ICT applications can represent an important ingredient and stimulus for a wider organisational and strategic innovation by user firms. In this sense, due to the specificity of problems, the heterogeneity of firms, and the localised nature of the innovation process, the role of local computer service suppliers is vital.

Local providers can act as transfer assistants of innovations from the places where technologies are developed (i.e. University, multinational software providers, etc.) and the small local users. The most critical part of the work of such KIBS companies is to make the user needs explicit, to match such needs to the available technical solutions, and to assist the customers in the process of implementation and use. The key role of such companies is to promote organisational and strategic innovations that are carried by ICT applications. It is for this reason that the deep knowledge of the users, the capability to manage supplier-client relations, and the specialisation in specific industries or sectors appear to be even more important than the technical capability itself.

Lastly, since knowledge co-production is a key element of any service delivery, there is a sort of virtuous (but also vicious) circle that links KIBS companies with their clients: only a mutual commitment can make the technology transfer efforts successful. On the contrary, for example when a bad climate lessens the willingness to innovate of clients, even the presence of a "strong" KIBS sector may be not sufficient to trigger innovation.

References

Amesse, F. and Cohendet, P. (2001). Technology transfer revisited from the perspective of the knowledge-based economy, *Research Policy,* 30, pp. 1459-1478.

Bolisani, E. and Scarso, E. (2004). Knowledge-intensive transfer of innovation: electronic commerce and small business, International Journal of Networking and Virtual Organisations, 2, pp. 335-352.

Bozeman, B. (2000). Technology transfer and public policy: a review of research and theory, *Research Policy,* 29, pp. 627-655.

Den Hertog, P. (2000). Knowledge-intensive Business Services as co-producers of innovation, International Journal of Innovation Management, 4, pp. 481-528.

Drejer, I. and Vinding, A.L. (2003). Collaboration between manufacturing firms and knowledge intensive services – The importance of geographical location, DRUID Summer Conference, Copenhagen/Elsinore, June 12-14.

Egan, E.A. (2001). Application Districts: An Emerging Spatial Form in the Computer Software Industry, *Journal of Comparative Policy Analysis: Research and Practice,* 2, pp. 321-344.

Eurostat (2006). European Business: Facts and Figures – Data 1995-2004, Office for Official Publications of the European Communities, Luxembourg

Gorman, M.E. (2002). Types of Knowledge and Their Roles in Technology Transfer, *Journal of Technology Transfer,* 27, pp. 219-231.

Howells, J. (2000). Computer Services: The Dynamics of a Knowledge-Intensive Sector, in Andersen, B. et al. (eds.), Knowledge and Innovation in the New Service Economy, Elgar, Cheltenham, pp. 121-141.

Jones, C. (1994). Globalisation of software supply and demand, *Software Engineering Journal,* 9, pp. 235-243.

Kuusisto, J. and Meyer, M. (2003). Insights into services and innovation in the knowledge-intensive economy, Technology Review, 134.

Miles, I. (2005). Knowledge intensive business services: prospects and policies, Foresight, 7, pp. 39-63.

Miles, I. et al. (1995). Knowledge-Intensive Business Services: Users, Carriers and Sources of Innovation, EIMS Publication, n. 15.

Muller, E. and Zenker, A. (2001). Business services as actors of knowledge transformation: the role of KIBS in regional and national innovation systems, *Research Policy,* 30, pp. 1501-1516.

Rodriguez, M. and Camacho, J.A. (2004). The role of services in the European national innovation systems: are they 'real diffusers'?, DRUID Summer Conference, Elsinore, 14-16 June.

Thomi, W. and Böhn, T. (2003). Knowledge Intensive Business Services in Regional Systems of Innovations. Initial Results from the Case of Southeast Finland, 43rd European Congress of the Regional Science Association, August 27-30.

Williams, F. and Gibson, D.V. (eds.) (1990). Technology transfer: a communication perspective, Sage, Newbury Park, Ca.

Wong, P.K. and Singh, A. (2004). The pattern of Innovation in the Knowledge-intensive Business Services Sector of Singapore, Singapore Management Review, 26, pp. 21-44.

Wood, P., (2006). The regional significance of knowledge-intensive services in Europe, Innovation, 19, pp. 51-66.

CHAPTER 10

TOWARDS DIGITAL INTEGRATION: PLATFORM THINKING IN THE FASHION BUSINESS

Finn Kehlet Schou

Department of Architecture & Industrial Design, Aalborg University, Denmark

Customers are demanding lower prices, a larger range of products and faster product innovation. Despite steady streams of advanced technologies developed to support product development many companies find it difficult to achieve the full potential of these as practical advices on selecting and implementing new technology are rare (Boer and Krabbendam, 1998). On the basis of case studies within the highly competitive fashion business the aim of this paper is to present strategic as well as practical aspects on how a company can improve time to market of styling objects (optical frames).

1. Introduction

Fashion, a business area built on imagination, dreams, a subtle understanding of market movements - and not least, hard work. Being a part of the fashion-business, the optical frame designers must be able to deliver their share of new (and interesting) products to the market.

Since it has proved difficult to radically improve the functional qualities of the products, focus has been on mastering aesthetic aspects – *styling*. This to attract attention to a highly competitive market and gain acknowledgement as important players by retailers, individual opticians, or optical chains that to a great extent believe in high product pricing based on customer experienced value. Although frames are intricate products, production prices of optical frames are fairly low because much of the work is "made in China". For a long time, outsourcing has concentrated in this part of the world and accepted in the business as a good practice of handling materials and production processes in combination with an affordable workforce. Expectations of high return on investment (ROI) are often followed by increased competition, and few business areas are as tight as the optical frame design business. Still, history has shown that even minor optical frame design companies (at least for a while) are able to place themselves in pole position by launching the right product to the market at the right time. Be it a lucky punch or strategic understanding.

As in any other business, the distinction between success and failure is a razor-thin margin. Much has been tried to reduce numbers of the latter by amongst others IT investments - primarily in 2D CAD and secondly in 3D CAD. The latter often in terms of AutoCad, which has been the default software in the optical frame design business for years. However, as many of the designers in this business area are opticians by trade – and thus autodidact industrial designers - they have little or no prior experience in handling CAD at the operational level. As reported by Boer and Krabbendam (1998), many have found it difficult to achieve the full potential of these technologies. This is because practical advice that is helpful to companies and managements in their decision-making when selecting and implementing new technology is rare.

In this paper, through presentation of case studies within the optical frame design business from France; Alain Mikli (AM) and Denmark; Lindberg A/S (LG) and Pro Design International A/S (PDI), information is presented on when and how to use digital platforms to increase competitiveness. In the following, three core competences related to the design of optical frames, as found by the case studies, are presented and the CAD-strategies behind them are discussed.

2. Core Competences Related to Design of Optical Frames

Naturally, the objective of optical frames is to enable the end user to see. However, according to some optical frame designers it is "evenly important to be seen" (Alain Mikli, 2004). According to others, optical frames must be discreet (if possible, non-visible) to "avoid taking focus from the face" (Lindberg A/S, 2004) or "aiming at obtaining clean and simple lines – the unmistaken look of Danish Design" (Pro Design International A/S, 2004). Though different in values and mission, there is no doubt that styling is in focus by all of the optical frame designers. From a strategic point of view, there were three core competencies among the companies investigated that all related to styling - the creation of visual attributes of artefacts (Figure 1). Each of these can stand alone, or in combination with one or more of the others to enhance the competitiveness:

- Mastering Technical Innovation
- Mastering Materials
- Mastering giving Shape (White Body)

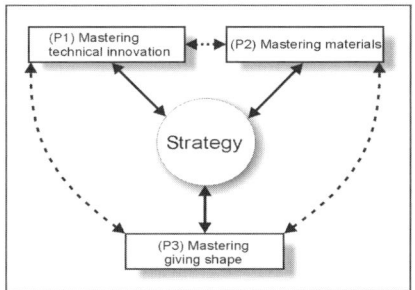

Figure 1. Core competences related to strategy.

2.1. *Mastering technical innovation*

For the STARCK EYES® collection, manager and designer Alain Mikli was collaborating with industrial designer Philippe Starck. One of the crucial components was the development of a flexible hinge named the Bio-Link. Even though Philip Starck frequently uses 3D CAD (e.g. Rhino) in his design studio, no use of CAD (neither 2D nor 3D CAD) was used for this part of the project. All drawings where carried out by hand. Perspective hand drawings are made in approximate size 1:1 in order to give an impression of the overall product style. To be able to evaluate visually the very small parts they, as is customary in the optical frame design business, are illustrated in oversize (10:1). Technical drawings are carried out by use of a drafting machine combined with several curve templates. Likewise to evaluate mechanical movements, oversize prototypes are produced. After conclusion of the technical construction, styling options are explored. In drawings, several visual platforms are created to explore the design, e.g. frames for sport, men, women, and children. Finally, prototypes are made from the technical drawings. From a strategic point of view (time to market) it is interesting to notice that despite high constructional complexity the design process is kept manual: paper born + physical models. Contrary to the digital design process, it is difficult and time consuming to re-use the prior knowledge already gathered in future projects. It is also interesting to notice the efforts taken to hide intricate technical solutions.

In contrast to hiding technical innovations discreetly inside the walls of the frame by Mikli / Starck, the Danish company, Lindberg has another strategy. Being a part of their styling paradigm, innovative technical constructions are used to create the distinct looks of the Lindberg frames. A style that was derived from a wish to design optical frames assembled without screws, rivets or soldering (as stated by the official website). Each technical design is used as the basis for a styling platform as illustrated by the "AIR, Rim" frame (Figure 2).

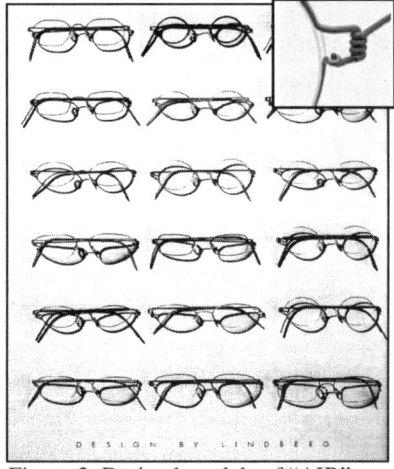

Figure 2. Derived models of "AIR" from one technical innovative detailed design.

In contrast to autodidact designers, many of the Lindberg Design Team members do have educational backgrounds in design (candidates from the ID department of the School of Architecture in Aarhus, Denmark). In terms of digital design tools, AutoCAD (2D + 3D) is preferred - as by many of their fellow designers (and manufacturers) in the optical frame design business. However, as many Lindberg products are based on thin plate terminology, the selection of AutoCad as a design/constructional tool causes problems to the designers. Presently, AutoCad does not support unfolding or other such features that are closely related to the construction of thin plate designs. As the company has its own production plants, Lindberg designers must be able to calculate and make constructions that can actually be produced in thin plate. For example, by creating laser cuttings and foldings themselves without software support. A task that sounds easy, but in reality is difficult and time consuming to perform. As such, the styling paradigm of Lindberg requires supplementary digital tools (e.g. 3rd party plug-ins to AutoCad), a radical change in software and design methodologies (e.g. selection of a feature based software like Solid Works), or an open minded management that spends time on extended calculations and (manual) drawing in combination with the creation of physical models. In a recent Danish newspaper (Erhvervs-Bladet, 2005) the managing director of Lindberg explains that to be able to compete, they have had to move much of the production from Denmark to the Philippines, and that many of their most fashionable competitors have moved their product development in the same direction. In other words, focus has been set on low cost manpower rather than investment in a technology lift that could enhance design productivity. In particular, by a highly integrated digital design process – from CAD to rapid prototyping and re-use of constructional ideas for future projects.

2.2. Mastering materials

It is evident from the case studies that the handling of materials divides the waters entirely. Entirely homogeneous materials or materials with a strict geometrical internal structure can be easily created and one can simulate the optical frame design in 3D CAD. Whereas heterogeneous materials (Figure 3), such as most acetate sheets today, are impossible to handle in 3D. This leaves physical prototypes as the only choice. Quoting Alain Mikli, "…as is with any natural material, acetate has its advantages and disadvantages. I have always wanted to work with the advantages of this material and forget or hide its disadvantages…" and when discussing use of CAD "…In the workshop there are no computers, only the work of human hands. I never make virtual 3D models on a computer; they would be meaningless. Every step by hand gives the frames soul and fills them with emotion…" (Mikli, 2004).

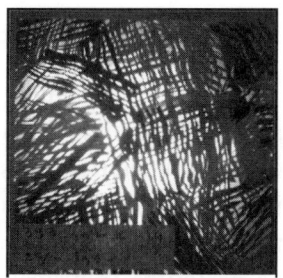

Figure 3. Heterogeneous material.

From a strategic point of view, the benefits of emphasizing the use of heterogeneous materials (e.g. acetate) in Mikli´s designs are obvious. As mentioned in the introduction, much of the optical frame production is carried out by companies in the Far East. This also applies to the production of prototypes. Having a workshop for the design and creation of physical prototypes of his own, Alain Mikli is able to explore the material and exploit it to its full potential. Quoting Alain Mikli, "everything revolves around passion, mistakes and improvisation: that is what makes the product so fantastic". And when discussing internal collaboration, communication, and exploitation of physical prototypes, "what I really like is to convey my passion for the product to my team who makes 300 models a year, but just 200 of them will be launched on the market!", (Mikli, 2004). In contrast his competitors who use a digital design process must base product development on their own imagination, knowledge of existing products on the market (which can inspire new products and even be an important part of the documentation for them), skills, experience, and ultimately on the prototyping facilities of their professional counterparts in, primarily, the Far East.

2.3. *Mastering giving shape (White Body)*

The matter of "technical innovation" exemplifies the separation of constructional details from the process of giving shape. Likewise is the handling of materials. As such, the last core competence related to styling, "giving shape", is regarded as a "White Body". The terminology - an expression derived from the Automobile design business - means that a prototype (physical or digital) is displayed without any material qualities (for example, surface textures, colouring, and so on) to enable visual evaluation of the mere body without interfering visual information. The prime objective of "giving shape in white body" is to create the looks of the "front", which designers in this business area define by the frame surrounding the spectacle lenses. Many optical frame designers regard this as the most important part of the design as this is the "eye catcher" to which all other components are visually related. Depending on expected geometric complexity of the front, different methodologies are used by Pro Design International A/S. For frames that are basically 2-dimensional (e.g. thin plate, titan frames) the design process is straightforward - a digital version of the classical, paper born design process. Line drawings in orthographical projections are made using AutoCad. Primarily, the front view is presented as the top view, and sides include complex visual elements (double curvatures) that are time consuming to create perfectly, and which add very little information.

From a strategic point of view, this process ensures a maximum of creative freedom and makes innovatively shaped products a reachable goal. Furthermore, as the digital design process enables re-use and communication of information to a much higher degree than the manual process, time to market is increased. In contrast, to the straightforward process described above is the process of giving shape to geometrically complex fronts. As in the first process mentioned, the line drawing of the front, in orthographical projection (front view) is made in AutoCad. Secondly, the top view is created. Thirdly, a 13 step procedure is followed where the most time consuming steps are the redrawing of 2D lines and addition of details like bevels or facets to all sides of the frame. 2D AutoCad line drawings are imported as DWG files into Solid Works (3D, MCAD). All lines are in segments and cannot be used to process 3D features. To enable extrusion and cutting in 3D, spline lines are drawn using the original AutoCad lines as background.

From the 2 views in orthographic projection the 3-dimensional solid body is constructed. Details like bevels or facets are applied to all sides to finish the design of the front. The critical aspects in this process are:

1) Low integration, i.e. minimal re-use of digital information between AutoCad and Solid Works (MCAD);
2) Although the result looks correct, it is not because the front must fit two parabolic curvatures (the spectacle lenses), but it does not. Being imprecise, the digital representation of the front is unusable for production of a mould or rapid prototyping (milling, STL, Z-print and so on) and last but not least;
3) Even though created digitally the result is rigid. Re-designing, e.g. manipulation of the style in 2D and 3D, is a time consuming task to perform.

This leads to the conclusion that one should create a platform for styling objects, based on 3D CAD, white body terminology. First aspects of the selection and implementation in CAD are discussed below.

3. Towards a Digital Platform

As is evident from the case of Pro Design International A/S, selection of CAD plays a major role. For 2D, AutoCad has been selected due to its default status in the optical frame design business. The unarticulated argument is that though manufacturers use this software it must be suitable for us too. Seen as a stand alone tool, AutoCad serves its purpose as it enables (technical) drawing. The designer, however, could just as well have selected a 2D software like Corel Draw for graphic illustration. First of all, the user-interface and the tools for manipulating line drawings are much more intuitive than the ones in AutoCad. This implies that Corel Draw users will have a quicker learning curve than AutoCad users - not least for autodidact industrial designers. Secondly, Corel Draw has import and export facilities for splines. In other words, time consuming redrawing of complex 2D curves, e.g. in Solid Works, would be history. As for selection of 3D CAD, there are basically two types of software to consider, MCAD (like Solid Works) and CAID (like Rhino). The latter is supposed to be superior in terms of easy creation of organic objects. However, as recent tests conducted by the author of this paper indicate, it seems that differences between the two in terms of modelling capabilities are diminishing. In other words, any of these would be suitable for a digital platform. As discussed in Schou and Nielsen (2004), differences remain in terms of different user interfaces, feature management, and the ability of MCAD to make technical drawings directly in 3D (copy and paste).

4. Creation of a Digital Platform

As previously mentioned, a digital platform for useful creation and (not least) manipulation of optical frames can be based on either of two 3D CAD systems: MCAD or CAID. In the following is a description, adapted from Song *et al.* (2004), that fits both systems. To facilitate the design process and reduce time, only half of the front is created in the beginning. Hence, the other half is mirrored. The initial process of creating this digital platform has two steps:

1) One parabola, the size and shape of standard spectacle lens is created.
2) Half of the (nose) bridge is built in terms of a curved line that is connected to the surface of the parabola (these steps are to be performed only once).
3) Imported from 2D and transferred to the curved surface - or created directly in 3D space, one outline (the top) of the frame is designed on the curved surface of the parabola.
4) The bottom line of the frame is created in the same way as in Step 3. Next, all lines are placed in 3D.
5) Sections (that represent the surface geometry) are placed along the lines. A sweep along path is executed. Now one half of the (solid) front is designed.
6) A mirror function is executed. Two identical half parts of the front are displayed.
7) The bridge is designed as a (curved) line between the two objects. To make a solid bridge Step 3 is used.

What has been gained? A digital styling platform (White Body) has been created that is believed to enhance the design process in many ways for some optical frame designers:

- The surfaces of the frame are defined by internal sections that are easy to manipulate in 2D or 3D. As for the latter this enables instant previews of experiments and results of manipulations.
- The digital 3D-model is accurate and can as such be used for rapid prototyping and manufacturing (milling of acetate frames or moulds for the manufacture of metal frames).
- Even though a model is geometrically complex, it can be documented in 2D technical drawings (guide, spline lines in combination with sectional cuts). Although this feature may sound outdated, it is still very valuable today because some manufacturers

have limited access to (sophisticated) 3D CAD systems and/or others lack the necessary knowledge to handle this kind of technology.

5. Conclusion

Since it has proved difficult in the business of optical frame design to radically improve the functional qualities of the products, styling of objects is very much in focus. To improve the succes rate (especially to launch the right product to the market at the right time), it has been found strategically valuable to focus on a digital styling platform, which is based on a white body terminology, in the product development process. A template of such a digital styling platform has been presented in this paper. Open for future investigation is to see how this is implemented in practice and evaluate its pros and cons. This will be carried out in an upcoming research project.

References

Boer, H. and Krabbendam, K. (1998). Organising for market-oriented manufacturing, Twente University, The Netherlands, 1.
Erhvervs-Bladet, 4.02.2005. Article on Lindberg outsourcing.
Lindberg, Catalogue (2003).
Lindberg official Website (2004, 2005).
Mikli, Alain (2004). Made in Passion, profile book + DVD.
Pugh, Stuart (1996). Creating Innovative Products Using Total Design. Addison-Wesley Publishing Company, 157, 209-273.
Schou, F. and Nielsen, S.H. (2004). Managing of total design: Use of computer tools in support of innovation, design and product development, IAMOT 2004, 4-5.
Song, Y., Vergeest, J.S.M. & Spanjaard, S. (2004). Fitting and manipulating freeform shapes by extendable freeform templates. Proceeding of the Shape Modeling International 2004, IEE 2004, 1-2, 4, 6-8.
Tjalve, Eskild (1983). Systematisk udformning af industriprodukter, Akademisk Forlag, 30-57.
Ulrich, Karl T. & Eppinger, Steven D. (2003). Product design and development, McGraw-Hill/Irwin, 40 - 41, 163 – 165, 180 -183, 200, 255 – 260.

CHAPTER 11

VALUE-CENTRIC E-GOVERNMENT: THE CASE OF DUBAI MUNICIPALITY

Habib Talhami and Mohammed Arif
The British University in Dubai, Dubai

In the past five years e-government initiatives have gained tremendous momentum and as a result governments worldwide are actively pursuing them. The implementation methodologies currently available view the implementation process as a progressive technological build-up resulting eventually in cost savings. It is important, however, that these methodologies take into consideration value-addition as early as possible. This paper presents the case of Dubai Municipality (DM) which has taken value-centric steps that have led to successful implementations and high acceptance rate. These steps are: 1) Developing effectiveness measures to track and measure value added; 2) Identifying high value services to both customers and the government; 3) Selecting services that are highly visible as well as less complex to implement first from a list of high value services; 4) Developing services through an all inclusive process in which the stakeholders are both involved and contribute to defining the final product; and 5) Providing training and support after implementation. These value-centric steps are generic and can be incorporated in any implementation process to add value that uses technology effectively to enhance service quality and reduce costs.

1. Introduction

E-government is gaining popularity worldwide. Virtual government is an ambitious goal that some researchers have been discussing for a few years now (Fountain, 2001). However, in the long run, the eventual success or failure of e-government will depend on the value it adds to citizens' lives, its government services as well as the cost savings that can be achieved. Some of the values e-government can add are: 1) 24 hour and seven day accessibility; 2) Active citizen participation (Wimmer, 2002); 3) Open government (an essential component of e-democracy); 4) Public access to information (Doty and Erdelez, 2002); 5) Avoidance of physical trips to government offices; and 6) The government can avoid maintenance of brick and mortar type facilities to handle citizen services (Kaylor *et al.*, 2001). The literature is replete with success factors for implementing successful e-government systems. Some of these success factors are: 1) To ensure competence in the required technologies (Borins, 2002); 2) Educating citizens about the value of e-government (Jaeger and Thompson, 2003); and 3) Development of both methods and performance indicators. There are several challenges associated with e-government that the General Accounting Office (GAO) report identifies as: "1) sustaining committed executive leadership; 2) building effective e-government business cases; 3) maintaining a citizen focus; 4) protecting personal privacy; 5) implementing appropriate security controls; 6) maintaining electronic records; 7) maintaining a robust technical infrastructure; 8) addressing IT human capital concerns; and 9) ensuring uniform service to the public" (GAO, 2001).

Taking into account the success factors and environment constraints, value has to be planned into an e-government project. Benefit to customers and government itself can be enhanced if value is planned properly into the project right from the beginning. It is critical to evaluate the extent to which e-government initiatives have to be implemented in order to harness maximum value from them. Knowing when to act is as important as knowing when not to act in e-government (Salem, 2003). This paper presents some ways of planning values into an e-government project through a case study of Dubai Municipality (DM). The rest of this paper is divided into four sections. The next section reviews IT and e-government implementation frameworks in the literature. Section 3 presents the DM case study. Section 4 analyzes the implementation and discusses the results; it is followed by Section 5 – the conclusion of this paper.When a molecular ion captures an electron, the molecular analogues of radiative and dielectronic recombination are in principle possible, however, they are usually completely overshadowed by a process which is far more effective than any of the atomic processes.

2. Implementation Methodologies

Layne and Lee (2001), presented a four stage model for implementation of e-government (see Figure 1).

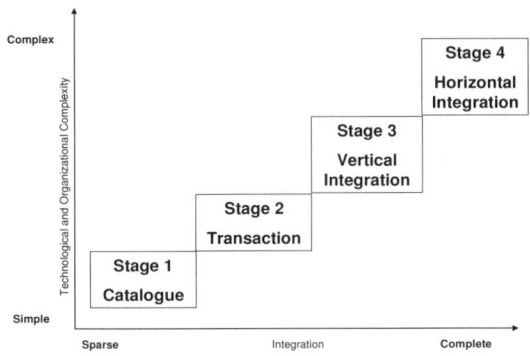

Figure 1. Four stage implementation model for e-government (Source: Layne and Lee, 2001).

As depicted in the model (Figure 1) the first stage of e-government implementation is the cataloging phase. At this stage government agencies announce their web presence through uploading a website. The website includes basic features that provide citizens with information and limited functionalities (such as form downloads). The second stage is the transaction stage. At this stage transactional capabilities are added to the website. A customer can submit a form, a request, or make a payment online. The third stage is the vertical integration and it involves the complete integration of back end as well as front end functions of the government for a certain process. The final stage is the horizontal integration stage. At this stage integration across departments and agencies is achieved. The case presented in this paper is at stage two for most of the functions that have been implemented so far.

A similar implementation model is presented by Hiller and Balenger (2001). They present a five stage e-government implementation strategy.

The first stage is the so-called one way communication, similar to the cataloging phase in the Layne and Lee (2001) model. The second stage is the two-way communication model. In this stage two way communications between the government and citizens take place without any financial transactions, which are still carried out through physical visits to government offices. The third stage is the financial transaction stage. At this stage, along with two way communication, financial transactions take place over the web. The fourth stage is the horizontal and vertical integration steps. At this stage, both vertical integration of back-office processes and horizontal integration across the organization takes place. The last stage is political participation stage. At this stage major exercises like elections can be conducted online.

Vriens and Achterbergh (2004) provide another three stage framework for e-government implementation. The first stage is to generate an e-government application portfolio. In this stage, the level of automation and the type of e-government services are identified. The second stage is the infrastructural requirements gap analysis phase. In this phase changes to technical, organizational, and infrastructure are identified. The third stage is the project definition and categorization stage. At this stage projects are categorized based on operational implications and other factors like political pressures.

All of these models are flexible enough to be implemented in most situations. However, as Layne and Lee (2001) pointed out most of the e-government initiatives are chaotic and become unmanageable in a very short time. As opposed to recent speculation in the literature (Moon, 2002), it seems that government institutions worldwide are far from the maturity, cost savings, revenue generating, or acceptable downsizing targets. Until now the status quo has been rather that of mild encouragement for internet technologies to reinvigorate local government (Musso *et al.*, 2000). One of the major reasons for this failure has been the technology driven implementation methodologies for e-government. Out of the three methodologies described above, the first two are technology driven. Stage 1 is a simple website; stage 2 builds up on the first stage and incorporates a two way communication through electronic data interchange and emails. Stage 3 and 4 involve implementation of an enterprise-wide system (e.g. ERP) and integration of the current system within the ERP. For the Hiller and Balenger (2001) model, the fifth stage involves adding technical features like encryption and authentication, chat-rooms, more sophisticated interfaces and a seamlessly integrated system. As is apparent from the description above, these methodologies are technology driven – the system moves from one stage to the next progressively incorporating more advanced technology. However, the success of an IT project depends on the attention paid to the processes to be automated, more than the technology used to implement it (Arif *et al.*, 2005).

The Vriens and Achterbergh (2004) approach involves two overall planning steps. However, in these two planning steps, the implementation of services is prioritized based on cost implications and political pressure. The approach emphasizes cost savings rather than adding value. The objective of this paper is to highlight some steps that can be taken prior to and during implementation in order that value is inherently embedded and constantly tracked both during as well as after the implementation. The following section presents a case study of DM, in which several steps were taken in the planning and implementation to increase the value-added to both citizens and the municipality. The case study was developed through interviews with DM officials and a review of the literature in the public domain.

3. Dubai Municipality – A Case Study

E-government initiatives in Dubai were started in 2001, through the directive of His Highness Sheikh Mohammed Bin Rashid Al Maktoum. The dual mission of this initiative was "to ease the lives of people and businesses interacting with government, and contribute to the establishment of Dubai as a leading economic hub". There were 27 government departments in Dubai and all of them were required to move 90% of their services online by the year 2007. One of these departments was the Dubai Municipality (DM). So far these initiatives are way ahead of its other gulf neighbors (Kostopoulos, 2004).

Dubai Municipality is regarded as one of the largest establishments in Dubai, in terms of the number of people it employs, the volume of services it provides to the public, and the projects it carries out.

The municipality was established in the 1940s with three employees, housed in a one room office. However, it was in 1965 that the Dubai municipality came into being officially. The municipality kept up a steady growth since its inception and now employs more than 11600 people.

In 2001, the Dubai Municipality embarked on a major e-government initiative that was triggered by a wider government initiative to automate all governmental functions. The vision for the Municipality e-services as defined by HH Sheikh Mohammed Bin Rashid Al Maktoum was:

"Use E-government solution as the primary delivering channel to provide a single, easy, integrated, and reliable means of access to Municipal information and services in order to continuously improve the quality of services provided for the residents, businesses, and partners, reduce internal operational overheads, enhance revenues, and promote Dubai's image as a commercial and tourism hub in the Gulf region."

This vision for e-government gave priority to transform the municipality into a customer-oriented, agile, accountable entity, and move away from the existing public bureaucratic organization. This was a major paradigm shift and required detailed planning on part of top management.

3.1. *Defining stakeholders and benefits*

The first step towards implementation taken by DM was to define the stakeholders in this project, and the benefit that each one of them was supposed to achieve through this initiative. The three stakeholders identified were: 1) Dubai Government; 2) Dubai Municipality; and 3) Customers. Benefits to these stakeholders were subsequently identified through brainstorming sessions, surveys, and focus groups. All the benefits to the stakeholders are summarized in Table 1.

As can be seen, the emphasis of this initiative was both on value to the government and the customers. For any e-government initiative to be successful it is important that emphasis be paid to both the government's perspective as well as the customer's perspective (Vassilakis *et al.*, 2003; Devadoss *et al.*, 2002; Stratford and Stratford, 2000).

Table 1. Stakeholders and benefits for DM e-government.

Dubai Government	Dubai Municipality	Customers
Achieve a balanced economy	Improve customer service	Quick and easy access 24/7
Attract large corporation	Improve processes	Prompt resolution of complaints
Compete with regional governments	Reduce costs	Promote electronic exchange
Reduce costs	Increase revenue	Reduce physical traffic
Generate enough revenues to become self financing		Transparency of internal procedures
		Faster processing of transactions

3.2. *Implementation plan development*

The DM offers more than 900 services. However, transforming all those services to e-government format is difficult, takes time, and needs resources. Therefore, a small sample from within these services had to be selected. The DM management decided to use the following 5-steps to prioritize its municipal services.

Step 1: Review the DM's Business plan - This step involved reviewing the 5-year business plan developed by the DM's senior management to understand the key business objectives and growth targets of the municipality.

Step 2: Identify the high value services which need to be e-government – enabled: This step involved identifying those Municipal services that would deliver the maximum value benefits to the DM and its customers if e-government enabled.

Step 3: Collect information & statistics about the various Municipal services: This step involved collecting operational statistics and information about each Municipal service to assist in the prioritization process. Such information included transaction volume, customer type, service type, and the number of DM departments involved in delivering the services.

Step 4: Prioritize the implementation of high value services – This step involved defining when to implement each of the high value services identified earlier. A structured analysis method is used based on service visibility and complexity.

Step 5: Validate and rationalize the results- This step involved using the experience of the DM project team and knowledge of DM's customers and operations to further verify the results of the analysis.

3.3. *Identifying high value services*

The second step in the 5-step methodology was to identify high value services. Identifying high value services involved reviewing the various Municipal services before selecting those services that once e-government enabled would deliver the maximum value to the DM and its customers. A service value questionnaire was designed by the e-government team and later used by department directors to evaluate different services. The questionnaire explored how e-government can add value to DM and its customers through a number of measures. These quantitative measures were categorized into two key groups summarized in Table 2.

Table 2. Values of services to DM and customers.

Values to DM	Values to Customer
Enhancing existing revenues	Minimizing the number of customer visits to DM premises
Setting up new revenue streams	Reducing the time required to request the services
Reducing cost of processing transactions	Reducing the time required to deliver a service
Delivering intangible benefits (i.e. Boosting the image of DM as a leading governmental organization)	Reducing fees and charges associated with the service
	Reducing the time spent by the customer to follow-up and track the progress of the requested services
	Reducing the time spent by the customer to file complaints, comments and suggestions

Based on the results of the questionnaire, services were classified into the following 4 major categories (see also Figure 2).

Category 1 – High DM value, High customer value: this group represented the high value services that should be web-enabled first (Marked by "1" in Figure 1). These services were expected to deliver key benefits to the DM in terms of increased revenues and cost reductions.

Also, these services were expected to deliver key benefits to DM's customers in terms of accessing Municipal services quickly, conveniently, and effectively. An example of a high value service is the "Service Directory" that is accessible over the Internet and describes the various DM Services, their target customers, and their key requirements. The directory service minimizes the customer service operational overhead across all departments by providing answers to customers' most frequently asked questions on-line. Furthermore, the service makes it easier and more efficient for customers to inquire about the different services, including their associated procedures and transactional requirements.

Category 2 – High DM value, Low customer value: this group represents one significant focus of governmental organizations that aspire to be self-financing or profitable. The services were expected to deliver key benefits to DM in terms of increased revenue and cost reductions. However, once the services are web-enabled, they will not necessarily deliver significant benefits to the customers. For example, the implementation of electronic procurement would enable the DM to communicate with suppliers electronically to acquire products faster. This would reduce the overall operational overhead of the purchasing department within the DM, yet the implementation of this advanced purchasing model would not have delivered any immediate and direct benefits to DM's customers.

Category 3 – Low DM value, High customer value: this group represents the main focus of non-profit governmental organizations – a focus exclusively on customer services. The services were expected to deliver key benefits to DM customers in terms of accessing municipal services quickly, conveniently and efficiently. However, enabling these web services would not have delivered any significant financial benefits to the DM in terms of increased revenue or reduced operational overhead. For example, "filing complaints" electronically over the internet delivers key benefits to the customer, but little financial value to DM.

Category 4 – Low DM value, Low customer value: this group of services provides little benefit and few advantages to the DM or its customers (shown by "4" in Figure 2). Therefore this was a low priority group of services for web-enablement.

Both Category 2 and Category 3 services were rated equally and it was decided that the prioritization would be done on a case by case basis. If it was difficult to determine the priority, then service visibility and complexity (described later) were used as tie breakers.

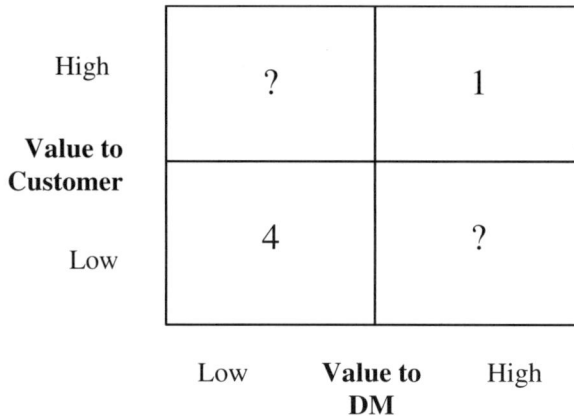

Figure 2. Step 1 in quick wins approach.

From the list of services developed until now, Category 1 services had to be prioritized to determine the timeline of their implementation. In addition, a methodology to break the tie between Category 2 and Category 3 had to be developed. Two attributes were discovered to prioritize the services.

Service visibility: This attribute described how significantly and extensively customers could potentially feel and experience the benefits achieved from enabling the web service. Services with a high volume of transactions and large customer base would be more visible to DM customers than other services with limited customer base. For example, the public health services provided for the residents of Dubai are more visible than housing services provided exclusively for UAE nationals.

Service Complexity: This attribute described how easily the service could be web-enabled. This depends on a number of factors such as the degree of existing automation, number of DM departments involved, number of external parties involved, and the number customer documents processed. For example, issuing of a "No Objection" certificate is more difficult to transform into an E-Service than issuing of a "Public Health" certificate. This is due to the large number of customer documents and DM departments involved in delivering the service.

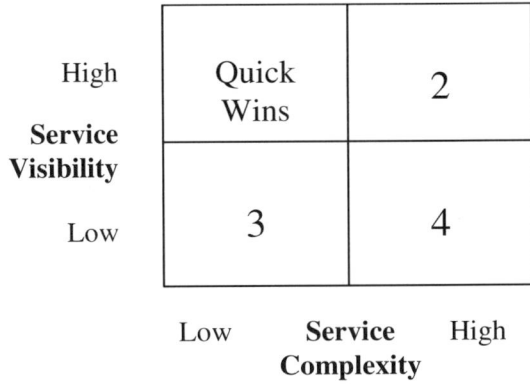

Figure 3. Step 2 in quick wins approach.

The analysis showed that high value services can be grouped into four major categories (see Figure 3):

1. *Category 1 (Quick Wins): high visibility, low complexity*: this group of high value services should be implemented first because these services can easily be transformed into e-services, which deliver highly visible benefits to customers of the DM.
2. *Category 2: high visibility, high complexity*: this is the second group of high value services to be implemented.
3. *Category 3: low visibility, low complexity*: this is the third group of high value services to be implemented because although these services can be easily web-enabled, their visibility is low.
4. *Category 4: low visibility, high complexity*: this is the final group of high value services to be implemented owing to the extra effort to enable them and their low visibility.

3.4. Designing the selected services

Once a service was selected for web-enablement, a 12 step product development process depicted in Figure 4 was followed.

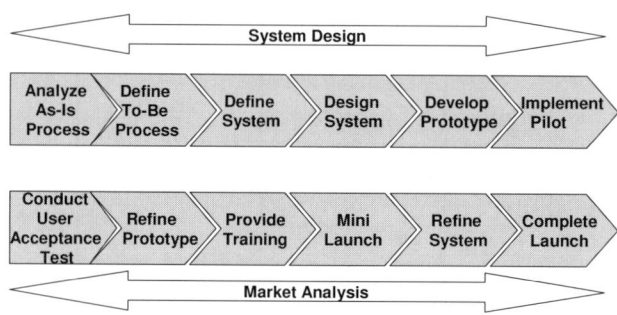

Figure 4. Service design process.

The first six steps are the system design steps and the subsequent six steps are the market analysis steps. The first system design step analyzes the current processes in an organization. Based on the needs of the organization, both redundancies and steps adding no value to the process were eliminated, and then a new process proposed. Based on the definition of the to-be process, system requirements are defined, and the defined system is designed. Once the system design is finalized, work begins on developing a prototype. In brainstorming sessions, the prototype is presented to representatives from the DM and its customers. Once the results of the acceptance tests are analyzed, the prototype is refined. Free training is offered to employees and customers at DM facilities. Following the training, the service is introduced at local level through kiosks placed at DM facilities. The objective of this so-called mini launch is to widen the base of customers who will test the system and provide feedback. Final modifications are made, which are based on the feedback received from the mini launch, before the new system launched.

This is a very comprehensive process. In the first six steps internal vision, strategies and future business goals are incorporated, and in the later six steps the quality of the service is refined through user tests. For all the services launched so far DM has maintained a 24 hour phone service to guide users through the process. Thus a system that adds value to both DM and the customer is obtained.

3.5. *Measuring the performance*

DM implemented three types of performance measures to manage the implementation of e-government and ensure the realization of tangible business benefits. These measures are

Progress Measures: involved establishing a set of quantitative indicators to measure the progress delivered in implementing e-government. These measures assess the progress achieved in transforming DM operations into E-government. The following quantitative measures were used:

- Percentage of customers (for each customer type) using e-services.
- Percentage of Municipal services transformed into eservices.
- Percentage of Municipal service information published over e-government.
- Percentage of transactions of each service executed electronically.
- Percentage of transactions for each customer type executed electronically.

DM Value Measures: involved establishing a set of quantitative indicators to measure the benefits gained by DM through using e-government. These measures aim to assess the advantages and benefits gained by DM through using e-Services. The following quantitative measures were used:

- Percentage of reduction in the cost of services (average cost of processing per transaction).
- Percentage of increase in revenue for each service.
- Number of new e-services revenue streams for each service.
- Total e-Service revenue achieved.
- Total number of transactions executed per staff member for a fixed period of time.

Customer Value Measures: involved establishing a set of quantitative indicators to measure the benefits gained by DM's customers through e-government. These measures aimed to assess the advantages and benefits gained by DM's customers through using e-Services. The following quantitative measures were used:

- Reduction in the number of visits to the DM premises for each service.
- Reduction in the number of documents required for submission for each services.
- Reduction in the time and effort required to request a service.
- Reduction in the time and effort required to track pending transactions for each service.
- Reduction in the time and effort required to acquire a service.
- Reduction in the time and effort required to file in and follow up complaints and suggestions.
- Reduction in the fees charged for each service.
- Reduction in the time and effort required to obtain information for each service.

In the initial phase, a select number of measures are being tracked (these are documented towards the end of the next section). However, all the DM value and the customer value measures will be tracked from 2007 onwards.

4. Implementation Factors Adding Value

Even though the implementation of e-government is still work in progress, the results so far are promising. The implementation process has been very value-centric and this has contributed significantly to its success. Some of the ways in which value has been added are:

4.1. *Value through defining effectiveness measures*

As has been described in the previous section, there are three categories of measures: 1) Progress tracking measures; 2) DM Value measures; and

3) Customer value measures. The first measure has been helpful in tracking the progress of the project so far. The last two measures will be helpful when the implementation is complete and the value added to DM and customers can be tracked. Having a clear indication of the effectiveness measures prior to implementation has provided an effective roadmap for implementation.

4.2. *Value through managing risks*

The implementation plan for the DM has been quite comprehensive and hence it can be argued that the well planned approach pays dividends. A great deal of planning and analysis has been performed in choosing services, prioritizing them and then making them customer driven. Service statistics and analysis tools such as service value questionnaires, service value analysis, and service implementation priority analysis reduce the risk of failure. Both customers and DM management provide valuable opinions and feed back for analysis, which helps tremendously in arriving at the right set of services to be web-enabled.

4.3. *Value through customer-centered implementation*

Inventory of all the services to be web-enabled were divided into four major categories (see Figure 2). Services that were high value to customers and DM were then selected. The timeline of implementation was then determined through a prioritization technique where they were further divided into four categories based on their implementation complexity and visibility (see Figure 3). Services that were highly visible, but less complex to implement were categorized as high priority or quick wins and implemented first. This two stage methodology has resulted in high value and less complex services being implemented first, thus establishing a highly visible web presence immediately.

4.4. *Value through a comprehensive product development process*

Once the services were identified and prioritized a twelve step process was performed for each of the services (see Figure 4). Through these twelve steps the DM ensures that both the DM and its customers' interests are safeguarded as well as high value services for the DM, which are very acceptable to its customers.

DM's emphasis on value has started showing some preliminary results. Not all the data being currently tracked is available in the public domain. However, the limited data available demonstrates that DM has been successful at transforming its vision into high value e-government services. By the end of 2003 the DM had launched 304 e-government services (54 transactional, 250 informational). The Number of e-government services launched by DM constituted 34% of all Internet services by the end of 2003. Compared to all other government departments in Dubai (304), the DM successfully launched the largest number of Internet services. The average number of monthly transactions has increased from 16,000 in 2003 to 45,000 in 2004, an increase of 181%. So far 750,000 online transactions have been performed and DM has 10,000 registered users. Currently the online usage of automated services is an impressive 90% of all services. In a recent study, Melitski *et al.* (2005) ranked a set of 84 municipality websites worldwide on the following five major criteria: 1) Privacy and security; 2) Usability; 3) Content; 4) Service delivery; and 5) Citizen participation. The DM is ranked 18th overall and 10th for privacy and security.

This paper has so far summarized the DM case, where value was considered as an integral part of e-Government implementation. The following section presents the main findings of this research

5. Summary and Concluding Remarks

E-government has been gaining popularity worldwide. However, the long term survival of this phenomenon is dependent upon all the stakeholders realizing the value in using this alternate channel for performing government operations. On the one hand current e-government implementation methodologies are technology centric, (Layne and Lee, 2001; Hiller and Balenger, 2001) in which implementation is performed by progressively building on the available technology. On the other hand they are cost centric (Vriens and

Achterbergh, 2004), in which services are inventoried and analyzed – a gap analysis using technological and organizational changes as key drivers – to ascertain the least expensive services for implementation. Even though these methodologies are very open ended and flexible, research into value incorporation for both governments and customers is sparse. Moreover, the emphasis of value as a major driver in the implementation process or existing methodologies is severely lacking. This research has documented the case of the Dubai Municipality; in particular attention was paid to addition of value for both customers and the DM during the implementation process. These steps can be incorporated in any implementation methodology and will contribute to its success. These steps include: 1) Developing effectiveness measures to track and measure value added; 2) Identifying high value services to both customers and the government; 3) Choosing services from amongst the high value services that are highly visible as well as less complex to implement first; 4) Developing services through an all inclusive process in which the stakeholders are involved with defining the final product; and 5) Providing training and support after implementation.

Even though the implementation is still work in progress, the results obtained so far have been encouraging. The user traffic has seen an immense growth, the number of transactions being performed has grown and DM now has 10,000 registered users. Despite the recent beginning of E-government initiatives, the progress until now is not only impressive but also the rated best e-government initiative in the Middle East (Kostopoulos, 2004), and among one of the best in the world (Melitski *et al.*, 2005). They offer a quality of service comparable to, and in some cases better than, cities like New York, Seoul, Sydney, Toronto, and Paris, where e-government initiatives are already long established (Melitski, 2005). The main reason for this success has been the value-centric approach highlighted throughout this research.

References

Arif, M., Kulonda, D.J., Jones, J.I., and Proctor, M.P., (2005), "Enterprise Information Systems: Technology First or Process First?", *Business Process Management Journal,* 11(1), 5-21.

Borins, S. (2002). On the frontiers of electronic governance: A report on the United States and Canada. *International Review of Administrative Sciences*, 68, 199-211.

Devadoss, P.R., Pan, S.L., and Huang, J.C., (2002), "Structural Analysis of e-government Initiatives: A Case of SCO", Decision Support Systems, 34, 253-269.

Doty, P. and Erdelez, S, (2002), "Information micro-practices in Texas rural courts: methods and issues for e-government", *Government Information Quarterly*, 19, 369-387.

Fountain, J., (2001), "Building the Virtual State: Information Technology and Institutional Change, Washington DC, Brookings Institution.

General Accounting Office, (2001), "Electronic Government: Challenges Must be Addressed With Effective Leadership and Management", GAO-01-959T, July 11, pp. 1-2.

Hiller, J., and Belanger, F., (2001), "Privacy Strategies for Electronic Government. Series. Arlington, VA: PricewaterHouseCooper Endowment for Business of the Government.

Jaeger, P.T. and Thompson, K.M. (2003), "e-government Around the World: Lessons, Challenges, and Future Directions", Government Information Quarterly, 20, 389-394.

Kaylor, C., Deshazo, R., Van Eck, D., (2001), "Gauging E-government: A Report on Implementing Services Among American Cities", *Government Information Quarterly*, 18, 293-307.

Kostopoulos, G.K., e-government in the Arabian Gulf: A Vision Towards Reality, *Electronic Government – An International Journal*, 1(3), 293-299.

Melitski, J., Holzer, M., Kim, S.T., Kim, C.G., and Rho, S.Y., (2005), "Digital Government Worldwide: An e-government Assessment of Municipal Websites", *International Journal of Electronic Government Research*, 1(1), 1-19.

Moon, J., (2002), "The Evolution of e-government Among Municipalities: Rhetoric or Reality?", Public Administration Review, July/August, 62(4), 424-433.

Musso, J., Weare, C., and Hale, M., (2000), "Designing Web Technologies for Local Governance Reform: Good Management or Good Democracy", Political Communication, 17(1), 1-19.

Salem, J.A. (2003), "Public and private sector interests in e-government: a look at the DOE's PubSCIENCE", Government Information Quarterly, 20, 13-27.

Stratford, J.S., and Stratford, J., (2000), "Computerized and Networked Government Information", *Journal of Government Information*, 27(3), 385-389.

Vassilakis, C., Laskaridis, G., Lepouras, G., Rouva, S., and Geogiadis, P., (2003), "A Framework for Managing the Lifecycle of Transactional e-government Services", Telematics and Informatics, 20, 315-329.

Wimmer, M.A., (2002), "A European perspective towards online one-stop government: the Egov project", *Electronic Commerce Research and Applications*, 1, 92-103.

CHAPTER 12

E-BUSINESS AND THE COMPANY STRATEGY: THE CASE OF THE CELTA AT GM BRAZIL

Sílvia Novaes Zilber

FEI University, Brazil

The Internet provides a global network infrastructure that is shifting business models, strategies and processes. Many authors reflect on the importance of incorporating e-business into the firm's global strategy. This paper deals with these issues in discussing the introduction of e-business activities by General Motors Brazil, specifically in connection with the launch of the Celta, an entry-level car designed to be sold on the Internet. A key to successful Internet strategies is the leadership shown by senior management. Technological demands may also conflict with the successful implementation of e-business initiatives, requiring greater interaction between the CEO and CIO. The importance of integration between employees on the business side and in information technology (IT) is highlighted in the context of GM Brazil's strategic objective to increase its market share for lower-price cars.

1. Introduction

The Internet has been a driver of change in business relationships. Electronic business enables consumers to interact directly with corporate information systems through the public infrastructure of the Web. The interconnectivity and interactive nature of the Internet make it a unique medium in a strategic context that differs from previous applications of information technology in business communications. For example, the electronic data interchange (EDI), which in its traditional form is based on rigid standards of information exchange over private networks between pre-existing business partners. Persuasive evidence has described the strategic use of information resources in organizations; information systems are strategic to the extent that they support a firm's business strategy.

Strategy can be defined as a "quest to match a firm's resources and capabilities to the opportunities and risks created by its external environment" (Grant 1991). During the 1980s the dominant Porterian *competitive forces* approach (Porter 1980) emphasized the relation of a company to its external environment, while the more recent resource-based view (RBV) highlights the need to consider a firm's internal resources and capabilities. Barney (1991) adds that for resources to create sustained competitive advantage they must be valuable, rare, imperfectly imitable, and not strategically substitutable.

Strategic use of Information Systems (IS) and related information technologies, such as Internet, can impact on organizational-level variables such as entry barriers, suppliers and customers, industry rivalry, search and switching costs, and intra and inter-organizational efficiency (Porter and Miller, 1985; Bakos and Treacy, 1986; Mahmood and Soon, 1991; Barua *et al.*, 2004).

In the context described above, Internet and e-business create competitive advantage. This paper discusses the specific case of an automotive manufacturer that used the Internet to achieve a strategic goal.

In fact, increased attention has been given to the alignment of IS and business strategies to create competitive advantage. It has been argued that lack of such alignment is the reason why many businesses fail to realize value from investments in IT (Henderson and Venkatraman, 1993).

According to Barua *et al.* (2004) there is no academic literature on the economic payoffs for Internet based business initiatives, but there is a rich body of literature on IT productivity and business value. In this research, we address the results of e-business adoption within the domain of Internet enabled business initiatives.

In the past, the firm had to deal with a fragmented customer and supplier base, and often incurred high costs when expanding its customer base. In contrast, Internet technologies have a significantly different impact on customer reach and richness of communication. Any customer with access to the Internet is able to gather information interactively, possibly customize and order products, check order status, and seek online advice.

When General Motors Brazil decided to increase its share of the market for entry-level cars, it opted for e-business as a means of making this possible: selling cars directly to consumers via the Internet in what is commonly called B2C, or business-to-consumer marketing.

This article has the objective of evaluating the applicability of a model of e-business operation in a company that has long operated in the physical world ("brick-and-mortar" company) yet has implemented business activities using the Internet.

The questions that arose from this objective were: *How was the e-business structured? How has electronic business addressed a market need? And, has it been consistent with the strategic choices for levering business?*

The case chosen to investigate these questions was the launching of the Celta car by GM Brazil: this company was able to opt for direct marketing of its new entry-level car model, the Celta, without dealers or other intermediaries for two reasons. First, it had a flexible manufacturing facility with customized assembly lines that were configured directly in accordance with online preferences from the end consumer. The second key facilitator of direct sales was the Internet.

2. Literature Review

Contemporary organizations are aligning Information Systems (IS) and business strategies to improve organizational performance (Kearns and Lederer, 2000). According to Turban (2000), e-business is the delivery of information, products and services, or payments via telephone lines, computer networks or other electronic means. Cunningham (2001) defines e-business as commercial transactions conducted over public or private networks, including public and private transactions that use the Internet as the means of implementation. These transactions include funds transfers, online exchanges, auctions, product and service distribution, supply chain activities, and integrated corporate networks.

According to Barua (2004) any customer with access to the Internet is able to gather information interactively regardless of time and location, (possibly) customize and order products/services, change orders dynamically, check order status, and seek online advice. According to a study by Freeland & Stirton (2000) for The Boston Consulting Group, in theory many established companies with solid experience in the "real world" or traditional "brick-and-mortar" companies are well positioned to succeed in e-commerce. They have critical assets, such as strong brands, customer relationships, and logistic systems, which provide them with an edge over startup competitors. In practice, however, traditional bricks-and-mortar firms will not be able to exploit these assets unless they are effectively organized for e-business. Indeed, for large firms the most difficult challenges of e-business are not so much strategic as organizational. According to Applegate (2001), a business model describes succinctly how the business is structured, which people are needed for that business and what roles they perform. The description of a firm's e-business activities in its business model facilitates analysis of its business structure and of the roles people play in it.

The model proposed by Applegate (2001) was used for this study, a model that consists of three components: business concept, capabilities, and value. Chesbrough (2001) stresses that the creation of a business model differs from the conventional notion of developing a strategy because a business model is more than an attempt to hypothesize an exploratory initiative within a given market: it is a fully worked out and well-defined plan of action. Most large corporations have not one but many e-business initiatives. In such cases a business model can be useful to coordinate e-business activities, set out their overall goals, and serve as a framework for evaluating their results. In particular, to measure whether the results match the goals set in the model. According to Oliveira (2001) the organizational structure describes the arrangement and grouping of activities and resources to achieve a companies established goals and results.

When a firm decides to implement e-business as an additional activity it will have an impact on existing processes and operations. Freeland & Stirton (2000) also report that the most difficult challenges are organizational rather than strategic in their analysis of the Boston

Consulting Group study of the organization of e-business. The success of a company's e-business strategy depends on its ability to organize appropriately, these authors conclude. In many situations a sound strategy founders on organizational problems. Several authors (Lientz 2001, Tapscott 1998, Plant 2000, Robert & Racine 2001, Kalakota 1999, Turban 2000) stress the importance of integrating the firm's objectives and global strategy with the e-business design adopted. Success in e-business requires that investment in technological infrastructure for e-business be linked to the organization's plans, strategies and tactics. As for the question of governance, Freeland & Stirton (2000) suggest the creation of a small e-commerce center with decision-making authority within the organizational structure. The authors reported on hundreds of large firms in several sectors and found a tendency to create a small yet powerful central unit that coordinates e-commerce. This unit is led by a senior executive who reports directly to the CEO. Barua *et al.*'s (2004) model of business value for Internet enabled business suggests that performance is ultimately judged by traditional financial performance measures such as revenue per employee, gross profit margin, return on assets, and return on invested capital.

3. Research Methodology

The present study can be considered as having an exploratory nature due to the contemporary character of the phenomenon studied and because of the limited amount of academic knowledge accumulated on the topic to date. Therefore, resulting from the genre of the research undertaken, there is no concern here in establishing relations between dependent and independent variables in order to prove or disprove pre-determined hypotheses.

The present research was based on a case study method that focused on the launching of the Celta, an entry-level car produced by GM Brazil and studied in depth by Zilber (2002). According to Yin (1990) the case study is the preferred strategy when the questions are presented in the form of "How" or "Why," which is the present case, where the questions are: "How is the e-business structured?" and "Why was a given model adopted for the e-business operations?"

It was decided to adopt the business model proposed by Applegate (2001) for studying the issues proposed in this study. This model consists of three components: business concept, capabilities, and value. The categories of analysis for each component of the model are shown below:

An organization's ***business concept*** defines its: Market opportunities; Products and services offered; Competitive dynamics; Strategy for capturing a dominant position; Strategic options for evolving the business.

An organization's ***capabilities*** are built and delivered through its: People and partners; Organization and culture; Operating model; Marketing/sales model; Management model; Business development model; Infrastructure model.

Value is measured by: Benefits returned to all stakeholders; Benefits returned to the firm; Market share and performance; Brand and reputation; Financial performance.

The factors selected from this model of study were:

- Within the *"**Business Concept**"* component: *Market opportunities, Competitive dynamics,* and *Strategic options to lever business.*
- Within the *"**Capabilities**"* component: *Organizational structure* and a *Marketing/ sales model.*

The case chosen to be studied in depth was that of General Motors Brazil and, more specifically, its launching of the Celta, an entry-level and lower-priced automobile. This choice was made, in the first place, because this car was sold entirely over the Internet, which constitutes a case of the use of the Internet in a strategic way for carrying out the company's business.

Managers and directors of GM whose activities had been affected by e-business operations were interviewed. They were the people responsible for IT areas in the company, as well as for the areas of marketing, purchasing, sales and the so called e-business area. These interviews were based on a semi-structured guideline, and drawn up with the purpose of obtaining replies to the questions being studied.

4. Case Presentation: The Launching of the Celta by GM Brazil

4.1. *The role of the company's global strategy in defining the e-business model*

There are two major blocks of e-business operations at GM Brazil: a) B2C operations, which involve direct sale of cars on the Internet, including the case described in this article, the launching of the Celta (a lower-priced car released only through the Internet); and b) B2B operations, which involve the company's purchasing area and its relationships with suppliers.

There is no business coordinator for all e-business operations. Therefore, B2C activities are coordinated by an area subordinated to the Marketing Department, while B2B operations are subordinated to the purchasing area, with no specific coordination. It is interesting to note that, right from the conception phase, the sale of the Celta on the Internet was an idea generated by the company's senior marketing board, with backing from upper management. An organizational structure dedicated to this project was created that continued to operate even after conclusion of the project by providing maintenance to the routine e-business operations related to B2C. It was also responsible for generating new ideas. But the B2B operations were not centralized in a single dedicated organizational structure. They were inserted into the existing organizational structure. Perhaps for this reason, they failed to attain the same financial results as the B2C. Therefore, a dedicated e-business management for B2C operations, which was subordinated to the director of marketing, was setup. Under this management is both staff (a total of approximately ten people) and technical areas.

On the IT side, there is an e-commerce management that is also fully dedicated to these activities. The e-business initiatives exercised in the area of purchasing (B2B) do not have the volume of business generated by the B2C area. The business model used was a *value creation model* based on business-to-consumer (B2C) retailing and using metrics for results and infrastructure suited to e-business. B2C means direct sales via the Internet. Results are to be measured using a specific set of metrics for e-commerce. The model also calls for investment in e-business infrastructure in the form of close links between the company's information technology (IT) department and the business units involved in direct marketing of entry-level cars via the Internet.

The project triggered changes in the company's organizational structure, including hiring of new personnel and the creation of new units. It is worth noting the clear evidence of a link between company strategy and the formulation of the e-business project. The idea of selling cars via the Internet arose from the strategic goal of increasing market share in the entry-level segment. Owing to GM Brazil's investment of more than US$800 million in the Celta, it had a significant confidence in the Brazilian market over several years. The aim was to improve GM's positioning against its competitors and in particular to challenge its main competitor, the market leader in the entry-level segment.

The success of the new car would also mean the difference between the success and failure of a strategy established in 1992. At that time GM Brazil decided to expand its offerings to include products for all segments from entry-level cars to heavy-duty trucks. The strategy proved successful. GM's market share in Brazil rose from 21% to 25%. The problem was that the strategy could not be sustained without a significant presence in the market for small or entry-level cars, which accounted for 70% of all car sales in Brazil. The aim of the Celta project was to increase GM's competitiveness in this segment. The price would be higher than that of its competitor, but GM expected superior design and technology to offset the higher price.

In addition to superior technology, GM focused on using a low-cost production process rather than an inexpensive design. This was achieved by implementing a modular plant in consortium with suppliers and flexible production methods to turn out 120,000 units per year. GM's modular consortium has 17 suppliers strategically located in the same plant and connected in real time. This enables suppliers, for example, to know just when their products are needed on the assembly line.

The purpose of this short introduction before our detailed presentation of the B2C model developed by GM Brazil is to highlight the connection between the carmaker's e-business initiative and its very clear strategy of growing market share using technology, which enables product sales directly to the consumer.

When the project began at GM (around 2000), there was a worldwide movement in favor of giving priority to e-business and GM Brazil's parent company created a new structure called "e-GM." Thus the initiative of direct sales via the Internet matched the parent company's expectations. A clear directive came from headquarters, indicating that subsidiaries should "invest in e-business projects." At the same time, the Brazilian subsidiary adopted for a strategy of market leadership in the entry-level car category.

The marketing director of GM Brazil then had the idea to combine both of the policy directives - investment in e-business worldwide and growing market share in entry-level cars in Brazil - with a new plant in Brazil that made flexible manufacturing possible. The initial idea was to use the "e-shop" system already put in place by the parent company in the U.S. This system assisted the buyer in configuring the product, calculated the price, and told the buyer where the product was located. However, it did not enable direct sales via the Internet.

The marketing department in Brazil was bolder. It created a website that completed the process by adding direct sales to configuration and pricing. Thus GM Brazil created a structure dedicated to developing this project of direct marketing via the Internet.

The next section describes the business model for direct sales of entry-level cars to consumers via the Internet developed by GM Brazil's marketing department in collaboration with IT.

4.2. *GM Brazil's process for selling the "Celta" online*

The business model for the B2C project to sell the Celta online had four pillars:

- Consumers were to be billed directly and not through dealerships
- Pricing was to be unified for the entire country (prices may now differ depending on the region)
- Rapid delivery
- Given the need to develop IT systems for this strategy, it was decided to use the Internet. The expected volume of customer interactions was large, justifying an automated and integrated

process. A concrete benefit would be direct knowledge of consumer wants and needs.

Dealerships were involved only in final delivery of the car to the purchaser. The selling process was direct to the final consumer but delivery remained in the hands of dealers in order to avoid channel conflict. The system adopted by GM Brazil includes rather than excludes dealers: the consumer buys a car and selects the dealer who will deliver it. Dealers receive a delivery fee, which is smaller than the commission on traditional sales but may be commercially more attractive because the dealer is not required to hold inventory. However, dealers need to have a good costing system to realize that the new procedure is more profitable.

The system presents a number of advantages for GM:

1) GM saves 5% of the commission normally paid to dealers, and transforms this saving into a discount on the price for the consumer;
2) Producing cars for delivery by dealers means holding inventory, and inventory equals cost. Producing a direct sale to consumers reduces inventory and hence cost;
3) Direct contact with consumers enables GM to obtain customer information, which can be used to improve CRM. The type of information relates to color, model and accessories. The information is stored in a customer database and provides business intelligence for use in future transactions; and
4) Direct marketing leads to better knowledge of customers' wants and needs, thus it reduces the number of models and options required as well as facilitating production, which can be tailored to demand.

Advantages for consumers: The consumer pays less and gets faster delivery.

Dealership's involvement in online sales:

- Customers with Internet access make choices online, save the configuration, and go to the dealer to pick up a car.

- Customers without Internet access go to a dealership to use the sales site, where they can configure the car they want to buy. For the dealer the benefit is selling cars without needing to have them physically on the premises.

4.3. *Possible problems that could arise from this sales system*

- Resistance on the part of the consumer public, used to "trying out" the product on dealers premises, where they can get into the car and check its equipment, take test drives and have other types of direct personal contact.
- Resistance on the part of the dealers, who would receive a lower commission for delivering cars sold on the Internet than if they made the sales themselves.
- Since consumers can configure certain items according to their needs and preferences, the factory could run into problems if it were not flexible enough, and this could cause delays in delivery.

The results obtained with the launching of the Celta went far beyond expectations: sales exceeded initial estimations and the company is thinking of selling more products directly to consumers in this way. The site has to be updated every time a new product is included. There are also logistic issues that have to be analyzed carefully before new models are put on the market. Thus, this e-business model is intrinsically related to the company's strategy in the sense that it is directly linked to its core-business.

In 2002, 30% of GM Brazil's total sales were carried out online. In 2001, sales of the Celta via the Internet accounted for almost half of the revenue generated by B2C overall in Brazil.

4.4. *Organizational structure for online car sales*

The marketing director's idea was not only approved by the executive committee of GM Brazil's parent company, but put together with a global budget that resulted in a world pilot of e-commerce and direct online marketing. The world's first direct sales e-commerce site for entry-level cars went live in September 2000, selling the car Celta. The e-business area needs information from other parts of the company such as: brand or product management, pricing, vehicle distribution (which cars can be delivered where, deciding how many cars are sent to each distribution center), billing and accounts, and sales. E-business personnel are constantly working to familiarize themselves with the day-to-day operations of other areas and make or collect suggestions on new projects.

Figure 1 below shows the structure of the e-business group (B2C) at GM Brazil in the early 2002:

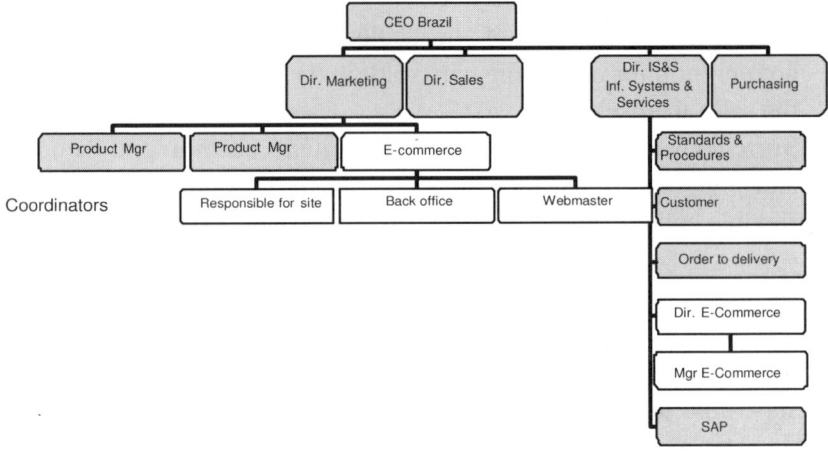

Figure 1. Organizational structure of GM Brazil and e-business areas – early 2002 (Source: Developed by the author).

Although the e-business manager has a certain amount of decision making authority, in certain cases issues are resolved by the head of department, director, board and so on, depending on the type of decision required.

The e-business marketing director was already a GM employee. He was formerly a marketing manager and was reallocated to this position. The e-business director in the IT area was hired from outside with a specific mandate to implement the online selling project. In 2001 the e-business director acquired control of sales support in general, and not just Internet sales support.

A new structure was set up to both increase the hierarchical status of e-business and focus on the online selling of the Celta. This new structure originated from the idea of selling the new entry-level model directly to consumers via the Internet. The fact that there was a centralized structure for B2C operations allowed the information coming in from the content providers to be analyzed. It was then utilized for: keeping up-to-date information, allowing periodic input into this information, updating the sales site, and communication with consumers.

5. Conclusions

The main objective of this article was to analyze the implementation of e-business operations in a company that has long operated in the physical world. In particular by using an e-business model that took into account the following dimensions:

- In regard to the ***business concept*** component: business opportunities, competitive dynamics and strategic options;
- In regard to the ***Capabilities*** component: Organizational structure and a Marketing/sales model.

The segment studied was the automotive industry, which in terms of its dynamics is highly competitive. All of its players are looking for innovations and opportunities to give them competitive advantages over their competitors. The segment of small and entry-level cars is responsible for 70% of car sales in Brazil, and GM Brazil had no significant slice in this segment. There was therefore an attractive opportunity for growth and the company went looking for strategic options that would lever its sales in this segment.

GM was able to opt for direct marketing of its new entry-level model, the Celta, without dealers or other intermediaries, for two reasons. First, it had a flexible manufacturing facility with assembly lines configured to allow small modifications in accordance with configurations selected online by consumers. The second key facilitator of direct sales was the Internet.

It was a clear innovation in the definition of "core competence" of the assembly plant. To put this direct sales initiative into practice, one of the company's strategic options was to present a clear link between e-business initiatives and corporate strategy. The aim of the B2C project was to increase sales of entry-level cars, furthermore many authors (Kearns and Lederer, 2000; Henderson and Venkatraman, 1993; Chan and Huff, 1993) report that this strategic alignment increases the company's competitive advantage.

The *competitive dynamics* of this segment show that price is a fundamental factor in consumers' final decisions. The use of the Internet as a selling tool enabled a reduction in the price of the Celta - through a reduction in costs (reduction in the IPI Tax because of direct sale, and reduction in the fees paid to dealers) - and proved itself as a new form of obtaining competitive advantage.

Besides the price factor, the possibility that the final consumer could configure the car before delivery proved to be an extra factor that adds value. It is important to emphasize that this type of configuration was only possible due to the existence of an innovative manufacturing model at the plant, which employs a flexible production process.

Top management involvement was total, with the original idea coming from the marketing director of GM Brazil. The project was successful in financial and marketing terms, exceeding expectations. This success was apparently due to the combination of an appropriate business model and a highly committed top management.

Partnering with dealers played a fundamental role by avoiding channel conflict. The role allotted to dealers was to deliver cars purchased on the Web. As noted by Weil & Vitale (2001), one way of avoiding channel conflict is to realign the distributor's functions.

Porter (2001) suggests that Internet has a complementary role, in the sense that it should reinforce the competitive advantage of the "real" part of the business. When this is the case, results are satisfactory. This conclusion undoubtedly applies to the case of GM Brazil, in which the e-business initiative was in line with a clear strategic objective, i.e. to increase market share in the entry-level segment. This was the first project in the world to sell cars directly online, stimulating both global acclaim and recognition by the respective parent company.

In terms of *strategic options used to lever business*, the company made an innovative use of the Internet. In fact, linking consumers to manufacturers for direct marketing of passenger cars would have been practically impossible without the Internet. In addition, the Internet suits the strategies that the automotive manufacturers are currently pursuing, according to a study by Santos (2001) on: differentiation (for example, the Internet facilitates differentiation in services), associations and alliances (facilitated by online interconnectivity) and, above all, by geographic expansion (the Internet is a global network).

The initiative made a significant difference, as reflected in the organizational structure utilized. The company created (and has kept in place), especially for this project, an IT unit totally dedicated to supporting direct sales of this entry-level model via the Internet, thus guaranteeing adequate IT infrastructure. On the business side, a totally dedicated structure was also created to guarantee the necessary functionalities and meet the needs of customers, dealers, and manufacturing.

6. Contributions

The following aspects of the Celta case can be tallied up as contributions to understanding the mechanisms involved in implementing e-business operations in a company that until then had only operated in the physical world:

- The fundamental role of the company's global strategy in defining the model to be adopted for e-business.
- Involvement of top management in providing the needed resources.
- Use of an adequate organizational structure in both the conception and implementation stages of the matrix structure-project, with central coordination of the e-business operations by one person responsible for IT and another for the business area.
- The existence of an assembly plant with a modular consortium and flexible production that allows real-time alterations in the production line so that customers can customize cars over the Internet.

- According to Amit and Zott (2001) value is created by e-business through the way in which transactions are carried out. In this sense, the company was efficient in using the Internet as a means of facilitating the traffic of information between manufacturer, customer and dealer (for delivery of the car). A high level retention was attained by allowing customers to customize the details of their cars over the Internet as well as putting customers into direct contact with manufacturers.
- The innovative use of the Internet to let customers configure car models was already implemented all over the world, however the final step of purchase for the complete processing of the sale over the Internet was the new innovation in the case of the Celta in Brazil.
- GM showcased its own use of the Internet as a marketing tool.
- The following can be considered the main difficulties involved:
  - Centralization of all e-business operations under a single coordinator. For example, B2B operations have a decentralized coordination, and post-sale operations do not use the same structure. Results obtained through these operations were not as successful as those attained through sales over the Internet.
  - Measurement of other e-business operations: direct sale over the Internet generates revenue, whereas other operations that consist purely of relationships are harder to measure.
  - Competitors have since launched their own direct sales sites. Therefore, GM is no longer the sole reference for this innovation.

References

Applegate, Lynda M. (2001). Emerging E-Business Models: Lessons from the Field, Harvard Business School,
  http://monkey.icu.ac.kr/sslab/course/ICE720/data/subjectPresentation/EmergingE-BusinessModel.pdf
  [www.stuart.iit.edu/courses/ecom530/fall2001/bmodels2.pdf]
Bakos, J.Y. & Treacy, M.E. (1986) Information Technology and corporate strategy: a research perspective MIS Quarterly 10 (2) , 107-119.
Barney, J. (1991) Firm Resources and Sustained Competitive Advantage, *Journal of Management* 17(1): 99-120.

Barua, A., Konana, P., Whinston, A.B. & Yin, F. (2004) Assessing Internet Enabled Business Value: An Exploratory Investigation, MIS Quarterly, crec.mccombs.utexas.edu http://crec.mccombs.utexas.edu/works/articles/barua_konana_whinston_yin_01311.pdf

Chesbrough, Henry & Rosenbloom, Richard S. (2001). The Role of the Business Model in Capturing Value from Innovation: Evidence from Xerox Corporation's Technology Spinoff Companies, www.hbs.edu/dor/papers2/0001/01-002.pdf

Cunningham, Michael J. (2001). B2B business-to-business: como implementar estratégias de e-commerce entre empresas, Rio de Janeiro, Ed. Campus.

Freeland, Grant D. e Stirton, Scott (2000) Organizing for E-commerce, discussion paper, The Boston Consulting Group Inc., www.bcg.com/publications/files/organizing%20ECommerce%20Apr%2000.pdf

Grant, R. M. (1991), The Resource-Based Theory of Competitive Advantage: Implications for Strategy Formulation, *California Management Review*(Spring): 114-135.

Henderson, J.C. & Venkatraman, N. (1993) Strategic Alignment: leveraging information technology for transforming organizations. IBM Systems Journal. 32 (1) 4-16.

Kalakota, R. & Robinson, M. (1999). E-Business-Roadmap for Success, Addison Wesley Longman, Inc, Massachusetts.

Kearns, G.S. & Lederer, A.L. (2000) The effect of strategic alignment on the use of IS-based resources for competitive advantage, *Journal of Strategic Information Systems,* 2000 - staffs.ac.uk http://web.staffs.ac.uk/~bstpjw/ebusiness/downloads/stman04.pdf.

Kraemer, K.L. & Dedrick, J. (2002) Strategic use of the Internet and e-commerce: *Cisco Systems, Journal of Strategic Information Systems,* 2002 , 5-29. http://www.doai.uom.gr/mai/Course_Material/E_Commerce/folinas/7.pdf.

Lientz, Bennet P. & Rea, Kathrin P. (2001). Transform Your Business Into E, California, Academic Press.

Mahmood, M.A. & Soon, S.K. (1991) A comprehensive model for measuring the potencial impact of information technology on organizational strategic variables. *Decision Sciences* 22 (4), 869-897.

Oliveira, Djalma de Pinho Rebouças de. (2001). Sistemas, organização e métodos: uma abordagem gerencial, Atlas, São Paulo.

Plant, Robert (2000). E-Commerce Formulation of Strategy, NJ , Prentice Hall.

Porter, M. E. (1980) Competitive Strategy: Techniques for Analyzing Industries and Competitors, New York: The Free Press.

Porter, M. E. (2001). Strategy and the Internet, *Harvard Business Review,* pp. 63-78, March.

Porter, M.E. & Millar, V.E. (1985). How information gives you competitive advantage *Harvard Business Review,* 149-160.

Robert, Michael & Racine, Bernard (2001). E-Strategy Pure & Simple, New York, McGraw-Hill.
Santos, Angela M. Medeiros M. (2001). Reestruturação da Indústria Automobilística na América do Sul, BNDES Setorial, Rio de Janeiro, n. 14, September, www.bndes.gov.br/conhecimento/bnset/set1403.pdf
Slack, N. & Lewis, M. A. (2001) Operations Strategy: FT Prentice Hall.
Tapscott, Don, Lowy, Alex, & Ticoll, David (2000). Plano de Ação para uma Economia Digital, Makron Books do Brasil Ltda.
Turban, Efraim et al (2000). Electronic Commerce: a Managerial Perspective, New Jersey, Prentice-Hall Inc.
Weil, Peter & Vitale, Michael R. (2001). Place to Space – migrating to e-business models, Harvard Business School Publishing Corporation, USA.
Zilber, Silvia Novaes (2002). Fatores Críticos para o desenho e Implantação de e-business por empresas tradicionais, PhD thesis, Faculdade de Economia , Administração e Contabilidade (FEA), Universidade de São Paulo (supervisor Prof. Eduardo Vasconcellos).

SECTION II

BUSINESS ORGANIZATION

CHAPTER 13

BUSINESS MODEL DESIGN AND EVOLUTION

Michael Weiss* and Daniel Amyot**

*School of Computer Science, Carleton University, Canada;
**School of Information Technology and Engineering,
University of Ottawa, Canada

In today's rapidly evolving world, companies need to constantly adjust their business models to changes in their environment. A good approach to evolving business models strikes a balance between capitalizing on new opportunities, and preserving investments in existing business processes. In this chapter, we argue that the User Requirements Notation (URN) provides such an approach. URN supports the modeling and analysis of user requirements in the form of goals and scenarios. Goals can be used to model high-level business (as well as system-level) objectives, and scenarios to describe the business processes to meet those goals. The approach is lightweight, and allows the quick evaluation of business model alternatives. Business models are represented in terms of actors and their dependencies, which correspond to value flows between the actors. Those value flows can subsequently be refined into business process activities. The approach gives business managers a tool for the systematic and incremental evolution of business model alternatives for their organizations. It allows them to model the strategic options available to them, and the conditions for their successful application.

1. Introduction

The objectives of this chapter are to:

(1) Introduce a lightweight approach for evaluating business model alternatives;

(2) Using a short example, demonstrate how the approach allows business managers to model the strategic options available to them, and the conditions for when they apply.

The focus of this chapter is on the early stages of business model design given a set of business objectives and informal requirements. We describe how a business model can be represented in terms of actors, the goals of those actors, and the dependencies between actors in achieving those goals. The dependencies indicate value flows between the actors, which can subsequently be refined into business process activities. In this chapter, we do not discuss those later refinement stages, but refer the reader to our recent work on business process modeling using URN (Weiss and Amyot, 2005).

The chapter first provides a short introduction to URN, illustrates business model design using goals, and introduces the supply chain management case study. It then discusses business model evolution and describes how URN allows stakeholders (such as a manufacturer) to experiment with different business model alternatives from their perspective. A brief overview of related work and conclusions follows.

2. User Requirements Notation

The purpose of URN is to support, in a semi-formal and lightweight manner, the modeling and analysis of user requirements in the form of goals and scenarios. URN has many concepts that are relevant for business process modeling, such as behavior, structure, goals, and non-functional requirements. URN combines two complementary notations.

The *Goal-Oriented Requirements Language* (GRL) is described in (URN Focus Group, 2003) and summarized in Figure 1. GRL captures business or system goals, alternative means of achieving goals, and the rationale for goals and alternatives. The notation is especially good for the modeling of non-functional requirements. It provides a higher, strategic level of modeling of the current system and its future evolution.

GRL originates from the Non-Functional Requirements (NFR) and *i** frameworks (Chung *et al.*, 2000), and supports multiple types of diagrams. Actor diagrams are used to model the *strategic dependencies* between actors, as well as the internal goals of individual actors. Rationale diagrams are used to compare architectural alternatives. They allow us to model the impact of each alternative on high-level business or system goals.

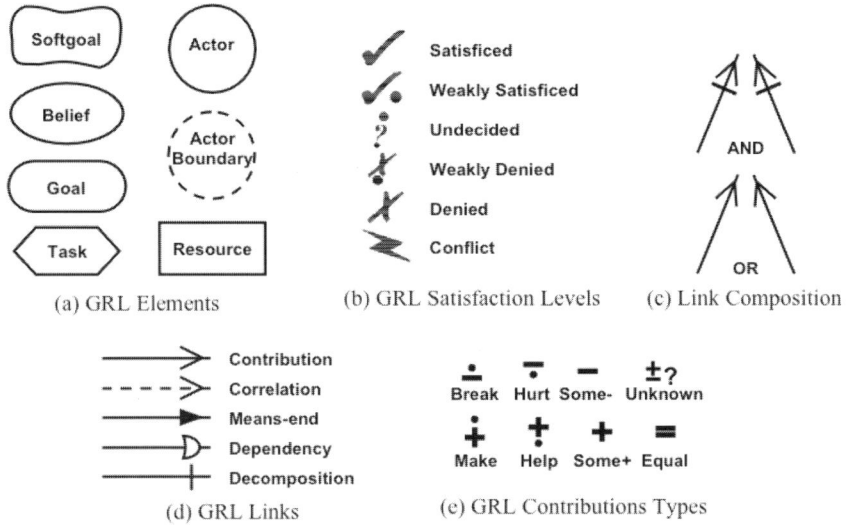

Figure 1. Elements of the goal-oriented requirements language.

The second part of URN is the *Use Case Map* (UCM) notation, described in (URN Focus Group, 2003b). This notation was first proposed to capture emerging behavioral scenarios during the high-level design of distributed object-oriented reactive systems (Buhr, 1998). It was later found to be an appropriate notation for describing operational requirements and services. A UCM model depicts *scenarios* as causal flows of responsibilities that can be superimposed on underlying structures of components.

As noted earlier, our emphasis in this chapter is on modeling strategic options and the conditions on applying them. For this reason, we will not discuss the refinement of GRL models into UCMs in this chapter. However, interested readers are referred to the URN tutorial by (Amyot, 2003) for a general overview on the use of UCMs in URN, and to our work on business process modeling with URN (Weiss and Amyot, 2005).

3. Business Model Design

In this section we focus on modeling the current business (its evolution is discussed in the next section). We also introduce the supply chain management case study.

We adopt the definition of an *(e-)business model* from (Weill and Vitale, 2001) as a set of participants and the flows between them. The participants include the company whose business model we are describing, its customers, suppliers, and allies or intermediaries. Value is created in the form of information, product, and money flows between the participants. At present, we do not represent the type of value flow in our GRL models (they are expressed in abstract terms as dependencies between actors).

Figure 2 shows a GRL actor diagram for a manufacturer that sells to stock via warehouses and retailers. This model represents each participant in the business model (consumer, retailer, warehouse, and manufacturer) as an actor, and indicates their dependencies. Thus, for example, the Consumer depends on the Sales Support provided by the Retailer, whereas the Retailer relies on the Consumer to Receive Payment. The half-moon symbol indicates the direction of the dependency.

Given the dominant role that intermediaries (warehouse and, in particular, the retailer) play, we will also refer to this business model as the **R** (Retailer) strategy. In this business model, the retailer controls the customer relationship. On the demand-side, customers benefit because the retailers offer a one-stop shopping portal, and support them during product selection. On the supply-side, the retailer gives warehouses access to consumers, and to high-volume sales. The manufacturer benefits from this arrangement as it gains in market share, and demand fluctuations are buffered by the warehouse.

In addition to dependencies, actor diagrams can also show the internal goals of a particular actor. Here, the main actor of interest, the Manufacturer, is expanded (its boundary is shown as a dotted circle partially under to the actor) to reveal its internal goals. There are two tasks (hexagons) that the manufacturer performs, Sell via intermediary and Build to stock. The Sell via an intermediary task is decomposed into three soft goals (where soft goals, shown as clouds, are goals that can never be fully satisfied).

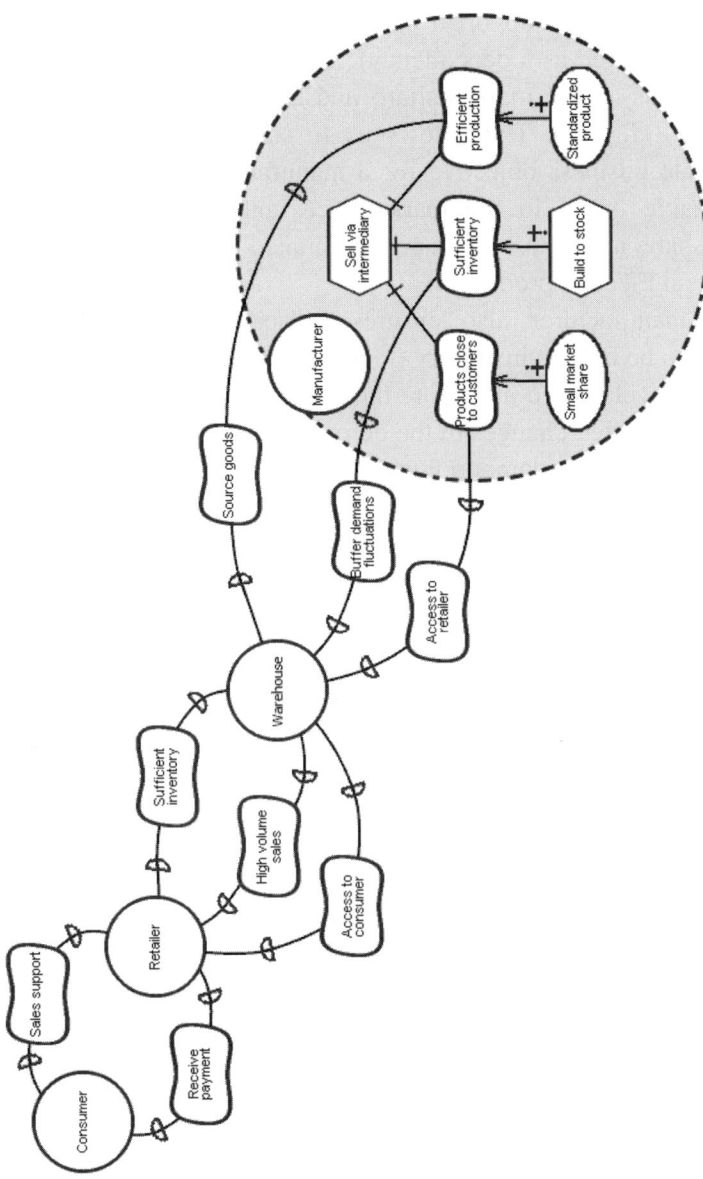

Figure 2. GRL actor diagram for the **R** strategy (sell to stock).

Tasks, goals, and soft goals can be recursively refined via such decomposition. As shown in the diagram, the manufacturer wants to position its Products as close as possible to its customers. This positioning requires support from warehouses for an effective Access to retailers (modeled as a dependency). This goal is also guarded with two preconditions (Small market share and Standardized product) modeled as beliefs (ellipses). This allows us to state that the goal is only an appropriate business objective for a manufacturer who does not have a recognizable brand in the marketplace, and a correspondingly large market share, and who offers undifferentiated products, and is thus likely to focus on Efficient production.

The manufacturer also ensures Sufficient inventory by building products to be held in inventory (modeled as the task Build to stock). The inventory levels try to anticipate the market demand. However, as there can be unexpected changes in the demand, the manufacturer relies on the warehouse to Buffer demand fluctuations. Preconditions for this business model are modeled as beliefs and connected to other model elements through *make* contributions. Therefore, the *levers for evolving this business model* are strategic moves that increase the market share or make the product more differentiated.

4. Business Model Evolution

This section discusses business model evolution and describes how URN allows stakeholders to experiment with different business model alternatives.

Consider the strategic options for evolving the current business model implied by the actor diagram in Figure 2. The levers for evolution are changes to either one or both of the two preconditions, Small market share, and Standardized product. Both options also result in increasing control over the customer relationship as they are applied. The possible evolutions of the business model are summarized in Figure 3.

The arrows indicate the evolution between these business models, and the labels on the arrows characterize the nature of the transition between the models. For example, the transitions from **R** to **W**, and **R** to **WR**, are both about increasing market share. However, in the former the manufacturer keeps selling a standardized product, whereas in the latter, it can offer a differentiated product. It is the warehouse that assembles the customized product. In both options, the warehouse (**W**) keeps control of order processing.

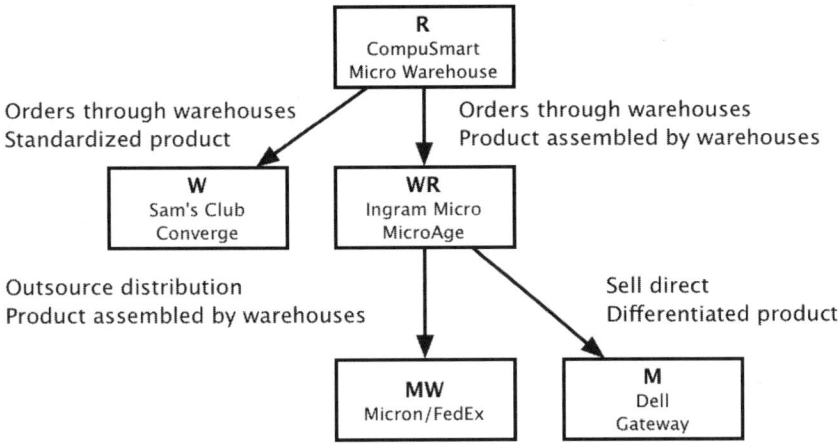

Figure 3. Possible evolutions of the current business model.

The manufacturer could increase its market share by partnering with a warehouse. In the **W** strategy, the warehouse now owns the relationship with the customer, and its implications for the manufacturer are in many ways similar to the **R** strategy. However, higher revenue can be expected due to the shorter supply chain. The manufacturer also keeps selling a standardized product. In the **WR** strategy, the warehouse assumes additional responsibilities such as (partial) product assembly. The main difference from the **W** strategy is that the manufacturer can now (via the warehouse) offer a customized product, and can strengthen its market position against competitors who do not.

Of greater interest to the manufacturer, however, should be the third option (**MW**). In this strategy the manufacturer is in the driver's seat. It sells its products directly to the customer, but, in part to share revenue risks, and in part to leverage the distribution experience of a warehouse partner, it outsources distribution to a warehouse. Traditional shipping service providers such as Micron's partner FedEx have developed additional capabilities to manage the inventories of their clients.

The most evolved of these strategies (**M**), however, is to assemble all key responsibilities (order processing, inventory management, and production) within the manufacturer. Note that this does not necessarily imply that the manufacturer handles the physical product, but refers to the control the manufacturer exerts over the information flow in the supply chain. The manufacturer could manage a virtual value chain.

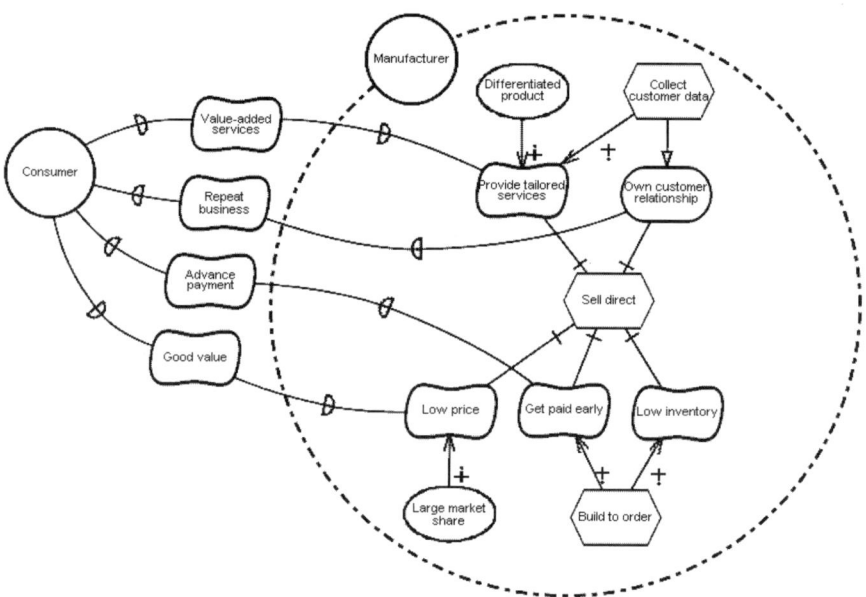

Figure 4. GRL actor diagram for the **M** strategy (sell-to-order).

The impact of choosing any of these alternatives can be analyzed within an actor diagram. The GRL model for the **M** strategy is shown in Figure 4. It shows that selling direct via an internal warehouse allows the manufacturer to benefit: Provide tailored services; Collect customer data; achieve high rates of Repeat business; and sell at a Low price, while realizing a high margin. The latter is the result of assembling a product only upon receipt of a firm order (Build to order), efficiencies in inventory levels (Low inventory), as well as the float gained from Advance payment. However, the **M** strategy can only be adopted if two preconditions are met: (1) the manufacturer already has a Large market share, and (2) it can offer a Differentiated product.

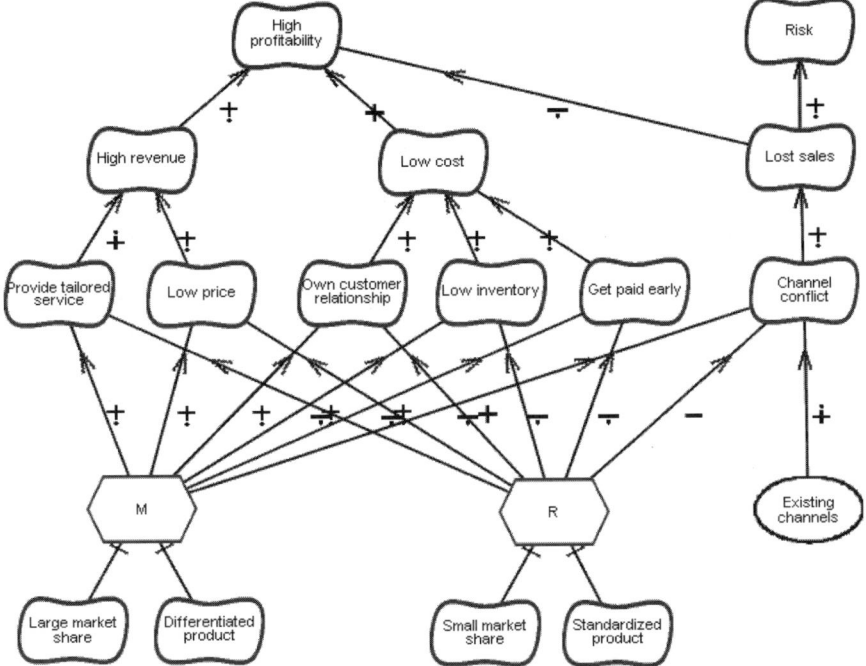

Figure 5. GRL rationale diagram for comparing the **R** and **M** strategies.

To compare the business model alternatives, we use a GRL rationale diagram. For reasons of space, we will only consider the two extreme strategies, **M** and **R**. Figure 5 summarizes the impact of choosing the business objectives: High Profitability and Low Risk. High Profitability can be achieved by increasing revenue (High revenue), or reducing cost (Low cost). The contributing factors of these goals are the five subgoals of Sell direct that can be seen in Figure 4. However, the model also indicates a key obstacle for evolving quickly from the **R** to the **M** strategy for manufacturers with existing resellers: Channel conflict, which (on opting for **M**) results in Lost sales.

For a full understanding of the implications of each business model alternative on the underlying business processes, we also need to look at the UCM scenario models at the next level of refinement. In order to support our goal of protecting the investment and organization already made in existing business processes, we do not want the business processes to change significantly as we evolve our business model. In related work (Weiss and Amyot, 2005), we present evidence that we can use the *same* scenario to describe *different* business models at the level of the business architecture. This property of UCM models lays the basis for the incremental evolution of the business model.

5. Related Work

The application of use-case driven design to business process re-engineering has been proposed by (Jacobson *et al.*, 1995). However, the use case approach has a number of well-known disadvantages that can be averted by using UCMs to model the early requirements of a business process. Use-case driven approaches seldom provide notions of modeling design goals and linking them to other design artifacts, as in URN.

Conceptual value modeling or *e3-value* (Gordijn and Akkermans, 2003) provides a means to evaluate the feasibility of an e-business model focusing on the creation, exchange, and consumption of objects (i.e., the revenue streams) in a multi-actor network. *e3-value* uses scenarios to model causal flows. It also provides a means for performing value-based trade-offs. However, unlike as in URN, value is mainly expressed in monetary terms; other non-functional goals cannot be modeled directly.

The Strategy-oriented Alignment in Requirements Engineering approach (Bleistein *et al.*, 2004) uses GRL to link requirements for strategic-level e-business systems to business strategy and to document recurring patterns of best business practices. They explore goal modeling to provide traceability and alignment between strategic levels (business model and business strategy) as well as tactical and operational ones (business process model and system requirements). This work however does not address how goals are converted to operational requirements.

6. Summary of the Approach

Our approach for business model design and evolution can be summarized as follows:

(1) Model the current business by capturing its business objectives and informal requirements in terms of actors and their strategic dependencies in an actor diagram.
(2) Document the internal goals for the actor of interest, and of any other actors when it furthers understanding of current business situations.
(3) Model the preconditions associated with the actor diagram. Changes that can be made to these provide the strategic options for evolving the business model.
(4) Explore the application of those strategic options in a series of business model alternatives (which can be related to one another in a line of evolution).
(5) Compare alternatives by assessing their implications using a rationale diagram. Include in the model the preconditions that drive each business model.
(6) Perform a qualitative evaluation[1] of the rationale diagram to determine the best next stage of the evolution of your business model as well as any associated risks.

As discussed earlier, for a full understanding of the implications of each business model alternative, we also need to look at the refinement of the GRL models into UCM models. The above steps only cover modeling of the *strategic options*. A full picture of the business situation emerges only from a combination of both approaches.

7. Conclusion

The approach described in this chapter gives business managers a conceptual tool for the systematic and incremental evolution of business model alternatives for their organizations. We have illustrated the approach with a supply chain management case study. However, its capabilities are in no way limited to supply chain scenarios. Evidence of that is the variety of applications to which URN has been put previously as summarized in (Amyot, 2003). One objective for future work is to

---

[1] The OME tool (Yu and Liu, 2000), used to generate the GRL models in this chapter, and the jUCMNav tool (Roy *et al.*, 2006) both support qualitative evaluations of models.

analyze various other business situations such as the (expected) evolution of the wireless payment industry.

References

Amyot, D. (2003). Introduction to the User Requirements Notation: Learning by Example. Computer Networks, 42(3), 285-301.

Bleistein, S.J., A. Aurum, K. Cox and P.K. Ray (2004). Strategy-Oriented Alignment in Requirements Engineering: Linking Business Strategy to Requirements of e-Business Systems using the SOARE Approach. *Journal of Research and Practice in Information Technology,* 36(4), 259-276.

Buhr, R.J.A. (1998). Use Case Maps as Architectural Entities for Complex Systems. *IEEE Transactions on Software Engineering,* 24(12), 1131-1155.

Chung, L., B.A. Nixon, E. Yu and J. Mylopoulos (2000). Non-Functional Requirements in Software Engineering. Kluwer Academic Publishers.

Gordijn, J. and J. Akkermans (2003). Value-based Requirements Engineering: Exploring Innovative e-Commerce Ideas. *Requirements Engineering Journal,* 8(1), 114-135.

Jacobson, I., M. Ericsson and A. Jacobson (1995). The Object Advantage: Business Process Reengineering with Object Technology. Addison-Wesley.

Roy, J.-F., Kealey, J. and Amyot, D. (2006). Towards Integrated Tool Support for the User Requirements Notation. Fifth Workshop on System Analysis and Modelling (SAM'06), Germany, May 2006. http://jucmnav.softwareengineering.ca/

URN Focus Group (2003a). Draft Rec. Z.151 – Goal-oriented Requirement Language (GRL). http://www.UseCaseMaps.org/urn.

URN Focus Group (2003b). Draft Rec. Z.152 – Use Case Map Notation (UCM). http://www.UseCaseMaps.org/urn.

Weill, P. and M. Vitale (2001). Place to Space. Harvard Business School Press.

Weiss, M. and D. Amyot (2005). Business Process Modeling with URN. International *Journal on E-Business Research,* 1(3), 63-90, July-September, 2005.

Yu, E. and L. Liu (2000). Organization Modelling Environment (OME). http://www.cs.toronto.edu/km/ome.

CHAPTER 14

THE ENTREPRENEUR HAS "SOLD OUT": AN EXPLORATORY STUDY OF THE SALE OF HIGH-TECH COMPANIES TO OFF-SHORE BUYERS

Sally Davenport

Victoria Management School, Victoria University of Wellington, Wellington, New Zealand

The sale of high-tech New Zealand companies to offshore buyers has attracted much criticism as it is assumed that the sale is bad for New Zealand. This paper explores the drivers that contribute to the sale by the founder entrepreneurs of ten companies since 1996. The literature on entrepreneur/founder sale tends to focus on the lack of skills of the entrepreneur to take the company to the next stage of growth as the key factor in their departure. The case studies show that the driving forces are much more diverse. The analysis suggests that the sale to offshore buyers may be a necessary growth transition that enables companies to resource the next phase of growth. Labeling this the "growing global" transition, it is argued that such sales are external validation of New Zealand's technology and companies and that the entrepreneurs and companies need to be supported through such transitions.

1. Introduction

"Peter Maire - The Navman founder takes out our number five spot after wallowing in the role of poster-child for "Kiwi companies done good" (and enjoying the odd government grant to boot) before becoming strangely reclusive and passing over the majority stake to US giant Brunswick" (Top 5 Prima Donnas, Unlimited, 2003).

"The sale to Brunswick highlights a familiar trend among New Zealand's most successful IT companies – they get plucked up by

overseas investors" (Navman Sells out to NYSE-listed Giant Brunswick for $56m, Griffin, New Zealand Herald, June 25, 2003).

In 2003 Peter Maire sold 70% of his award-winning marine products company Navman (formerly Talon Technology) to US-based Brunswick, owners of Mercury products amongst other brands. With Brunswick's size and credibility behind him, Maire has taken Navman from a $50 million turnover company a few years ago to an estimated $240 million turnover company in 2004 undertaking development projects with such diverse global leaders as SONY and Timex. Maire still smarts at the treatment he received when he sold the company:

> *"When we sold 70% it was all over the newspapers that another New Zealand company got "stolen". It was all bad news, doom and gloom. In the last 12 months we added another 230 people and we just spent $5m on a new R&D centre and we're growing like crazy."*

It has become almost legend in policy and management circles that one of NZ's major "problems" is that we are not able to retain promising high-tech companies in New Zealand (NZ). Essentially the premise of the legend is that before the companies have the chance to grow to a reasonable size they are sold by the founder/entrepreneur to overseas buyers and thus any potential to significantly contribute to the economy, in direct or indirect ways, is lost for good. Examples of such companies that have been sold off-shore usually include Allflex, PDL, Fletcher Paper, ALAC, Deltec, MAS Technology, Interlock, Marshall Software, Jade as well as Navman. No doubt there have been many other, less high profile, sales such as those carried out through private equity deals (Stride, 2004).

The offshore sale is almost invariably talked about in the media in negative tones. Not only is the company sold, but the removal of all assets, including intellectual property, offshore is assumed to happen in short order. In many ways, this legend is akin to a company level "brain drain". The legend also includes connotations that this is a very selfish act on behalf of the founder who is realising his/her equity for life-style and early retirement reasons.

But what is really going on when a company is sold offshore? The paper will explore this important company transition - the sale of the company by the founder/entrepreneur to a foreign buyer - that we have identified as being a significant evolutionary step of salience to NZ's companies and which marks a turning point in the company's life-cycle.

2. Literature

As background to this research, we look at several literatures: the entrepreneurial life-cycle including founder departure; the literature about company growth and internationalisation of companies from small country environments; and the literature on foreign direct investment, to investigate the motives of offshore buyers in acquiring NZ companies.

Entrepreneurship is concerned with the "discovery and exploitation of profitable opportunities" by entrepreneurs and thus, as a field of research, it tends to concentrate on either the "front end" of the individual's entrepreneurial life-cycle or on the performance of entrepreneur-led SMEs (Shane and Venkataraman, 2000; Shane, 2003). There appears to be less interest in entrepreneur departure and, when this is studied, the framing of the transition is that the company has outgrown the entrepreneur's skills so that founder departure is a function of size and age of the company or that the entrepreneur must change to be able to lead an institutional growth phase (Willard *et al.*, 1992; Boeker and Karichalil, 2002; Swiercz and Lydon, 2002). Investigating the transition in more depth, Boeker & Karichalil proposed that founder departure was more likely: in companies experiencing high (or low) growth; in companies with more concentrated ownership (but not if it is the founder that has most ownership) and in companies with more external board members. Conversely, founders were less likely to depart when: the founder worked (or presumably had a strong hand in) research and development functions; when the founder had a great deal of industry experience; and when the founder was also the founding CEO. Whilst we may find that many of these propositions apply to NZ companies, that fact that the company has outgrown the entrepreneur's skills is less likely to be a factor. Virtually all of the companies had already grown from the entrepreneur-led "start-up" stage to an internationalised company still managed by the founding entrepreneur and thus the entrepreneur had already "transitioned" through the skills gap phase to be a "professional manager" (Boeker and Karichalil, 2002).

The transition may also be driven by strategies for company growth and particularly for growth in international markets. As Chetty and Campbell-Hunt (2003) state, "scaling up to a global scale from a base as small as the NZ market is particularly overwhelming". Traditional internationalisation theory includes several stages models of internationalisation whereby companies are assumed to follow a step-by-step process with increasing commitment to international markets. The Uppsala model, for example, suggests four stages: no regular export activities; export via independent agents; creation of an offshore sales subsidiary; and overseas production facilities (Johanson and Vahlne, 1990). There have been many criticisms of these models and evidence that refutes either such ordered progression or that one universal theory can be proposed (Coviello and McAuley, 1999; Chetty and Campbell-Hunt, 2003, 2004). In the case of NZ companies, Chetty and Campbell-Hunt (2003) found that there are at least two distinctive patterns – one where companies adopt a regional approach as an attempt to manage controlled growth, the other where the strategy is truly global in terms of depth of offshore involvement.

Lastly we are interested in the buyer's side of the purchase. Theoretical approaches to FDI have centred upon the impacts of foreign direct investment (FDI) on the host country's economy which depend not only on the characteristics of the host country but also on how the multinational enterprise (MNE) chooses to operate through its affiliate. Most studies of FDI have focused on the immediate impacts on capital, technology and employment with little emphasis on the "second round" effects such as the longer-term impact on local industry (Enderwick, 1998). Recent work has shown that FDI has been directly responsible for facilitating the upgrading of local (NZ) capability (Scott-Kennel and Enderwick, 2001).

The focus on high-tech companies introduces an added component to the FDI mix. Studies of the globalization of R&D (as one source of FDI) suggest that there are two drivers for MNE access to the host country (Howells, 1990). The first is as a "demand" or "market control" strategy in which customising technology is an input to understanding local markets, whilst the second is a "supply-side" strategy involving access to new technologies and capabilities necessary for competitive advantage. The supply-side perspective generates an additional motivation for foreign buyers-the "tapping" into local pools of scientific and technical talent (Howells, 1990; Florida, 1997).

3. The New Zealand Context

New Zealand has 277,842 private sector enterprises but is unusual in that it is dependent on a very small number of companies for its export earnings (MED, 2003). Estimates range from 33 companies accounting for 80% of earnings to 133 companies accounting for 60% of merchandise trade (Bowen *et al.*, 2003). In turn, whilst the majority of NZ's private sector enterprises export only up to 10% of their products or services, NZ has a disproportionate number of companies that derive at least 75% of their revenue from exports – about 1500 in total and they are typically producers of quality, niche products sold to the global market (Heeringa, 2004).

Foreign ownership of NZ companies has been increasing in recent years, particularly of larger companies. Through the latter 1990s, NZ was one of the most heavily dependent (developed) counties on FDI as a source of fixed capital formation (Scott-Kennel and Enderwick, 2001). In 2003, there were 5450 NZ businesses with 50% or more foreign ownership and 15% of all full-time workers were employed by these businesses. As these figures indicate, foreign ownership of NZ companies and/or their "loss" overseas is not new (Glaxo originally started in NZ) nor novel. It can been argued that it is another "baby boomer" phenomenon (Rotherham, 2003) in that those entrepreneurs who founded their businesses during the 1960s and 1970s began reaching 65 around the year 2000 and "the bulk of them will be looking to ease back between now and 2020" (Stride, 2004). It is this "easing back" with the reaped rewards that underpins the selfish connotations, yet there are other ways for withdrawal from a company which would allow the entrepreneur to do so without bearing the brunt of recriminations (Stride, 2004; Peart, 2004).

4. The Research Project

This research forms part of the Competitive Advantage New Zealand (CANZ) project which uses in-depth historiographic case research to explore growth transitions in NZ companies (Eisenhardt, 1989; Goodman and Kruger, 1988). For this exploratory study, ten high-tech NZ companies sold to offshore buyers between 1996 and 2004 were chosen for study (Table 1). All of the companies in this study had survived the (often very rapid and intense) "going global" transition

Table 1. A selection of sales of New Zealand companies by the founder/entrepreneur.

Company	Founded	By	Known For	Size at Sale by Revenue (Staff)	Sold	To (Home Country)
ADIS International	Late 1960s	Graeme Avery	Pharmacovaluation, medical research publishing	$45m (225 in NZ, 400 worldwide)	1996	Wolter-Kluwer (Netherlands)
Switchtec	1985	Dennis Chapman	Power supplies for telecoms	$40m	1997	DTR Power System/ Invensys (UK)
Binary Research		Murray Haszard	'Ghost' disk cloning software		1998	Symantec (US)
MAS Technology	1976	Neville Jordan	Microwave telecom/ electronics	$100m (240)	1999	DMC* (US)
Holliday Group	1990	Phil Holliday	WAP applications	$2-3m (30)	2000	Itouch (UK)
Deltec		Peter Graham	'Teletilt' remote antennae technology	$34m (80)	2001	Andrew Corp (US)
Interlock	1961	Stuart Young	Innovative window and door hardware	$60m (420)	2001	Assa Abloy (Sweden)
Marshall Software	1994	Spin off of Design. Tech	Content security software	$12m (57)	2002	NetIQ (US)
Navman	1988	Peter Maire	Satellite navigation systems	(350)	2003	Brunswick (US)
Jade	1978	Gil Simpson	Software environ. (LINC, JADE)	$36.8m (330)	2004	ICap and LLC (US)

* Officially a "pooling" of shares as MAS Technology was a publicly listed company.

(Chetty and Campbell-Hunt, 2003) to establish a presence in a range of international markets and were exporting between 50 and 100% of their product at the times of sale. Scoping interviews were held with a selection of the entrepreneurs and other primary (eg. notes from seminar presentations by some of the entrepreneurs) and secondary data (especially the considerable number of media analyses of the sales) for all of the companies were gathered.

4.1. *Entrepreneurial factors*

It is the entrepreneurial factors that underpin the selfish connotations. Certainly a number of the entrepreneurs would fit the "baby boomer easing back" mode looking for a change after many years, often decades, of devotion to building their companies. Taking a break immediately after the sale was important for some (usually after years of not taking holidays) but realising the value and passing on the risk, after decades of continually reinvesting in the company were also drivers. Alan Wilkinson, one of the founders of Marshall Software, was a typical case:

"We sold because we were offered an excellent deal allowing us to finally walk away from the huge risk we have carried for five years. Essentially, we were continually playing double or quit with our entire stake... We had to reinvest every cent to fund our growth to remain viable and achieve a significant market share and reputation. All this reinvestment is taxed heavily. Consequently, the shareholders had taken nothing out for 5 years. Instead they had large outstanding loans to the company. This had purchased 5% of the world market. We were aiming for market leadership. That prospect was another five years of total reinvestment and continuing risk."

Even so, many of the entrepreneurs wanted to continue in business, and went on to set up venture capital companies or to get involved with other businesses, either as a mentor or to try growing another business (often called serial entrepreneurship). Of this "letting go" stage, Jordan, who founded MAS Technology as well as venture capital company Endeavour Capital, stated:

> *"It's too much fun doing business. I had seen venture capital in the States and it's quite hands-on. I wanted to be involved with business and science and technology in New Zealand – hands-on."*

In fact, some entrepreneurs never intend to leave the company at all (eg. Maire, Simpson, Holliday) but just see this transition as yet another step in growing the company. Others did so only after a significant period, as many of the offshore buyers want to retain the entrepreneur's skills and included an "earn-out" clause whereby the buyer pays a certain value only if revenue holds up for (say) another two years. The last factor then captures the fact that the entrepreneurs, particularly those that were intending to stay with the company, usually wished to see their companies grow to major players in the world but released that to do this was impossible under their current business model.

Together these entrepreneurial drivers lead to four factors contributing to the potential sale, albeit at this stage not necessarily to an offshore buyer:

1. Entrepreneur's desire for change, to do something different;
2. Entrepreneur's desire to reap reward for years of "hard graft";
3. Serial entrepreneur's desire mentor/grow other businesses;
4. Entrepreneur's desire to see company grown to next stage globally.

4.2. Company factors

Whilst most of these companies were doing very well in exporting, the entrepreneurs almost all talked of internationalization strategies that would see their companies grow to the next stage on the world markets. However, they were frustrated by the company's lack of resources to enable this. In 2001, Maire labeled this as "phase three" for his company:

> *"We're an ant. We're one grain of sand on the 90-mile beach of the world market. To survive, we've got to go to the market as fast and as large as possible. It's time for an all-out worldwide strategy."*

More recently, after the sale, Maire expanded on the barriers Navman had faced:

"We got to around $50m in sales a couple of years ago and we started getting very frustrated about our growth. Although we looked at $50 M like a reasonable player, $50M is $30M US, which is a very small company in the US. What we couldn't see was how to generate real growth because we couldn't get credibility in the world market. We had great IP but all the companies we were going to get business from would look at our technology and say "that's absolutely fantastic" but they'd say "look unfortunately you are way too risky because you are way too small. We can't afford to really deal with a little company like that – if you want to license your IP to us, maybe we could look at something like that but we can't let you build product for us."

This last comment from Maire also captures another of the company factors – that these NZ companies were relatively small on a world scale and they were unable to attract major customers because of the perceived risk of contracting with a small, relatively unknown NZ company. Jim Donovan, CEO of Deltec at the time of its sale concurred regarding Deltec's major customers such as Motorola and Nokia who were saying:

"We like your product, but you've really got to be global if you're going to be a partner for us. We can't keep dealing with you NZ. You've got to be in the US and EU markets with product available."

Labeling this decision point as *"do it now or don't bother"*, Donovan explained that the expectations of these potential customers included sales, servicing, support, logistics and manufacturing of a scale that was orders of magnitude larger than Deltec's NZ operations. This would need significant investment (estimated at $10m minimum in 2000) and risk and he doubted that the then shareholders *"had the appetite for the game"*, leading to the decision to sell.

As indicated above by the interest from major MNEs in these companies, the last company factor centres on the attractiveness of the technology, the R&D capabilities and/or the market position of the companies' products. Continuing with Deltec, Donovan commented that Andrew Corporation got "sick" of hearing from their customers *"have you got a product to compete with those Deltec guys?"* Assa Abloy took note of Interlock when they started to make their presence felt in the Australian market against Assa Abloy subsidiary Lockwood Locks. Navman's attractiveness is centred more on its ability to develop innovative uses for existing product concepts. Thanks to its creative, skilled engineers, it costs Navman about 50% less than its competitors to design, develop and build GPS products. This capability is very attractive to MNEs as the General Manager of Navman's customer, Intel Australasia, indicated:

> *"In a globalising world, multinationals get even bigger and less nimble. Smart ones recognise this trap and voraciously seek out innovators and entrepreneurs who can enhance their businesses."*

Extracting drivers from the company perspective on the potential sale leads to four more factors:

1. Company has reached a perceived resource barrier (and/or shareholder reluctance) to further international growth strategies;
2. The company does not have enough credibility or reputation internationally to take on large projects;
3. The company does not have enough manufacturing capability to compete with major international competitors;
4. The company has technology, development capability or a market position that is attractive to potential buyers.

4.3. *Buyer factors*

The sale to an offshore buyer was not a decision made quickly. The entrepreneurs and shareholders had usually considered carefully all the available options but, in general, local options did not provide either the immediate financial rewards or the ability to take the company to the next stage. As Marshall Software's Wilkinson indicated:

"What other options did we have for diversifying our investment and reducing our risk? We could have sold a portion of the company to a VC or listed on the sharemarket. But when we looked we found little interest. Those who were interested did not have a compatible culture or attractive expertise. The immediate money on offer was small and the potential impact on our management was large and not attractive. Listing would have increased our compliance costs and bureaucracy, taking critical focus away from our core business challenges."

Capital acquisition through initial public offerings (IPO) had figured in the plans of several of the companies. Jordan had taken MAS Technology to the NASDAQ exchange rather than to the local exchange, prior to the sale to DMC. Jade had planned to list locally in 2003 but had postponed the listing due to difficult market conditions after the unsettling world events of 2001/2002. Maire had explored a wider range of options such as floats in NZ and abroad and further injections of venture capital. Each met some needs but not all. Offshore buyers, such as $2 billion Assa Abloy conglomerate or the $1.3 billion Brunswick Corporation, certainly offer the capital to take the company through the next phase of international growth. In a typical response, Maire says:

"The deal with Brunswick will allow Navman to continue its international expansion in the marine industry as well as improve its competitiveness in other related markets. Navman will continue to design, manufacture and introduce new GPS technology-based products from New Zealand. It's a case of business as usual for Navman, except that business will be faster, bigger and better."

Although capital is important for growth, it is not the only asset that these MNE offshore buyers bring to the table. They also offer the backing of their reputations and brands, their market knowledge and distribution resources. The Mercury Marine division of Brunswick, for example, owned a huge list of popular boat brands and had 5000 distribution outlets that became accessible to Navman. Many of the companies that had successfully entered some regions but the offshore owner gave access to other markets. Typically the NZ companies were well established in Australia and Asia and even parts of Europe, but few had "cracked" the North or South American markets. This phase of growth enabled by the sale, therefore, would also be marked by a potential expansion of market access.

The MNE is also reaping resources in return, but there was usually a "good fit" between the NZ company and the MNE buyer in terms of overlap of expertise and technology/product. NetIQ, for example, wanted Marshal's technology as it was a "hot niche in the software market" but also because it would "strengthen" NetIQ's total offering. They intended giving the products more exposure under the NetIQ Marshal Solutions brand and focus on large North American customers as this was a market into which Marshal had not made "inroads". Andrew Corporation, intended to integrate Deltec's Teletilt technology in their own existing radio frequency subsystem product portfolio enabling much wider penetration of the technology. At the time of the Deltec sale, founder and chairman, Graham said:

> "The Andrew deal means our Teletilt technology can really take its rightful place in the global mobile telecommunications market. We didn't have the scale or market reach to do it on our own, but Andrew does. We share similar vision and values, and this deal means an expansive future for our sales and development staff."

This last quotation also illustrates the last of the identified themes – the intent of the buyer had to be in line with the vision of the entrepreneur for the company. Whilst in some cases this did happen a period after the sale, intent to remove all operations from NZ was not acceptable to either the entrepreneur or current shareholders. Not all sales involved retaining all current staff or operations but the core of the technology and capability that had made the NZ company attractive in the first place, was almost inevitably retained in NZ. In several cases the fact that the NZ company would evolve into an R&D or creative design arm of the MNE was an explicit buyer intention. MAS Technology was renamed Stratex Networks and became DMC's R&D wing and, even though it made up only a fifth of DMC, it quickly contributed about one third of DMC's profits. Binary Research's Auckland development team became the only full-scale Symantec development facility outside of North America. ITouch intended Holliday Group to be a centre of excellence for development of mobile office applications and hardware.

In summary, the four buyer factors for the sales are:

1. The lack of local buyer interest (including IPOs) of scale necessary for next growth stage;
2. The offshore buyer's resources, credibility, brands, market access and distribution channels;
3. The offshore buyer wishes to merge technology/product with own expertise and grow market;
4. The offshore buyer has intentions in line with the entrepreneur's vision including maintaining (the core of) the company in NZ.

5. Conclusion – The "Growing Global" Transition

Combining all these factors together underpin this NZ company transition. This transition occurs at the intersection of the entrepreneurial life-cycle at a point where the entrepreneur (may) want to take a different trajectory and the life-cycle of the company as it attempts to evolve from a "niche" international participant to become a major global player. The transition, which is labeled the "growing global" stage, builds upon the previous "gone global" transition that these companies had already successfully accomplished. In this "growing global" stage, the companies have generally reached a plateau in their ability to internationalize further without a step increase in resources of a range of types. The "right" offshore buyer appears to provide more of these resources than other possibilities such as IPOs or local venture capital.

What seems to be of particular importance, if this transition is to be negotiated successfully, is for the offshore buyer to be a "good fit" for the entrepreneur, the company and the NZ economy. As several commentators have noted, these sales are bringing great validation to these companies and to the country. Lane Finlay, of the Marine Export Group spoke in such terms about the Navman sale:

> *"It's terrific for New Zealand because Brunswick is recognizing NZ as a low-cost place to develop technologies. Think of what that can do if the word spreads. If Brunswick is successful other companies will come here looking for hotbeds of innovation."*

If this transition is yet another important phase for NZ companies to achieve, what of those that don't sell or raise the next huge tranche of resources through some other means? Are they limiting their growth? Tait Electronics is perhaps the most famous NZ company, and Angus Tait one of the most vocal New Zealanders, against such sales (off shore or through IPO). Tait developed the "Tait Charter", a company structure that ring-fences the business and prevents it being taken over. It has been observed that perhaps Jade missed a window of opportunity in the mid 1990s, that a partial float of the company then would have given the company the "financial grunt" necessary to promote the product internationally up against big names like SAP.

It is possible that what is being observed is a divergence in growth paths, like the "regional" versus "global" internationalisation strategy (Chetty and Campbell-Hunt, 2003). Those that chose not to sell (either offshore or through an IPO) are selecting what is perceived to be a less risky path that remains under the control of the entrepreneur. However, the unforeseen risk might be that it might miss a window of opportunity and only be able to sustain lesser levels of growth. This preliminary research suggests that an offshore sale and the resources that flow with it, enables faster "growing global" growth. Choosing not to sell does not mean that this level of growth will not happen but, like regional internationalisation strategies, may mean that growth happens much more slowly without those extra financial, reputational and distribution channel resources. The risk of not acquiring the resources provided by an offshore sale is that competitors will outpace a slow-growth company and market share will begin to decline, so in the long term such a decision may in fact risk the long-term sustainability of the company.

This study of the factors that influence the decision to sell a high-tech NZ company offshore suggests this transition needs to be re-framed towards a view that recognises that this "growing global" transition may be unavoidable, if not necessary, in the globalised markets in which companies must compete. The framing of the transition should be more positive than mere acceptance, however, but that these sales might be celebrated. Jordan argues that we should view the transition as important external validation of our technology and companies and that this is an important measure of economic progress, a *"proxy equivalent to the successful listing of companies"*. With this perspective, the concern should be if such sales were not occurring! Rollo Gillespie, former chairman of the Software Exporters Association, concludes:

> *"Each off-shore acquisition raises the credibility of all NZ technology companies in the eyes of the world. I say the more of them the better. I used to think it was sad, but now I think of it as sad only in the same way as when your last child finally leaves home."*

References

Boeker, W. and Karichalil, R., (2002). *Academy of Management Journal*, 45.
Bowen, E.; Haworth, N.; Wilson, N and H., (2003). *University of Auckland Business Rev.*, 5.
Campbell-Hunt, C., Brocklesby, J., Chetty, S., Corbett, L., Davenport, S., Jones ,D., and Walsh, (2001). P., *World Famous in New Zealand: How New Zealand's Leading Firms Became World-class Competitors* (Auckland University Press, Auckland).
Chetty, S. and Campbell-Hunt, C., (2003). *European Journal of Marketing*, 37.
Chetty, S. and Campbell-Hunt, C., (2003). *Int. Small Business Journal*, 21.
Chetty, S., Campbell-Hunt, S. and C., (2004). *Journal of International Marketing*, 12.
Coviello, N. and McAuley, A., (1999). *Management International Review*, 39.
Eisenhardt, K., (1989). Academy of Management Review, 14.
Enderwick, P., (1998). *Foreign Direct Investment: The New Zealand Experience* (Dunmore Press, Palmerston North).
Florida, R., (1997). *Research Policy*, 26.
Goodman, R. and Kruger, E., (1988). *Academy of Management Review*, 13.
Griffin, P., (2003). *New Zealand Herald*, June 25.
Heeringa, V., (2004).*Unlimited.*
Howells, J., (1990). *Regional Studies*, 24.
Johanson, J. and Vahlne, J-E., (1990). *International Marketing Review*, 7.
MED, (2003). *SMEs in New Zealand: Structure and Dynamics* (Ministry of Economic Development , Wellington).
Peart, M., (2004). *National Business Review*, 14 May.
Rotherham, F., (2003). *Unlimited.*

Scott-Kennel, J. and Enderwick, P., (2001).*The Academy of International Business Conference*, Sydney.

Shane, S., (2003). A General Theory of Entrepreneurship: The Individual-Opportunity Nexus (Edward Elgar, Cheltenham).

Shane, S. and Venkataraman, S., *Academy of Management Review*, 25 (2000).

Stride, N., (2004). *National Business Review*, 2 July

Swiercz, P. and Lydon, (2002).S. *Leadership & Organization Development J.*, 23.

Willard, G.; Kruegaer, D. and Fesser, H., (1992). *Journal of Business Venturing*, 7.

CHAPTER 15

TECHNOLOGY ACQUISITION THROUGH CONVERGENCE: THE ROLE OF DYNAMIC CAPABILITIES

Fredrik Hacklin, Christian Marxt, and Martin Inganäs
ETH Zurich, Switzerland

In recent cases of industrial dynamics and technological change, the acquisition of technologies is often not based on strategic choice, but can rather be regarded as a required operation in order to tackle risks in emerging phases of consolidation. In particular, the phenomenon of technological convergence is examined as a special case for acquisition of technologies. Introduced by a discussion of drivers for such a convergence, its implications on technology and innovation management practices are investigated. Special focus is laid onto the resulting impact in terms of business model convergence, where creative destruction might lead to severe disruptions in the competitive environment. Based on these reflections, two scenarios for acquisition approaches are introduced. In the first scenario, the convergence causes the current internal competencies to be merged with external ones, resulting in an emerging dominant design, from of which the firm holds a critical resource stake. In the second scenario, the firm's internal competencies remain outside the emerging dominant design. Especially in the latter scenario, the relevance of dynamic capabilities in managerial actions is underlined. The argumentation is illustrated by using the case of telecommunication industry actors in tackling convergence challenges, and in implementing practices for acquisition of technologies and related competencies.

1. Introduction

Technological convergence can be observed as an emerging discontinuous effect in a globalized industry. This effect is especially driven by the omnipresence of product components in a worldwide market, innovation opportunities based on an increasing amount of intersections and interfaces among technological solutions, business opportunities for establishing innovation collaborations, and in some cases, the customer need for full solution and service provisioning.

Particularly in the information and communication technologies (ICT) sector, the convergence of the telecommunication and computer industries has been a broadly observed trend (Borés et al., 2003; Duysters & Hagedoorn, 1997; Lee, 2003; Yoffie, 1997). In earlier stages of this ICT convergence, the belief was to offer innovative services by treating the various signals which flow in a digital network in the same manner (Sherif, 1998). More applied to daily applications, convergence represented the "merger of voice and data onto a unified network" (Edwards, 1999), or could be defined as "the ability of different network platforms to carry essentially similar kinds of services, or the coming together of consumer devices such as the telephone, television and personal computer" (European Commission, 1997). Generally, this convergence phenomenon can be classified into technology and business model convergence (Hacklin & Marxt, 2003), which not necessarily have to be interdependent trends (Freeman & Louçã, 2001). Similarly, the distinction can be made between technological and economical factors driving convergence through confluence, where technological factors consist in the evolution of communication and information technologies. On the other hand, the key economic factor can be seen in the worldwide liberalization of telecommunication markets (Borés et al., 2003), where convergence is enabled through deregulation (Karlsson, 1998; Steinbock, 2003). In many cases of current ICT industry environments, the effect of technological convergence can be observed as the collision of existing business models (Gartner Group, 2003; Pringle, 2003), i.e. the sudden inter-firm overlapping of existing technological solution concepts, causing an accelerated competitive environment. In other words, due to the rapid changes in this high-tech sector, current technological solutions and even entire business models can be rendered obsolete within a short period of time. This phenomenon can especially be observed in market structure evolvements in the wireless communications industry (Backholm & Hacklin, 2002; Camponovo & Pigneur, 2003). In general, the driving force of technological convergence can be seen in competition, where the need for new, convergence-based solutions arises in parallel to the evolution of enabling technologies (Edwards, 1999). This resulting business need causes a market pull towards convergence, in parallel to the technology push initiated by the technology development (Hacklin & Marxt, 2003). In a more long-term context, it is furthermore argued that technological convergence causes challenges for standardization organizations (Sherif, 1998) and can be responsible for reshaping entire markets (Achrol & Kotler, 1999).

Nevertheless, convergence of technologies is an effect not only occurring in recent industrial trends. The phenomenon of technological convergence was in literature for the first time observed in the industrialization process in the USA, and in particular, in the machine tool industry during 1840-1910, where apparently unrelated industries "from the point of view of the nature and uses of the final product became very closely related" (Rosenberg, 1963). As technological, and on a more generic level, industrial convergence in many cases implies a deconstruction of the value chain, existing products and solutions might get exposed to new competitive environments (Greenstein & Khanna, 1997; Li & Whalley, 2002; Sabat, 2002), if not even obsolete. This disruptive effect on competition can be differentiated by convergence in substitutes, being characterized by different firms developing "products with features of certain other product", and convergence in complements, occurring when "different firms develop products or subsystems within a standard bundle that can increasingly work together to form a larger system" (Greenstein & Khanna, 1997).

Furthermore, the emerging obsolescence and substitution of technologies has already been identified as a common phase in the technology evolution process (Betz, 1993), even though it was not originally and explicitly associated with technological or industrial convergence. Transferring this technology life cycle based view further to this case, technological convergence could be interpreted as the fusion of several incremental or sustainable technologies, which in their confluence achieve innovation with highly disruptive character (Hacklin et al., 2004b; Ireland et al., 2003). Examples for new business model potentials through converging technologies are illustrated in Figure 1.

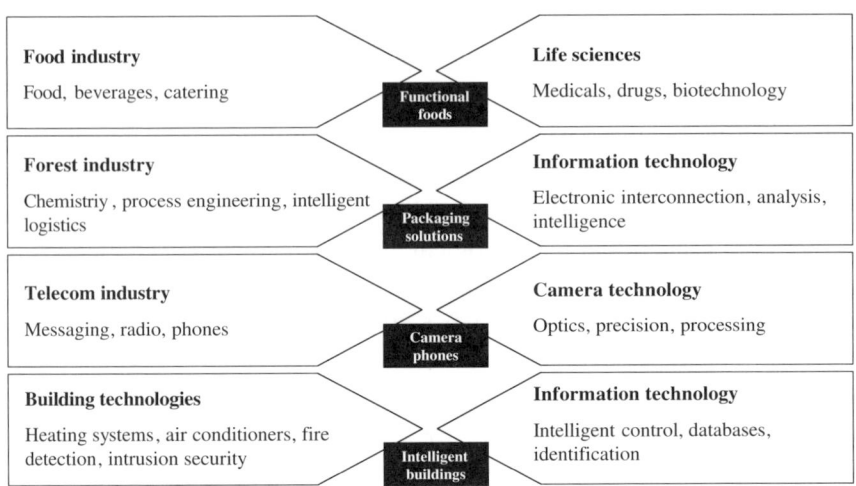

Figure 1. Examples for new business models through technological convergence.

2. Trajectory Lock-In and the Challenge for Competence Acquisition

In literature, convergence in ICT has so far been discussed mostly from a more technical perspective (Bohlin *et al.*, 2000; Edwards, 1999; Nikolaou *et al.*, 2002). Strategic aspects of such convergence, however, such as implications on business models (Duysters & Hagedoorn, 1997; Fransman, 2002; Olla & Patel, 2002; Sabat, 2002), on management practices (Backholm & Hacklin, 2002; Borés *et al.*, 2003; Duysters & Hagedoorn, 1997) and on approaches for fostering disruptive innovation through intersection (Johansson, 2004) can still be seen as rather rarely identified subjects. From an entrepreneurial and innovation management perspective, this issue deserves further attention in research, taking the technological convergence management challenge onto a generic level, developing theories not solely valid for the ICT industry. This identified challenge for current innovation management formulates a research need for developing strategic management tools, allowing entrepreneurial planning and technology management for sustaining the competitive advantage of actors in converging environments.

Special focus is laid onto the resulting impact in terms of business model convergence, where creative destruction might lead to severe disruptions in the competitive environment (Abernathy & Clark, 1985; Afuah & Tucci, 2003; Lei, 2000), posing a need to collaborate along the value chain (Afuah, 2001). The effect of business model convergence can be either technology or market driven, implying needs to open-up innovation activities and acquire future competencies along the value chain. Figure 2 attempts to depict both market and technology-driven convergence tendencies along the value chain framework. In particular, for involved players in converging environments, the decision to acquire competences and technologies from external sources does not necessarily come as a strategic choice; it is regarded rather as a required operation in order to avoid effects of coevolutionary lock-in as implied by the converging trajectories. In this context, technological convergence is regarded as a special case for technology acquisition, combining the challenges of managing technological change on the one hand, and tackling disruptive innovation on the other hand (Jones et al., 2001; Lambe & Spekman, 1997).

3. Capability-Based Management Approaches

In the dynamic industrial and market environment evoked by the convergence paradigm, firms' strategic responses also have to be built on flexibility. Therefore, the dynamic capabilities theory (Teece et al., 1997) is chosen as a reference model for considering the convergence from an acquisition perspective, aiming to align developed theories and models within the suggested "specific strategic and organizational processes like product development, alliancing, and strategic decision making that create value for firms within dynamic markets by manipulating resources in to new value-creating strategies" (Eisenhardt & Martin, 2000). The dynamic capabilities framework represents a rational strategic management approach for the challenge of converging industries, as it serves to "integrate, build, and reconfigure internal and external competencies to address rapidly changing environments" (Teece et al., 1997).

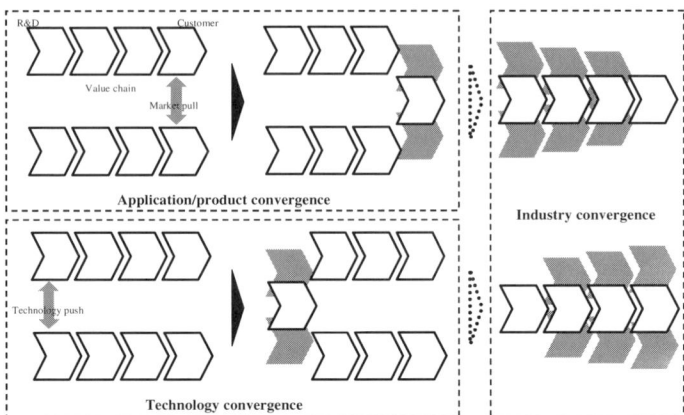

Figure 2. Sources of creative destruction through the convergence of value chains.

The proposed management approach based on dynamic capabilities can be characterized as threefold:

3.1. *Leveraging exogenous industry dynamics*

Being identified as one dimension within the dynamic capabilities framework, the issue of partnering and alliance formation is a crucial challenge. As convergence implies existing value chains to be deconstructed, the reconstruction and redefinition of value generation constructs has to be based on interfirm operations, i.e. one major management implication can be seen in an increased need for cooperation and network formation (Chesbrough & Teece, 2002; Duysters & Hagedoorn, 1997; Johannessen *et al.*, 1997; Rockenhäuser, 1999). It is considered as not sufficient, to solely develop an understanding of the dynamics of industrial value creation processes along a deconstructed and converging value chain. Furthermore the challenge is perceived in a successful focus on the entrepreneurial competence portfolios based on the underlying heterogeneous set of firms, representing key focal points of a value-oriented convergence strategy (since competencies and core competencies represent the basis for value creation and competitive advantage of a firm) (Rockenhäuser, 1999).

In such a converging industry scenario, management of technology-based projects in geographically distributed environments, independent of their hierarchical level of application, is a crucial but not trivial task in innovation. Conducted research has shown that approximately 50% of all technology-based companies, who have been involved in collaborative innovation projects, perceive the ventures with their partner as a failure (Littler *et al.*, 1995). However, current economic and technological development trends form the basis for a need of cooperation throughout the entire innovation process. On the suppliers' side on the one hand, issues of rationalization force these organizations to minimize their operations portfolio and to focus on their pure core competencies only. On the other hand, the customers' side claims a product range with high complexity, demanding the provisioning of full service solutions, implying an augmenting need for joining forces on the suppliers' side.

Based on these given circumstances, firms have designed new research and development (R&D) practices, including both internal organizational changes and the construction of complex networks to deal with growing outsourcing and various types of technological partnerships. This development is strongly supported by economic and social globalization, paving the way for worldwide competence complementation. Research on the specific characteristics of R&D cooperation in high-tech sectors exist (Miotti & Sachwald, 2003), and these observations, together with new studies strive to construct a network model. The research activity area of R&D networks *in general* shows a wide range of contributions (Birkinshaw, 2002; Gassmann & von Zedtwitz, 2003; Miotti & Sachwald, 2003). From an operationalized perspective, the assessment and selection of strategic innovation partners shall be mentioned in this context (Hacklin *et al.*, 2004a).

Combining the networked modeling approach with a classic value generation based view of firm activity (Amit & Zott, 2001; Porter, 1985), the basis for the *value network* construct is formed, serving to illustrate and explain the value generation and flow within a more complex business transaction framework than in a linear chain. A relatively small set of literature contributions exists, considering generic network-oriented value creation theories for understanding interorganizational firm operations exists (Christensen & Rosenbloom, 1995; Eisenhardt &

Schoonhoven, 1996; Zahn & Foschiani, 2002), but recently, articles especially covering this phenomenon's occurrence within the information and communication technologies (ICT) industry (Li & Whalley, 2002; Willcocks & Plant, 2003) have been published. From a convergence perspective, it is crucial to understand how interorganizational networks have to be formed, maintained and managed, in order to sustain competitive advantage and allow value creation.

3.2. Bundling of endogenous resources

As technological convergence can be characterized by the dilemma of continous reinvention (Christensen, 1997; Hacklin & Marxt, 2003), technology strategies and roadmaps have to be built in order to respond to short-term and long-term scenarios of convergence. In some cases, where current successfully positioned players will have to completely redefine their value proposition based on the obsolescence effect, a resource based view (RBV) is suggested to serve as foundation (Wernerfelt, 1984). Before drifting off to entirely new frontiers of business operations and models, a critical assessment of existing technology resources and their future applicability should be conducted. As dynamic capabilities are necessary, however not sufficient conditions for competitive advantage, it is argued that "dynamic capabilities can be used to enhance existing resource configurations in the pursuit of long-term competitive advantage" (Eisenhardt & Martin, 2000). On the other hand "RBV breaks down in high-velocity markets, where the strategic challenge consists of maintaining competitive advantage when the duration of that advantage is inherently unpredictable, where time is an essential aspect of strategy, and the dynamic capabilities that drive competitive advantage are themselves unstable processes that are challenging to sustain", which again motivates the combination (Eisenhardt & Martin, 2000). From a convergence perspective, it is therefore a question of how dynamically existing technological resources can be bundled into new solutions, quickly enough for responding to disruptive technology strategies implied by the industrial changes.

3.3. Bridging previously distinct knowledge bases

Also being identified as a strategic resource of a firm, therefore in theory even applicable to RBV, is the aspect of knowledge creation (Grant, 1996; Nonaka & Takeuchi, 1995). In rapidly changing environments, efficient knowledge management (KM) practices are needed for responding to the market dynamics and for sustaining value creation (Allee, 2003; Johannessen & Olsen, 2003; von Krogh *et al.*, 2000).

Special focus has to be laid onto interorganizational KM, which on one hand can be seen as a facilitator of strategic value networks. However, the knowledge asset also has to be covered, as in order to efficiently respond to convergence, interorganizational KM can serve to share and early identify emerging convergence trends, and therefore create competitive advantage.

4. Competence Aggregation Scenarios in Converging Environments

Based on the background as previously discussed, two scenarios for considering technological convergence as a special case of technology acquisition are introduced.

In the first scenario, the involved actor is regarded as a "convergence enabler", finding himself in a position where the convergence represents a potential for merging proprietary competencies with external ones into new business models and synergies. This scenario assumes that the involved actor remains in a position to control critical resources, envisioning a future stake of a dominant design. In such a setting, acquisition strategies would become proactive in terms of emerging new business models, rendering the actor to work on partnerships such as e.g. vertical integration. Also, the scope of foresight activities would include constant observation of external technology streams, with an evaluation of their potential intersections with own competencies. In summary, this proactive scenario could be summarized as "first acquisition, then convergence", referring to the active role in building the new business intersection.

On the other hand, a convergence follower scenario would refer to a rather passive role in terms of convergence, i.e. "first convergence, then acquisition". In such a scenario, proprietary competencies would remain outside the new dominant design, and the strategic response would consist of a rather resource-based redefinition of the value proposition and business model. In that sense, activities would be aligned against the convergence mainstream, committing onto other, previously unexploited competence areas.

The scenarios are summarized in Table 1. In either scenario, the bottom line of any managerial activities can be seen as based on the different dimensions of dynamic capabilities. In particular, these dimensions should be addressed by evaluating what technologies to commit on, what knowledge and competencies to keep and develop, and what alliances and networks to build and maintain.

The "convergence enabler"-scenario can be observed in the examined case company A, being active as an equipment and network manufacturer in the telecommunications sector. In order to receive a dominant stake in a new emerging service concept that was emerging through convergence, the company decided not to develop the service in-house, and thereby risking to compete against the emerging dominant design. Instead, company A decided to acquire the technologies through signing partnerships and initiating R&D collaboration with a broad variety of small companies, each of which targeting specific niche groups within the emerging service. Today, the company finds itself in a closer position to individual customers through the niche solution provider network, thereby covering a broader spectrum of specific customer demand.

The "convergence follower"-scenario can be observed in case company set B, representing a set of European incumbent telecommunication operators. These companies were challenged by their proprietary core competence set, consisting of analog telecommunication lines and circuit-switched systems, becoming obsolete as telecommunication systems started to converge with computer-oriented data communication. Whereas some incumbents managed to commit onto the new emerging packet-switched telecommunication paradigm at early stages (by performing activities as in the "convergence enabler"-case), the company set B failed to develop dominant designs in new emerging services, as they did not open up their innovation planning horizon beyond their present value chain, thereby missing opportunities to collaborate with firms representing new players in emerging business models. Also, in some cases, technologies or solutions were acquired only, leaving the required competence set out of control for the company set B.

Table 1. Classifying scenarios for convergence acquisition strategies.

Convergence Enabler	Convergence Follower
- First acquisition, then convergence - Envisioning new dominant design as confluence of proprietary and external competences - Work on partnerships, vertical integration - Integrate and roadmap competencies - Foresight on new external technology streams with their potential intersections	- First convergence, then acquisition - Own competences remain outside new dominant design - Commitment onto new, previously unexploited competencies - Work against convergence mainstream with proprietary business model - Foresight on new strategic orientation approaches based on resource-based business redefinition

5. Discussion and Outlook

In either case A and B, common areas of activities can be observed in all dimensions of dynamic capabilities, i.e. resources, competencies and alliances, illustrating the importance of considering all these aspects instead of e.g. sticking to technology roadmapping only. Both case companies A and B were evaluated through press surveys and individual interviews with representatives in this context.

This paper introduced the perspective of regarding the phenomenon of technological convergence as a special case for technology acquisition. On the one hand, an active role in management of convergence implies technologies and competencies to be acquired on an open basis, positioning the proprietary value proposition within a network of intersecting technology domains, securing a controlling stake in emerging business models. On the other hand, a passive role in terms of managing technological convergence implies technology acquisitions to be related to a profound revision and redefinition of the value proposition. In both cases, it is argued that the concept of dynamic capabilities serves as an essential framework for any strategic actions.

References

Abernathy, W.J., Clark, K.B. (1985), Innovation - Mapping the Winds of Creative Destruction. *Research Policy* 14(1): 3-22

Achrol, R.S., Kotler, P. (1999), Marketing in the network economy. *Journal of Marketing 63: 146-163*

Afuah, A. (2001), Dynamic boundaries of the firm: Are firms better off being vertically integrated in the face of a technological change? *Academy of Management Journal* 44(6): 1211-1228

Afuah, A., Tucci, C.L. (2003), A model of the internet as creative destroyer. *IEEE Transactions on Engineering Management* 50(4): 395-402

Allee, V. (2003), The future of knowledge: increasing prosperity through value networks. Butterworth-Heinemann: Amsterdam

Amit, R., Zott, C. (2001), Value creation in e-business. *Strategic Management Journal* 22(6-7): 493-520

Backholm, A., Hacklin, F. (2002), Estimating the 3G convergence effect on the future role of application-layer mobile middleware solutions: scenarios and strategies for business application providers in 3G and beyond. In Lu, W.W. (Ed.), Proceedings of the 2002 International Conference on Third Generation Wireless and Beyond (3Gwireless'02): 74-79. World Wireless Congress: San Francisco/Silicon Valley, CA, USA

Betz, F. (1993), Strategic Technology Management. McGraw-Hill: New York

Birkinshaw, J. (2002), Managing internal R&D networks in global firms - What sort of knowledge is involved? *Long Range Planning* 35(3): 245-267

Bohlin, E., Brodin, K., Lundgren, A., Thorngren, B. (2000), Convergence in Communications and Beyond. North-Holland

Borés, C., Saurina, C., Torres, R. (2003), Technological convergence: a strategic perspective. *Technovation* 23(1): 1-13

Camponovo, G., Pigneur, Y. (2003), Analyzing the m-business landscape. *Annals of Telecommunications* 58(1/2): 59-77

Chesbrough, H.W., Teece, D.J. (2002), Organizing for innovation: When is virtual virtuous? *Harvard Business Review* 80(8): 127-+

Christensen, C.M. (1997), The Innovator's Dilemma: When New Technologies Cause Great Firms to Fail. Harvard Business School Press: Boston

Christensen, C.M., Rosenbloom, R.S. (1995), Explaining the Attackers Advantage - Technological Paradigms, Organizational Dynamics, and the Value Network. Research Policy 24(2): 233-257

Duysters, G., Hagedoorn, J. (1997), Technological convergence in the IT industry: the role of strategic technology alliances and technological competencies, Research memorandum 97:15. Maastricht Economic Research Institute on Innovation and Technology (MERIT): Maastricht

Edwards, J. (1999), Convergence reshapes the networking Industry. Computer 32(5): 14-16

Eisenhardt, K.M., Martin, J.A. (2000), Dynamic capabilities: What are they? Strategic Management Journal 21(10-11): 1105-1121

Eisenhardt, K.M., Schoonhoven, C.B. (1996), Resource-based view of strategic alliance formation: Strategic and social effects in entrepreneurial firms. Organization Science 7(2): 136-150

European Commission (1997), Green Paper on the Convergence of the Telecommunications, Media and Information Technology Sectors, and the Implications for Regulation. Towards an Information Society Approach. European Commission: Brussels

Fransman, M. (2002), Mapping the evolving telecoms industry: the uses and shortcomings of the layer model. Telecommunications Policy 26(9-10): 473-483

Freeman, C., Louçã, F. (2001), As Time Goes By. From the Industrial Revolution to the Information Revolution. Oxford University Press: Oxford

Gartner Group (2003), Technology convergence driving business model collision: Teleconference presentation, 25 March

Gassmann, O., von Zedtwitz, M. (2003), Trends and determinants of managing virtual R&D teams. R & D Management 33(3): 243-262

Grant, R.M. (1996), Toward a knowledge-based theory of the firm. Strategic Management Journal 17: 109-122

Greenstein, S.M., Khanna, T. (1997), What does industry convergence mean? In Yoffie, D.B. (Ed.), Competing the age of digital convergence: 201-226. Harvard Business School Press: Boston/MA

Hacklin, F., Marxt, C. (2003), Assessing R&D management strategies for wireless applications in a converging environment. In Butler, J. (Ed.), Proceedings of the R&D Management Conference 2003 (RADMA). Blackwell Publishers: Manchester, England

Hacklin, F., Marxt, C., Fahrni, F. (2004a), Technology partner selection for collaborative innovation in production systems: a decision support system based approach., 13th International Working Seminar on Production Economics, Vol. 4: 115-124: Igls/Innsbruck, Austria

Hacklin, F., Raurich, V., Marxt, C. (2004b), How incremental innovation becomes disruptive: the case of technology convergence. In Xie, M., Durrani, T.S., Chang, H.K. (Eds.), Proceedings of the IEEE International Engineering Management Conference, Vol. 1: 32-36. IEEE Engineering Management Society: Singapore

Ireland, R.D., Hitt, M.A., Sirmon, D.G. (2003), A model of strategic entrepreneurship: The construct and its dimensions. Journal of Management 29(6): 963-989

Johannessen, J.A., Olsen, B. (2003), Knowledge management and sustainable competitive advantages: The impact of dynamic contextual training. International Journal of Information Management 23(4): 277-289

Johannessen, J.A., Olsen, B., Olaisen, J. (1997), Organizing for innovation. Long Range Planning 30(1): 96-109

Johansson, F. (2004), The Medici Effect: Breakthrough Insights at the Intersection of Ideas, Concepts, and Cultures. Harvard Business School Press: Boston

Jones, G.K., Lanctot, A., Teegen, H.J. (2001), Determinants and performance impacts of external technology acquisition. Journal of Business Venturing 16(3): 255-283

Karlsson, M. (1998), The liberalisation of telecommunications in Sweden : technology and regime change from the 1960s to 1993. Tema Univ.: Linköping

Lambe, C.J., Spekman, R.E. (1997), Alliances, external technology acquisition, and discontinuous technological change. Journal of Product Innovation Management 14(2): 102-116

Lee, G.K.-F. (2003), The Competitive Consequences of Technological Convergence in an Era of Innovations - Telephony Communications and Computer Networking, 1989-2001. Haas School of Business, University of California: Berkeley

Lei, D.T. (2000), Industry evolution and competence development: the imperatives of technological convergence. International Journal of Technology Management 19(7-8): 699-738

Li, F., Whalley, J. (2002), Deconstruction of the telecommunications industry: from value chains to value networks. Telecommunications Policy 26(9-10): 451-472

Littler, D., Leverick, F., Bruce, M. (1995), Factors affecting the process of collaborative product development: a study of UK manufacturers of information and communication technology products. Journal of Product Innovation Management 12(1): 16-32

Miotti, L., Sachwald, F. (2003), Co-operative R&D: why and with whom? An integrated framework of analysis. Research Policy 32(8): 1481-1499

Nikolaou, N.A., Vaxevanakis, K.G., Maniatis, S.I., Venieris, I.S., Zervos, N.A. (2002), Wireless convergence architecture: A case study using GSM and wireless LAN. Mobile Networks & Applications 7(4): 259-267

Nonaka, I., Takeuchi, H. (1995), The knowledge creating company: Oxford

Olla, P., Patel, N.V. (2002), A value chain model for mobile data service providers. Telecommunications Policy 26(9-10): 551-571

Porter, M.E. (1985), Competitive Advantage: Creating and Sustaining Superior Performance. The Free Press: New York

Pringle, D. (2003), Clash of the titans, The Wall Street Journal Europe: R1-R2

Rockenhäuser, J. (1999), Digitale Konvergenz und Kompetenzenmanagement. Deutscher Univ.-Verlag: Wiesbaden

Rosenberg, N. (1963), Technological change in the Machine-Tool Industry, 1840-1910. Journal of Economic History 23(4): 414-443

Sabat, H.K. (2002), The evolving mobile wireless value chain and market structure. Telecommunications Policy 26(9-10): 505-535

Sherif, M.H. (1998), Convergence: A new perspective for standards. Ieee Communications Magazine 36(1): 110-111

Steinbock, D. (2003), Globalization of wireless value system: from geographic to strategic advantages. Telecommunications Policy 27(3-4): 207-235

Teece, D.J., Pisano, G., Shuen, A. (1997), Dynamic capabilities and strategic management. Strategic Management Journal 18(7): 509-533

von Krogh, G., Nonaka, I., Nishiguchi, T. (Eds.) (2000), Knowledge creation: a source of value. Macmillan Press: Houndmills

Wernerfelt, B. (1984), A resource-based view of the firm. Strategic Management Journal 5(2): 171-180

Willcocks, L.P., Plant, R. (2003), How corporations e-source: From business technology projects to value networks. Information Systems Frontiers 5(2): 175-193

Yoffie, D.B. (1997), Competing the age of digital convergence. Harvard Business School Press: Boston/MA

Zahn, E., Foschiani, S. (2002), Wertgenerierung in Netzwerken. In Albach, H., Kaluza, B., Kersten, W. (Eds.), Wertschöpfungsmanagement als Kernkompetenz. Gabler Verlag: Wiesbaden

CHAPTER 16

THE PROCESS FOR ALIGNING PROJECT MANAGEMENT AND BUSINESS STRATEGY: AN EMPIRICAL STUDY

Sabin Srivannaboon and Dragan Z. Milosevic

Portland State University, USA

This study focuses on the process for aligning project management with business strategy. We conducted a case study research with a total of 9 projects in 7 organizations across industries, leading to a theoretical framework for such the alignment process. We found that there are processes at the strategic level to interpret the business strategy into the context of project management. These processes include strategic planning and project portfolio management. A project is then selected into the project portfolio to fulfill business needs, and it is executed through a project life cycle process. As the project progresses, a stage gate is a major mechanism to ensure the quality of the alignment throughout its life cycle.

1. Introduction

Recognition of the strategic importance of project management (PM) in the corporate world is rapidly accelerating. One reason for this may be a strong belief by business leaders that aligning PM with business strategy can significantly enhance the achievement of organizational goals, strategies, and performance. However, empirical literature that offers advice on how to achieve this alignment is scanty. Many companies are suffering from misaligned projects and a lack of a systematic approach to align PM with the business strategy. In fact, PM is not often recognized as a functional strategy or rarely perceived as a business process, although projects are the basic building blocks of organizational strategy in many companies (Cleland, 1999), making the achievement of a PM/business strategy alignment even more difficult. In particular, many projects are not chosen in a way that is supported by the business strategy (Meredith *et al.*, 2003). Accordingly, the projects may be later terminated or do not contribute to organizational goals and waste organizational resources. In many instances, organizations treat all projects in the same way, regardless of the business strategy chosen by the organization (Pinto & Covin, 1989; Shenhar, 2001). As a result, when the business strategy is translated onto the project level, its uniqueness (speed to market, superior product quality, etc.) may disappear. Understanding the PM/business strategy alignment may be a major challenge, one that practitioners must confront when they attempt to effectively manage their projects in today's competitive environment.

This study addresses one aspect of this under-researched area by thoroughly exploring the PM/business strategy alignment in terms of the process used for such an alignment. In particular, an empirically based theoretical framework is developed that highlights the mechanisms used to strengthen that alignment. We define the theoretical framework here as a set of well developed concepts, which include an integrated structure that can be used to describe the PM-business strategy alignment process.

2. Theoretical Background

To develop a theoretical framework for the process of aligning PM with business strategy, we examined multiple streams of related literature, streams that include business strategy, project management, and alignment literature.

2.1. *Business strategy*

Definitions of business strategy vary. For example, Chandler (1962) defined business strategy as the determination of the basic long-term goals of an enterprise, the adoption of courses of action, and the allocation of resources necessary to carry out those goals. Ansoff (1965) suggested that business strategies are rules for making decisions determined by product/market scope, growth vector, competitive advantage, and synergy. Miles and Snow (1978) argued that business strategy is a pattern or stream of major and minor decisions about an organization's possible future domain. Mintzberg (1987, 1994) offered multiple definitions for business strategy: plan, pattern, position, perspective, and ploy. Later, Lengnick-Hall and Wolff (1999) summarized strategy research streams into resource-based views of the firm's hypercompetition and high-velocity strategies, and ecosystem/ chaos theories. However, the focus on commonality found in these definitions is a better way deal with competition (Tse & Olsen, 1999) by means of creating competitive advantage (Hamel & Prahalad, 1989), which gives an organization a sustainable lead over its competitors for attracting customers and defending against competitive forces (Thompson & Strickland, 1995).

2.2. Project management

Project management (PM) is a specialized form of management, similar to other functional strategies, that is used to accomplish a series of business goals, strategies, and work tasks within a well-defined schedule and budget. The essence of PM is to support the execution of an organization's competitive strategy to deliver a desired outcome (i.e., fast time-to-market, high quality, low-cost products) (Milosevic, 2003).

As opposed to the traditional stereotype, the recent literature recognizes PM as a key business process (Morris & Jamieson, 2004). This view defines an organization as the process rather than the traditional functional or matrix form, and describes PM as one of the key business processes that enable companies to implement value delivery system.

2.3. Alignment literature

Research in the literature has examined the idea of alignment in various management areas. For example, a number of studies have discussed the alignment between tasks, policies, and practices (e.g., Boyer & McDermott, 1999; Kathuria & Davis, 2001); others have emphasized the relationship between alignment and performance in regards to organizational hierarchy: corporate, business, and function (e.g., Papke-Shields & Malhotra, 2001; Youndt, M.A, et al, 1996). The literature frequently mentions research and development (R&D), production, human resources, and information technology as functional strategies and used these as the variables to examine alignment vis-à-vis the business strategy. PM is similar to other of these functional strategies and therefore it too should be aligned with the business strategy (Harrison, 1992). However, traditional literature on aligning PM with the business strategy is vague. Most studies link business strategy with PM through project selection, viewing it as part of the alignment process (e.g., Baker, 1974; Bard, Balachandra & Kaufmann, 1988; Cooper, Edgett & Kleinschmidt, 1998a; Englund & Graham, 1999; Hartman, 2000).

Added to this is project portfolio management (PPM and also called pipeline management), another concept suggested in the literature to ensure the strategic alignment (Turner & Simister, 2000). Cooper, Edgett, and Kleinschmidt (1998b) define PPM as a dynamic decision process through which an organization can update and revise its list of active projects.

The organization's choice of business strategy is what drives their PPM process, the major purpose of which is to select and prioritize projects (Cooper, Edgett & Kleinschmidt, 1998b), balance projects (Archer & Ghasemzadeh, 1999; Cooper, Edgett & Kleinschmidt, 1998b), align projects with business strategy (Cooper, Edgett & Kleinschmidt, 1998b), manage rough-cut resource capacity (Harris & McKay, 1996; Wheelwright & Clark, 1992), and articulate empowerment boundaries for project and functional management (Harris & McKay, 1996).

Only recently have researchers started to explore the alignment of PM more thoroughly (e.g., Artto & Dietrich, 2004; Jamieson & Morris, 2004; Papke-Shields & Malhotra, 2001; Srivannaboon & Milosevic, 2004). For example, Jamieson and Morris (2004) suggest that most of the components comprising the strategic planning process – internal analysis, organizational structures, and control systems – have strong links to PM processes and activities. Thus, they strongly influence "intended" business strategies. Similarly, Artto and Dietrich (2004) suggest that an important managerial challenge of the project management/business strategy alignment is encouraging individuals to participate in using emerging strategies to create new ideas and renew existing strategies. These studies suggest a need for more research in this area; none, however, explicitly talks about the process used to align PM and business strategy cohesively and comprehensively.

3. Research Design

To accomplish our research, we integrated two main phases in the study: data gathering and data analysis. During data gathering (Phase 1), we conducted a literature review so as to understand the general research on aligning project management and business strategy. In parallel with our literature review, we researched case-studies over a ten-month period, studying the process of alignment in market-leading organizations through semi-structured interviews (ranging from 60 to 120 minutes per each interview). The interviewees were individuals holding key organizational positions, individuals such as senior managers, project managers, assistant project managers, team members – as well as a few customers – in order to obtain information from different perspectives (Boynton & Zmud, 1984).

In addition to the interviews, we reviewed related documents – meeting minutes, project descriptions, risk logs – to triangulate and validate our findings. To select the reviewed cases (companies, projects, and participants), we defined multiple criteria and identified the cases most relevant to such criteria as theoretical sampling, project frame of reference (projects completed in at least six month or under), and the PM experience of participants.

We classified projects in our sample into different types: including strategic projects (creating strategic positions in markets and businesses); extension projects (improving or upgrading an existing product); utility projects (acquiring and installing new equipment or software, implementing new methods or new processes, reorganization, re-engineering); and research and development projects (exploring future ideas, no specific product in mind).

Those projects were also labeled as external customers (external contract or consumers), internal customers (internal users or another department), or both. In addition, we evaluated success dimensions of projects, including project efficiency, impact on the customer, direct organizational success, and team leader and spirit.

In Phase 2/data analysis, we wrote case studies, 25-30 pages per case, based on the interviews and related documents. We sent these cases back to the companies for verification of factual accuracy. Then, we conducted within-case, cross-case, and content analyses. Altogether, we completed 8 case studies (Cases A to H) in 7 organizations with a total of 9 projects that differed in size, type, and complexity (42 interviews). In Phase 3, a panel of experts from academia and industry validated the essential findings.

4. Analyses and Results

In this section, we analyze the patterns of the processes used by the companies we surveyed to align project management and business strategy. In doing so, we discuss the similarities and the dissimilarities across all cases in order to generate a theoretical framework of the processes that organizations use to ensure the proper alignment. We performed content analysis to compare these cases and identify the patterns of the alignment process used across these cases.

The pattern we found revealed that organizations could divide the mechanisms used to align projects with business strategies into three levels: the strategic, the tactical, and the emergent strategic feedback. Each level contained distinct mechanisms to achieve alignment.

4.1. *Level 1 – Mediating process at the strategic level*

The general steps of the alignment process begin at the strategic level where the long-term business goals are defined and business directions are determined through a strategic plan (or "intended strategy" (Mintzberg, 1994)). The strategic plan, either formal or informal, was found that every sample company had a strategic plan; some used a formal plan, some used an informal one. In all but two cases, these plans were developed to reflect a 3-year planning horizon. One exception was

Case B, which at the time of our interview was a short-term plan (1-year horizon) that the company was actively expanding to a 3-year range. The other exception was Case G, which used an informal plan due to the nature of its business (construction). In some cases, roadmaps were included in the strategic plan as the guidance for the company's (or department's) future interests, such as a product roadmap (Cases A and B) and an information technology roadmap (Case D).

We also observed that the sample companies used a project portfolio process – again, some used a formal process, others used an informal one – as a mechanism for selecting the most valuable projects that would contribute to the organizational goals. To select such projects, and make them part of the portfolio, many companies matched their strategic goals with the project's contribution, with its strategic fit. In several cases, the term project portfolio was not recognized, but its project selection and prioritization functions were employed (Cases B, E, F, G, and H). In addition, two cases recognized the term project portfolio, but it was still an informal process (Cases C and D). Only Case A had a formal project portfolio management process and semi-annual portfolio reviews, where its functions included project selection and prioritization, risk balance, strategic alignment, and capacity management.

4.2. Level 2 – Mediating process at the project level

Once organizations select projects into the portfolio, they further plan in details and executed these through the project life-cycle phases. We refer to these mechanisms so as to ensure the proper alignment during the project life cycle as the mediating processes at the project level, which can be classified into the planning process and the monitoring process.

In the planning process, we found that the companies used varying mechanisms to ensure proper alignment. The most explicit planning mechanism used was in Case C: this company required that project managers identify the alignment link of their project plans and the goals in their strategic plan. This was accomplished through product definition and project definition, by linking these with the business goals outlined in the strategic plan. In other cases, this was implicitly accomplished through the development of the project plan, as based upon the objectives of the projects and the reason why these existed, such as achieving business goals.

We found that as project progress, most companies use common mechanisms to ensure these are properly aligned during execution, using mechanism such as project metrics, internal coordination mechanisms (i.e., project management office involvement and internal sign-off), customer involvement (sign-off), and stage gates. This last item, stage gates, is so important that we have separated it from this section to explain it separately as the mediating process at the emergent strategic feedback level.

4.3. Level 3 – Mediating process at the emergent strategic feedback level

Stage gates are points in the project life cycle where projects transition from stage to stage. The gates represent filters for the project status and provide opportunities for the project to be realigned to the requirements set by the project owner. In the sample companies, we observed such gates as milestone reviews for evaluating the project status (time, cost, performance). An exception to this observation is Case A.

This company covered staffing level and market shift considerations as additional concerns. When a project fails to meet a stage gate's requirements, the project team must adjust the project (if the owner has not killed the project), in accordance with the operating conditions of the project.

For certain instances the operating conditions of a project reveal significant changes resulting from internal or external factors. This reveals factors that may affect the overall success of the project because if the project manager fails to manage the changes, the operating conditions will impact on the deployment of the business strategy. In particular, the priorities under which the project is managed may change. For example, we found that one of the examined projects in Case A was considered an unsuccessful project by its project team and the company's upper management, even though the project was well aligned with the business strategy. Part of the reason for this perceived failure was that the project was committed to the wrong set of customers, which led to a poor product definition of the overall market. By the time the project was finished, the operating conditions of the project had changed (the market had shifted), and there was no longer a place for the product developed through this project. In this case, the stage gate failed to provide the organization with the information it needed to realign its process of managing the project to meet those changes. Once the problem was identified at a subsequent stage gate, the project team should have adjusted the product definition (as part of the project strategy). Unfortunately, the project team failed to identify the changes that were necessary to save this project in a timely manner. As a result, the team was not able to react to those changes effectively. To accommodate for this unsuccessful effort, the company later adjusted its stage gate reviews to cover market shift considerations.

The mechanism explained above is a feedback loop that emerges during project execution. It is a result that is not planned or intended but that emerges from a stream of managerial decisions through time. In other words, the operating conditions of reviewed projects are expected to support the company's business strategies by helping it adapt the business strategy and its competitive attributes to environmental changes. In essence, the combination of intended and emergent strategies is needed to align project management and business strategy.

5. A Theoretical Framework

Based on these three levels of the mediating process, a theoretical framework for the process of aligning PM and business strategy is constructed and depicted in Figure 1. Mediating processes are mechanisms to ensure that organizations create and maintain alignment between business strategy and PM. For the sake of illustrating the processes in general, we have used the traditional phases of the project life cycle, including conception, planning, execution, and closing, although in reality, each company has different project life cycle phases.

Firstly, to establish and maintain the processes used to achieve the alignment, mediating processes at the strategic level help organizations to interpret the business strategy in the context of project management.

Processes at the strategic level include strategic planning and project portfolio management. Projects are initiated and selected into the project portfolio (based on the high-level analysis) to fulfill business needs, and then they go through a standard life cycle during which project planning and project monitoring are used to ensure the quality of the alignment.

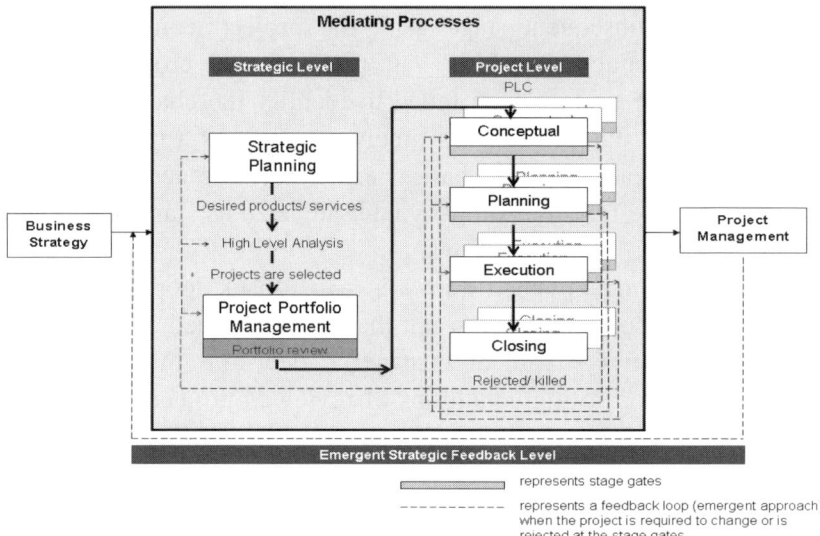

Figure 1. A theoretical framework for aligning PM with business strategy.

One of the major control mechanisms to ensure that projects remain in line with the initial expectations as they progress from one project phase to the next is the stage gate mechanism. This mediating process provides strategic feedback that can lead to what Mintzberg (1994) calls "emergent strategy."

Although cases have much in common, there were differences in the alignment process among the companies examined. These differences occurred at various points, but they all fit, within the framework depicted above, and include various details of (1) the mediating processes at the strategic level, (2) at the project level, and (3) at the emergent strategic feedback level.

6. Discussion

In this study, we explained an inductive logic process as a means to derive our mediating processes. The general process of developing these mediating processes was based on case study research, which heavily used within-case and cross-case analyses. The resulting framework can be used to explain the alignment process at the strategic level, the project level, and the corrective emergent feedback level. Our study expands on previous mostly anecdotal work by incorporating a rigorous theoretical approach into the proposed framework. Although Jameson and Morris (2004) identify strategic planning, portfolio management, and the emergent approach as important steps in the alignment process, which is also supported our research, they do not provide a framework. Further they do not position their research as a set of case studies or as a theoretical foundation. In addition, Turner and Simister (2000) argue, conceptually and without an empirical validation, that portfolio management is an important step in aligning projects with the business strategy.

In comparison with the existing literature, our framework contributes three elements:

- *Comprehensive*: The framework includes and relates all levels of participants in an organization and different levels of management processes into a coherent structured set, which describes the phenomenon of the PM/business strategy alignment in different situations.
- *Empirically established and validated*: The framework is based on a diverse set of companies and projects as well as real-world data. It also takes a multi-level view (no single-source bias), an approach that enabled us to develop a strong theoretical framework.

7. Research Limitations and Future Research

Although Eisenhardt (1989) argued that four-to-ten cases provide a sufficient range of measure and for analytic generalization, one major limitation in our study is the relatively small number of cases that we used to develop the framework (8 cases). In addition, this study may suffer from a bias of company management views. However, we were able to minimize any such bias by using multiple data sources and validating the findings with a panel of experts.

The research findings and limitations suggest the following directions for future research regarding project alignment. First, the alignment measurement methodology deserves an empirical study. If that is done in a comprehensive manner, it would be possible to standardize the measurement and create a framework for comparative studies of the alignment of various business strategy types and project types. That would enable further studies to determine the degree of alignment required under different circumstances to assure project and business success. Needed is a large sample study that focuses on the quantitative correlations of various strategy types and project management.

References

Ansoff, H.I. (1965). Corporate Strategy. New York: McGraw-Hill.
Archer, N., & Ghasemzadeh, F. (1999). An Integrated Framework for Project Portfolio Selection. *International Journal of Project Management*, 17(4): 207-216.
Artto, K.A., & Dietrich, P.H. (2004). Strategic Business Management through Multiple Projects in The Wiley Guide to Managing Projects, Eds. P.W.G. Morris and J.K. Pinto. Hoboken, NJ: John Wiley & Sons, Inc., 144-176.
Baker, N.R. (1974). R&D Project Selection Models: An Assessment. *IEEE Transactions on Engineering Management*, 21(4): 165-170.
Bard, J.F., Balachandra, R. & Kaufmann, P.E. (1988). An Interactive Approach to R&D Project Selection and Termination. *IEEE Transactions on Engineering Management*, 35(3): 139-146.
Boyer, K.K., & McDermott, C. (1999). Strategic Consensus in Operations Strategy. *Journal of Operations Management*, 17(3): 289-305.
Boynton, A.C., & Zmud, R.W. (1984). An Assessment of Critical Success Factors. *Sloan Management Review*, 25(4): 17-27.
Chandler, A. (1962). Strategy and Structure. Cambridge, MA: MIT Press.
Cleland, D.I. (1999). Project Management: Strategic Design and Implementation. 3rd ed. New York, NY: McGraw-Hill.
Cooper, R.G., Edgett, S.J. & Kleinschmidt, E.J. (1998a). Best Practices for Managing R&D Portfolios: *Research. Technology Management*, 41(4): 20-33.
Cooper, R.G., Edgett, S.J. & Kleinschmidt, E.J. (1998b). Portfolio Management for New Products. Reading, MA: Perseus Books.

Dubin, R. (1978). Theory Building. New York, NY: The Free Press.

Eisenhardt, K. (1989). Building Theories from Case Study Research. *Academy of Management Review*, 14(4): 532-550.

Englund, R.L., & Graham, R.J. (1999). From Experience: Linking Projects to Strategy. *The Journal of Product Innovation Management*, 16(1): 52-64.

Hamel, G., & Prahalad, C.K. (1989). Strategic Intent. *Harvard Business Review*, 67(2): 92-101.

Harris, J.R., & McKay, J.C. (1996). Optimizing Product Development through Pipeline Management, in The PDMAA Handbook of New Product Development. Ed. D.R. Rosenau. New Yorik: Wiley, 63-76.

Harrison, F.L. (1992). Advanced Project Management: A Structured Approach. 3rd ed. New York, NY: Halsted Press.

Hartman, F. (2000). Don't Park Your Brain Outside: A Practical Guide to Improving Shareholder Value with SMART Management. Newtown Square, PA: Project Management Institute.

Jamieson, A., & Morris, P.W.G. (2004). Moving from Corporate Strategy to Project Strategy, in The Wiley Guide to Managing Projects, Eds. P.W.G. Morris and J.K. Pinto. Hoboken, NJ: John Wiley & Sons, Inc., 117-205.

Kathuria, R., & Davis, E.B. (2001). Quality and Work Force Management Practices: The Managerial Performance Implication. Production and Operations Management, 10(4): 460-477.

Lengnick-Hall, C.A., & Wolff, J.A. (1999). Similarities and Contradictions in the Core Logic of Three Strategy Research Streams. *Strategic Management Journal*, 20(12): 1109-1132.

Meredith, J.R., & Mantel. Jr., S.J. (2003). Project Management: A Managerial Approach. 5th ed. New York, NY: John Wiley & Sons.

Miles, R.E., & Snow, C.C. (1978). Organizational Strategy, Structure and Process. New York: West.

Milosevic, D.Z. (2003). Project Management Toolbox: Tools and Techniques for the Practicing Project Manager. Hoboken, NJ: John Wiley & Sons.

Mintzberg, H. (1987). The Strategy Concept I: Five P's for Strategy. *California Management Review*, 30(1): 11-24.

Mintzberg, H. (1994). The Rise and Fall of Strategic Planning. New York, NY: The Free Press.

Papke-Shields, K.E., & Malhotra, M.K. (2001). Assessing the Impact of the Manufacturing Executive's Role on Business Performance Through Strategic Alignment. *Journal of Operations Management*, 19(1): 5-22.

Pinto, J.K., & Covin, J.G. (1989). Critical Factors in Project Implementation: A Comparison of Construction and R&D Projects. Technovation, 9(1): 49-62.

Shenhar, A.J. (2001). One Size Does Not Fit All Projects: Exploring Classical Contingency Domains. *Journal of the Institute for Operations Research and the Management Science*, 47(3): 394-414.

Srivannaboon, S., & Milosevic, D.Z. (2004). The Process of Translating Business Strategy in Project Actions in Innovations: Project Management Research 2004, Eds. D.P. Slevin, J.K. Pinto, and D.I. Cleland. Newtown Square, PA: Project Management Institute.

Thompson, A.A., & Strickland, A.J.I. (1995). Crafting and Implementing Strategy. Chicago, IL: Irwin.

Tse, E.C., & Olsen, M.D. (1999). Strategic Management,in The Handbook of Contemporary Management Research, Ed. B. Brotherton. New York, NY: John Wiley & Sons: 351-373.

Turner, J.R., & Simister, S. (2000). The Gower Handbook of Project Management. 3rd ed. Aldershot, UK: Gower.

Wheelwright, S.C., & Clark, K.B. (1992). Revolutionizing Product Development. New York, NY: The Free Press.

Youndt, M.A., et al. (1996). Human Resource Management, Manufacturing Strategy, and Firm Performance. *Academy of Management Journal,* 39(4), 836-866.

CHAPTER 17

RISK-SHARING PARTNERSHIPS WITH SUPPLIERS: THE CASE OF EMBRAER

Paulo Figueiredo*, Silveira Gutenberg**, and Roberto Sbragia**

*Boston University School of Management, USA;
**University of Sao Paulo, FEA-EAD, Brazil

Investing in new product development is a strategic option for companies that want to adapt to constant changes in customer preferences, anticipate new product releases of rival companies and/or respond to them, make use of technological opportunities and increase market share. This investment can be undertaken directly, through R&D, licensing of technologies or copying; however, there are other means to develop products based on cooperation between companies in the production chain, through partnerships.

Since the mid 1990s, the global aircraft industry has been creating new solutions for product development. Risk-sharing partnerships with suppliers began to be established in an attempt to reduce investments and, consequentially, the dependence on loans. Companies focused their development and manufacturing activities on specific and strategically interesting areas. The partners began not only to invest in tooling, engineering and infrastructure, but also to participate more directly in the projects, in the investments and design activities, acquiring rights to future sales income of products. This contractual modality, called risk-sharing partnership, is the focus of this study.

Specifically, this article analyzes the risk-sharing partnerships made by Embraer during projects for the ERJ-170/190 aircraft group. It also aims to justify these partnerships, considering the current global aircraft market conditions, evaluating the critical success factors, requirements and macro-economic conditions which supported the adoption of this new policy. Embraer is frequently studied and quoted as a successful example of a Brazilian business enterprise. This analysis may be a starting point to evaluate whether the business partnership model is useful to improve performance of Brazilian firms belonging to other industrial segments.

1. Introduction

A company has to deal with constant changes in customers' preferences, to anticipate new product releases of rival companies and/or respond to them, to make use of technological opportunities, and to increase market share. In order to address these challenges a strategic option to any firm is to invest in product development and exploit rival information for a better position in the market. This investment can be undertaken directly, through R&D, licensing of technologies or copying, for instance. However, there are other ways of developing products that involve cooperation of other firms in the process, through partnerships.

Since the mid 1990s the global aircraft industry found new solutions for product development and begun to establish risk-sharing partnerships with its suppliers, in an attempt to reduce investments and, consequentially, the dependence on loans, thereby diluting the risks associated to products. In turn, these companies focused their development and manufacturing activities on specific and strategically interesting areas, that is, the companies began to use the concept of core competences. This concept consists of a chain of capabilities and technologies which add value to the client, differentiating competitors and strengthening the company's competencies. (Hamel and Prahalad, 1990) and (Gingrich, 2003).

At first, this trend was reflected in a less vertical manufacture process, with an increased number of parts produced by thirds parties, not only responsible for supplying manpower but also for the material used and the manufacture process of the parts or pieces of the aircraft. Later, partners began to take over not only investments in tools, non routine engineering, and infrastructure but also to participate more directly in investments and project development, thereby acquiring rights to their future sales income. This contractual modality is called risk-sharing partnership (RSP) and is the object of the current study.

Specifically, this study analyzes the risk-sharing partnerships made by Embraer during projects for the ERJ-170/190 aircraft group, the so-called Embraer Regional Jets and places them in the global aircraft context. It evaluates the critical success factors, requirements, and macro-economic conditions that can support the adoption of a partnership policy. Embraer is frequently studied and quoted worldwide as an example of a successful Brazilian business enterprise. This analysis may be a starting point to evaluate whether the company's model of risk-sharing partnerships may be applicable to companies belonging to other industrial segments in order to improve their results.

The research methodology was the case study. This methodology is suitable for exploratory research insofar that it carries out an intensive analysis of a situation. This analysis allows for both an investigation of holistic nature and of the significant characteristics of the events, objects, and phenomena under study (Yin and Campbell, 1994). In view of the objective of the current research, the use of this methodology was justified.

Data for the study were collected from various sources. Initially, non structured interviews with the Director in charge of the academic liaisons at Embraer were conducted. These were followed by visits to the assembly lines. Studies from other researchers, progress reports, and presentations from Embraer were then accessed. Finally, reports from companies in the aircraft sector, which were available on the internet were checked.

Section 2 of this document includes a review of literature about risk-sharing partnerships, focused mainly on the aircraft sector. Section 3 presents some information on Embraer and analyses the risk-sharing partnerships in the program of the ERJ-170/190 family from a global perspective. Last section exhibits the conclusions and recommendations for future studies.

2. Foundations on Risk-Sharing Partnerships

Many explanations are found in literature to substantiate the entry of aircraft companies into risk-sharing partnerships. Most research focuses on strategic alliances. The purpose of this chapter is to provide a conceptual basis for an empirical investigation of risk-sharing partnerships (RSP), based upon work of other authors. Although an overview of the transaction cost theory is provided, it is not comprehensively described. There is not a single theory providing a complete basis for understanding the use of strategic alliances such as RSP, however, transaction cost theory (Williamson, 1975; 1985) offers theoretical explanations. The broader concept of strategic alliances is briefly explained, before analyzing this theory. At the end of this chapter, a description of RSP in the global aircraft market is made.

2.1. *Strategic alliances*

Seixas, Grave & Gimenez (2001) view a strategic alliance as a convenience game for the companies involved, which lasts as long as the parties are interested. They must have clear and well defined strategic intentions to establish trust and to bring about synergies, which should prove beneficial to the parties through a joining of forces. Some examples of strategic alliances are joint-ventures, operational collaborations, licensing and supplying agreements, and RSP.

Gordon (2003) mentions that a strategic alliance can significantly improve performance of an organization through joint actions. Klotzle (2002) has a different view compared to other authors by virtue of two facts: 1) alliances serve to facilitate the access of companies to the partners' valuable resources; 2) success of a strategic alliance relies on transfer of knowledge and capabilities during the partnership. With respect to the difficulty of knowledge transfer, Nyiri (2203) stresses that it is one of the critical factors in partnerships, which may cause differences between companies and regions. Furthermore, Oliveira (2003) discusses the fundamental role played by strategic alliances as accelerators of entrepreneurship and innovation. Liboni, Takahashi & Mauad (2004) emphasize that companies must develop mechanisms to achieve technological competencies that complement one another.

White (2001) affirms that strategic alliances have four possible objectives:

- Defensive – reduce the differentiations of competitors;
- Offensive/Optimizing – optimizing relationship with suppliers or partners to reduce costs. Sharing responsibilities, information and abilities;
- Cost Sharing – reduce R&D costs when investing in new technologies;
- Expand the Businesses – expand current markets and/or enter new markets.

For Liboni, Takahashi & Mauad (2004) the first objective is reduction of risk and uncertainties involved in the process. The other objectives are avoidance of investment in specific capabilities, new productive units, or access to new technologies and markets, as well as a search for complementarities in assets and competences. Complementarity of the alliance entails the existence of significant differences in terms of technology, product, market, qualifications and capabilities of the partner companies.

2.2. *Transaction cost theory and RSP*

Transaction cost theory attempts to explain why, in certain circumstances, hierarchical institutional structures may provide a more efficient means of governing economic transaction than markets. Alliances (in which RSP is included) are typically positioned as hybrid forms of organization, located somewhere between the arms' length contracts that characterize markets at one end of the spectrum and the complete equity control that characterizes "hierarchy" at the other (Jordan and Lowe, 2004). Generally speaking, the firm is said to economize on transaction costs because it internalizes transactions. Transaction cost theory focuses on the efficient choice between producing goods and services within the firm or purchasing them in the market. If the cost of purchasing goods or services is lower than the cost of internal production, the activity will be turned over to the market (Coase, 1937; Williamson, 1975, 1985).

If the transaction is a one shot enterprise, is highly certain in its terms, and does not require specific resources (an example of high specificity is a special machine or specially trained personnel), then the integrated firm enjoys no advantage relative to a simple market transaction. Actually, the three attributes - frequency of recurrence of transactions, uncertainty to which they are subject, and the degree to which they are supported by durable and specific investments (asset specificity) - all draw the balance in favor of the integrated firm (Koenig and Thietart, 1988).

There are, however, some limitations associated with a dichotomy classification scheme. Powell (1987) noted that "...analytical concepts such as markets and hierarchies may provide us with distorted lenses through which to analyze economic change. By looking at economic organizations as a choice between markets and contractual relations on one side, and at conscious planning within a firm on the other, we fail to see the enormous variety that cooperative arrangements can take...." Gomes Casseres (1996) similarly argues that "Economists and managers now realize that the old dichotomy between firms and markets no longer apply; perhaps it never did. Alliances fill the wide gap between these two extremes. They are a unique way to govern incomplete contracts between separate parties."

Cooperative agreements, such as risk-sharing partnerships, are governable structures that are intermediate to the market and the firm. This form of organization may be used when it is inappropriate for a firm to internalize an activity, but when at the same time there are high transaction costs associated with the market exchange. An explanation of the use of strategic alliances is implicit in transaction cost theory, although most of the body of theory does not specifically addresses cooperative government structures (Kogut, 1988).

2.3. *Risk-sharing partnerships*

In defining RSP it is important to stress that suppliers who invest in the development of parts or systems in a project are not always risk-sharing partners. A risk-sharing partnership with suppliers necessarily involves a participative sharing in the project, with rights to future sales income of products. The partner relies on the commercial success of the project to receive the total or part of his share for the activities and/or products delivered. The partner of a project is subordinate to the company that wrote the contract. It practices the activities of development and manufacture according to specified rules and acts jointly in an integrated manner with the manufacturer. As such, risk-sharing partnerships are different from joint ventures, contracts of technological cooperation, mergers and mere sourcing agreements between companies.

Risk-sharing partnerships can contribute to reduce the lead-times of projects. Stalk and Hout (1993) affirm that a decrease in the amount of time necessary to develop and launch new products, as well as a fast response to the client's orders, results in a competitive advantage. RSP can reduce the duration of projects because it enables parallel work, diminishes rework, synchronizes deadlines, and enhances the communication between suppliers and manufacturers. These techniques to enhance the innovative process were pointed out by Zirger and Hartley (1996).

As stated by Bernardes (2000), the design, marketing, logistics, distribution and trading, infrastructure activities, etc. are key elements of any entrepreneurial success. However, a sustainable competitive advantage is brought about not by repetitive manual work that does not add value and may be outsourced, but by technology and knowledge related to the development and manufacture processes. The recent strategy of Embraer and other companies in the sector of development and manufacture of aircraft reflects these changes. Many of these companies, although they are large size organizations with qualifications to carry out investments, faced a continuous increase in their costs for new product development, up to the point of being forced to devise a new business structure (Brown, 1998).

Difficulty in obtaining loans was another reason of this need for change. To make partnerships with the local and international suppliers was one of the approaches devised to face this situation. Constitution of international partnerships is also needed due to the lack of qualified suppliers in the country of origin. Many governments require that part of the aircraft production be made in their countries and by local companies, as a pre-requisite for the approval of contracts. China is an example of this. For Boeing to sell its 747 aircraft to the Chinese, it had to manufacture or assemble at least part of the product within this country (Pritchard, 2002). This kind of agreement has become increasingly complex, so that manufacturers such as Boeing and Airbus started to operate in innovative, decentralized, and global supplier networks that are not merely defined by cost criteria, quality, or logistics. Many of these outsourcing relationships were formulated in response to economic development priorities by governments of other countries, who control decisions of aircraft purchase for the domestic markets (Dixon, 1999).

The new techniques that emerged in the design and manufacturing activities increased the efficiency of the development process, cutting costs and reducing lead times (Krishnan, 2003). These techniques included simultaneous engineering, just in time (JIT), the use of programs like computer aided design (CAD), and computer aided manufacturing (CAM). New network communication technologies allowed rapid exchange of data and information even between enterprises on different continents, thereby becoming a competitive factor. They also permit a degree of cooperation which had been impossible, for development as well as for manufacture of products.

3. Risk-Sharing Partnerships in the Context of Embraer

Embraer is a Brazilian open capital enterprise, a manufacturer of aircraft, focused on the market segments of commercial, corporate, and military aviation. Table 1 presents the company's products, ranked in three segments. Embraer's families of regional jets place it among the four largest aircraft manufacturers in the world, having achieved an income over BU$3.9 in 2005. The firm backlog orders totaled US$ 10.4 billion on March 31 of 2006.

Table 1. Segments and families of aircraft.

Civil Commercial	Corporate	Military
ERJ 140, 145, 170, 175, 190, 195	Legacy	Super Tucano/ALX, AMX-T, BEM-145 AEW&C / RS/AGS / MP/ASW (P 99)

The company is prominent in the development and production of aircraft that operate in the regional aviation segment around the world, mainly with successful sales of the ERJ-145 with a capacity for about 50 seats. The segment for this model uses mostly medium size airplanes, the so-called commuters with 10 to 120 seats.

For Cassiolato (2002) the main thrust of Embraer's technological strategy was not the import of technology packages known as "black boxes", which were to be "opened", taken apart, optimized, and adapted to local conditions (according to the principle of reverse engineering). On the contrary, investments and efforts were directed towards acquiring competence through solid qualifications in basic and applied research, as well as continued training of human resources, apt to develop and design specific technological solutions.

In the mid 1990s Embraer started to pay close attention to strategic project activities such as development, systems engineering, materials, and integration. The company adopted the strategy to look for partners to manufacture parts and subsystems that correspond to the assembly of systems and kits, that is to say, to the off-load strategy. Thereby, investments in the sectors of machining and stamping were limited to partial modernization of existing equipment, in accordance with production requirements and also when suppliers were not available (Mendonça, 1997).

Currently, Embraer has an extremely modernized product development structure. In 1998 with the beginning of the development of a new family of regional jets ERJ-170/190, the company strived to act in the market of commuters, with a capacity of 70 to 118 seats. Development of the new program of the ERJ-170/190 family, that took about four years, and required around US$ 900 million of the company's own investments. Pre-launching took place seven years after starting production of the ERJ-145. It is important to note that the ERJ-145, although a lesser complex project also took four years, showing an improved company efficiency in the development of product families.

To meet its targets, the ERJ-170/190 program included modern management techniques of commercial aircraft development. The company has oriented its activities towards the generation of value, acting as a system integrator, mastering various technical phases and details of aircraft subsystems without, however, manufacturing them, maintaining the capability to combine and adapt according to project requirements. Therefore, Embraer sought to combine the complex technological issues associated to the demand, with a vision always oriented towards low costs, increase of income and solid return on investment. As an aircraft is comprised of more than 28 thousand parts, the capability to project and specify the product and harmoniously integrate components in various subsystems is a complex and difficult task. This activity is the core of Embraer's strategy. Linked with the marketing and technical services activities, it forms the core element of the company's competencies. It is this that permits Embraer to control its network of partners and its global supply chain, strengthening its commercial performance and competitive advantage, Cassiolato (2002).

Embraer established a well defined hierarchy regarding its suppliers. The firm coordinates a network structured on three levels in a decreasing order of importance. On the first level are the risk-sharing partners, those that take on financial risks in the projects, or better, multinationals that participate jointly in the design project and add technological value. The second level consists of suppliers that provide the systems, parts, components and services ordered by the company. The majority of these suppliers (98%) are companies from other countries. The companies at this level supply their equipment, avionics, components, etc., according to the specifications given by Embraer. An intense exchange of knowledge and technologies takes place between suppliers and the contracting company, however with less intensity than that taking place on the first level. This group may be subdivided. Some do not have rights to future sales income, just being paid for the supplied products and services. Others make significant investments in development such as: non-routine engineering, tooling and infrastructure. They also participate in the last part of the Joint Definition Phase and are therefore called risk suppliers. Different from risk-sharing partners, this last group is responsible for less complex and expensive components. On the third level are the outsourced suppliers, the companies and individuals that receive the raw material and the design from Embraer and sell their manpower services. Services outsourced by Embraer include project and system engineering services, machining and chemical treatments, as well as finishing and production services. Many of these companies are located near the plant (headquarters) in São José dos Campos and are directly subordinated to Embraer.

3.1. Risk-sharing partnerships in the ERJ-170/190 aircraft family

This section highlights major differences in the framework of partnerships compared to those made during the development of the ERJ-145, Embraer's first family of Regional Jets. Investment in the ERJ-170/190 program was a risky decision for a company that sought to participate in a market segment close to one in which Boeing and Airbus operate - that of airplanes for about 100 passengers. The development of the family of ERJ-170 and ERJ-190 jets, with 70 to 118 seats respectively, had an extensive participation of risk-sharing partners – a total of 16 - which include enterprises such as General Electric, supplier of the turbines, Honeywell manufacturer of the avionic systems, Gamesa, responsible for the empennage and rear fuselage units and the Hamilton Sundstrand, supplier of the tail cone, among various others, see Figure 1.

Another important consideration is the number of suppliers involved, which in comparison to the ERJ-145, was reduced from 400 to 40. The decrease was a strategic decision to better manage them, minimize costs, and improve product quality through collaboration with the best companies in the sector. The ERJ-170/190 program was developed according to a new philosophy – strategic partnerships. This philosophy brought about new dynamics for product development, aiming at commercial success so all could benefit from the results. This philosophy also mirrors the new competitive reality in the aircraft market – a new standard of "integrated organization" through networks of knowledge, development and technological innovation with participants that provide resources for project funding, as well as share the risks and uncertainties.

With implementation of the ERJ-170/190 program some changes in the partnership process were introduced. In addition to reduction in the number of suppliers, new forms of relationship evolved. For Cassiolato (2000), the American companies are loosing their share as Embraer's suppliers, retaining 57%. The Japanese now have 8%, the Europeans 27% and the remaining countries 8%. As for supplier participation in the manufacture of aircraft components, risk-sharing partners have 36% of the total shown below, international suppliers have 57% and Brazilians have a participation of 7%:

- 60% of equipments (engines, avionics, air conditioning systems);
- 34% of the metal structures (wings and careenage);
- 4% of the electrical components (cables, wires and systems) and mechanical systems (brakes, wheels); and
- 2% of the basic components (aluminum, titanium, Kevlar, carbon fibers).

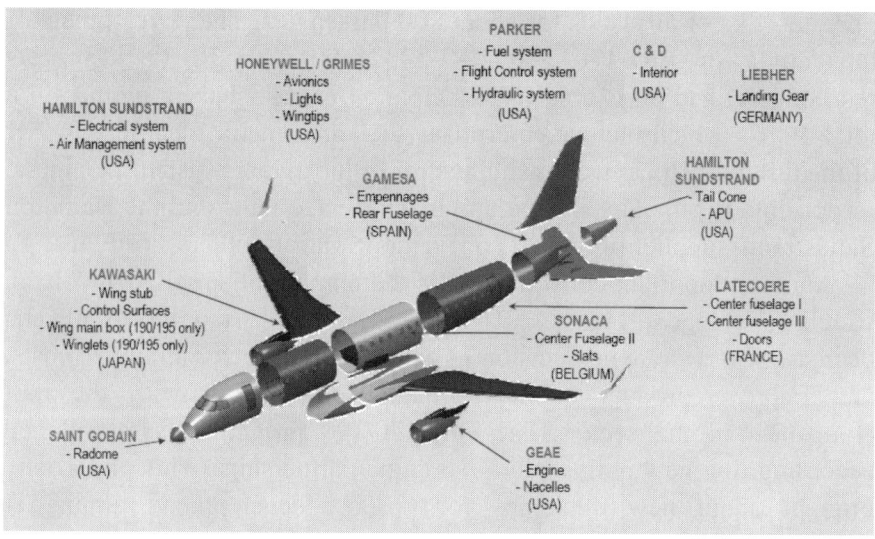

Figure 1. Risk-sharing partnerships of the ERJ – 170/190 program (Source: Embraer Company).

3.2. *Product development phases at Embraer*

The phases are in agreement with Embraer's Product Development System, based upon knowledge and organized on line with its risk-sharing partners for the development of the ERJ-170/190 aircraft family. Each of the project phases and the interactions with the risk-sharing partners are described below.

3.2.1. *Preliminary studies*

At this phase the target market is perceived and defined. For this purpose the terms and costs for conducting the preliminary studies, the business plan, as well as a definition of macro requisites and the basic design of the aircraft are set (Bernardes, 2000). Part of this phase is risk analysis in which the items that generate an impact on costs and terms are foreseen. Basically, studies that analyze the advantage of purchase or manufacture of parts and subsystems are carried out (make-or-buy decision). Therefore, the possible risk-sharing partners and suppliers are defined. During this phase, the company needs to evaluate the advantages and disadvantages of making a partnership or joint venture as opposed to keeping the manufacture and development activities in house in a vertical manner.

3.2.2. *Contact and selection of partners*

Through requests for proposals it was possible to establish which partners could qualify to meet the technical and commercial requisites and guarantee the quality standards. The strategy for selection of the international partner companies had, as guideline, three high level prerequisites to guarantee added value to the program. In other words it addressed the issues of: a) technical qualification; b) supply and integration of "technological packages"; and c) financial and investment structure.

In thesis, risk-sharing partners acted as first line suppliers, responsible for a significant part of the R&D and for the aggregation of a set of subsystems and components of the aircraft project to be supplied and integrated by Embraer, Oliveira & Bernardes (2002).

The risk-sharing partners of the ERJ- 170/190 family of jets were selected from a group of 85 potential companies. In a first selection 58 companies were pre-qualified and later only 16 were chosen from 7 countries located on four continents. For future projects it is hoped that the supplier and partner selection will be even more rigorous, because Embraer is achieving a stronger position to impose conditions on the network participants. Bernardes (2000) reports that for the assessment of the management of suppliers and partners, specific aspects were considered. For this purpose Embraer keeps a program of supplier follow-up supporting its activities. Some of these suppliers already adopted the JIT and Kanban philosophy for replacement of parts to reduce delivery and storage times, optimizing costs.

The capability to manage risks globally is a critical factor to keep alliances healthy. The criterion for choice of partners is one of the ingredients of Embraer's success. In its nationalization program Embraer hopes to produce 50% of the components in Brazil. In this context, risk-sharing partners view Brazil as a strategic place for investment, because of the country's tradition in this segment and the high level of education and knowledge in the area. Other appeals for investment in the country are low wages, political stability and industrial capability. Some of the risk-sharing partners have already invested in Brazil especially in the cluster of São José dos Campos, near Embraer. Among them are noteworthy the Kawasaki Heavy Industries (Japan), C&D Interiors (USA), Sonaca (Belgium) and the Pilkington Optronics (England).

3.2.3. *Joint definition phase*

In this phase more than 650 engineers, technicians and Embraer specialists and partners participated without having yet a breakdown of the project. Partnership contracts, that were signed prior to the Joint Definition Phase (JDP), defined the parts of the aircraft that would be manufactured and the investments needed in general terms only, nevertheless in a rigid manner. The outcome of the JDP was the production of complete electronic mock-up system of the aircraft or in other words, a mock up of all the structural, manufactured, system, and quality aspects of the project. Figure 1 depicts how the multidisciplinary and multicultural teams worked in the development of the ERJ program.

During the JDP, the activities were performed in the Embraer facilities. The performance requirements of each of the aircraft systems were specified as well as the requirements of functional and physical integration of the aircraft components. Partners' investments were made in development of parts and purchase of equipment.

For integrated and simultaneous product development, using multidisciplinary project teams, Embraer adopted the methodology called Integrated Product Development. As such, the project team was physically organized during JDP in the shape of an airplane – See Figure 1, in the Embraer facilities. The objective was to maximize interactions between the program's partner teams and allow all professionals to communicate better. In particular, professionals could understand the dynamics and goals of each group in order to carry out parallel activities. This organization is in agreement with the suggestion by Stalk and Hout (1993) who stressed the importance of a physical layout suited to reduce development time and improve product quality. According to the authors this layout must be flexible and organized by product. In addition, the teams responsible for one component must be kept as close together as possible.

Information Technology was mandatory for JDP success. Embraer supplied electronic equipment and exchange of information to make the work of the development team possible. The Virtual Reality Center (VRC) was of fundamental importance for this phase of the project, helping to integrate various aircraft subsystems, in addition to contributing towards customizing of products according to client requirements. Furthermore, simulations that permitted follow-up of the development were made available to future buyers of the aircraft. VRC also contributed to better marketing of the company/product because some of the sales were driven by presentation in virtual reality. Cassiolato (2002) stresses the importance of the VRC for the efficiency of decision making, identification of errors, problems in the design, etc. Embraer is noteworthy for investment in technologies that support the development process. Without Embraer's core competence in design, product specification and integration of components, JDP would not have been successful.

3.2.4. Development

In this phase of the program the detailed design of the airplane is accomplished. Conjoint work at the Embraer plant is concluded and the final definition of the aircraft is made. Engineers and technicians of the partners return to their home countries to finalize the project.

To facilitate communication, the management model adopted by Embraer is organized in an online network with its risk-sharing partners, an imperative fact for the successful development of the aircraft. Prior to the start of the project, few of the partner companies held the technological capability to use the powerful software for online connection. The software enabled a harmonious and standardized form for the creation and transmission of industrial design files for aircraft pieces and parts.

Technological advances gave the program substantial time savings and high quality in the aircraft development. According to Cassiolato (2000), approximately 18 months were gained. The prime responsible for this reduction of time between finishing of the design and certification is the VCR. It also contributed to a 50% reduction of the time to market by permitting a better evaluation of the aircraft by the certifying authorities. Utilization of an electronic mock-up system, instead of the traditional wood model, further contributed to process efficacy. Some other Information Technologies utilized in the project were the flight simulator, electronic data interchange and the CAD software in 3D - Catia®. This software eliminated the need for slow and expensive prototype constructions, virtually reproducing all functionalities, carrying out simulations and tests and detecting problems that the product might present, thereby minimizing the project risks.

4. Conclusions

The main objective of this study was to analyze the risk-sharing partnerships established by Embraer in the ERJ-170/190 aircraft family project. Some prerequisites and macro-economical conditions, which privileged the adoption of a policy of risk-sharing partnerships, were presented. This analysis may be used in various ways, particularly by the academic and entrepreneurial milieu. Scholars of productive systems may better understand the process of development and manufacture of products through risk-sharing partnerships and compare the solutions found by Embraer with those of other companies in this industry. The critical factors for success of the risk-sharing partnerships of Embraer may help, if this contractual modality may be used in other industrial or technology sectors, moreover it serves as an example for other aircraft companies.

One of the factors of success was the clear identification of the market needs. Especially in the regional aviation market and technological capabilities, having focused its actions on delivering products of high technology at competitive prices for its clients. Today, Embraer is a point of reference of competence and success for emerging high technology companies. The reasons that led the company to take this strategic decision of making risk-sharing partnerships are:

- Concentration of activities in core competencies;
- The need to reduce development time of its aircraft through shorter R&D cycles;
- Difficulty in finding government financing after the company's-privatization;
- Market opportunities achieved by means of partners in different countries.

As such, with its excellence in design, integration capability, and high aircraft technology, Embraer was able to attract partners in the global market who would take a stake in and invest in its projects. As for risk-sharing partners, technology transfers during their participation in the project represented a major advantage. Nevertheless, in the future, these partners may become indirect competitors of Embraer in areas where they acquired know-how.

Risk-sharing partnerships emerged as a crucial factor for Embraer's survival in the competitive market where it acts. Survival conditions in the aircraft industry may be tougher if companies are isolated, not very creative, not flexible enough in terms of associations and alliances, or slow to modernize its criteria and production methods. Based upon this study, some factors critical for the success of risk-sharing partnerships may be identified:

1. Capability for integration

Physical and simultaneous integration of various multidisciplinary and multicultural teams was decisive in some phases of the project, especially in the Joint Definition Phase. Embraer was able to coordinate the development process and establish a sequence of activities to save project time. Many of these complex development activities were performed in parallel. Usage of modern Information Technology tools such as extranet communication, computer aided design and manufacturing (CAD-CAM), Virtual Reality Center and electronic mock-up were crucial to assure integration of the project teams.

2. Mastering of key technologies

The fact that Embraer was able to establish the project's basic technical prerequisites permitted its autonomy as leader and enabled the choice of eventual partners. Concentration on core competences of design, materials engineering, system integration, project management, and the supply of technical support to clients ensures independence of the company's decisions and results in a sustainable competitive advantage.

3. Capability as a negotiator

The company was successful in making partnership contracts beneficial for both parties. Participation in the projects, investments, budgets for the activities of each partner, clauses related to terms to be met, quality and adjustment requirements to technical specifications, and the assignment of responsibilities in case of design or manufacture failures were all highly successful.

4. Post-sale partnership services

Embraer's partners also participate in client services, in the supply of replacement parts and services and even in training offered to clients. In a vertical production process, this activity may be offered only by the manufacturer, however in a shared design and manufacture process it is important that partner companies are available, as they specialize in specific technologies that the manufacturer does not.

For Embraer the most significant learning during the ERJ-145 project was management of contracts between companies, not advantages related to technologies it did not have. Another thing learned was how to achieve cost reduction of the subcontracted production processes. With the deverticalization process of production and balancing of its production plants, Embraer created conditions to reduce the price of its products. The strategy that oriented the partnership program definitely is focused on costs and financial engineering.

All this learning process led to an even more intense focus on the ERJ-170/190 program. Strategic partnerships were more integrated and complex. The project was carried out in co-design with partner companies. Another significant aspect was the technical requisites of the new partners, which were determined prior to beginning of the aircraft project, something that did not happen with the ERJ-145 project. Partnerships were made with large multinational companies, which made the aggregation of markets and distribution of development costs easier, thereby minimizing capital investments, enabling acquisition of business know-how and commercial and logistics infrastructure. To summarize, the major acquisitions in knowledge were:

- Development processes integrated with information systems and networks interconnecting clients, suppliers, and partners;
- Ability to integrate project teams from various countries in a shared physical environment and later, in a separate environment;
- Faster development of complex products, with shorter cycles;
- Experience in the offering of post-sales services together with other partners;
- Ability to negotiate strategic partnership contracts with other companies in the market;
- Integration and products sales and services on global level, increasing the opportunity to internationalize business/markets.

Examples of recent and punctual risk-sharing partnerships that were signed by other aircraft companies substantiate Embraer's position as a front line company in the local and international market. This type of contractual modality is a relatively new trend in the aircraft industry, which began to be a part of projects, mainly after the mid 1990s. It was precisely during this time period that the company began to develop the ERJ-145 family and carried out the changes in its production system and in R&D, which included the constitution of partnerships. Some companies in the sector, such as Boeing, are more conservative and resist adoption of partnerships in the same way as Embrear. However, considering Embrear's current success, it seems that risk-sharing partnerships will become standard in the future.

An interesting subject for analysis in future studies is the comparison of two distinct, seemingly opposing tendencies: a) that of concentration of companies through mergers; b) cooperation between companies through risk-sharing partnerships. During the Cold War, the great impact of government actions kept the aircraft sector in an artificial situation, with too many companies in activity. With the end of this policy, the sector became globalized and showed a tendency for concentration. The reduced number of large commercial aircraft manufacturing companies in the international market seems to signal that the process of mergers and acquisitions like that of Airbus in 2001 has reached a stabilizing stage. Companies have now begun to follow a strategy of verticalization and cooperation.

Another research possibility would be to assess the cost/benefit of using risk-sharing partnerships in other industrial and technological sectors. Besides assessing if these companies meet the critical factors previously cited, such analysis should also consider aspects such as scale, regional influences, investments, project complexity, macro-economical and political contexts, as well as the level of dependence on the partners.

It is questionable if the model adopted by Embraer can serve as benchmark and inspiration for other Brazilian companies. The company distinguishes itself by a strategic and innovative insight and by its system of developing personnel. With regard to the latter, in the building of its technological capabilities, excellence of human resources has been one of the decisive factors. Brazilian companies wishing to follow Embraer's model will have to invest in these managerial systems. The making of risk-sharing partnerships demands a whole set of managerial abilities. Should the companies lack them, they might be better off seeking foreign partners for the signing of joint-ventures. On the other hand, complexity of the projects is another important factor. It may not be interesting to establish partnerships in industries with low or medium complexity products, where there are not high risks involved and the projects do not require high investments.

References

Bernardes, R. (2000). Embraer – Elos entre estado e mercado. São Paulo: Ed. Hucitec.
Browm, R. E. (1998). http://www.wingsclub.org/speech3.html.
Cassiolato, J. E., Bernardes, R., Lastres, H. (2002). Innovation Systems in the South: a case study of Embraer in Brazil. Paper prepared for UNCTAD-DITE investment policy and capacity-building branch. New York and Geneva, United Nations.
Coase, R.H. (1937). The Nature of The Firm. Economica, 4, 386-405.
Dixon, M. (1999). State, Strategy, firm Strategy and Strategic Alliance: Evidence from United States-Asian Collaboration in Commercial Aircraft (Japan, China, Korea). Doctoral thesis, University of Pittsburgh.
Gomes-Casseres, B. (1996). The Alliance Revolution: The New Form of Business Rivalry. Cambridge, MA Harvard University Press.
Gordon, St. (2003). Computing infomation technology: the human side. USA: Idea Group Publishing.
Hamel, G. Prahalad, C. K. (1990). "The Core Competence of the Corporation", *Harvard Business Review,* Vol. 68, no. 3, May-June 1990, pp. 79-93.
Jordan, J., Lowe, J. (2004). Protecting Strategic Knowledge: Insights from collaborative Agreements in the Aerospace Sector. *Technology Analysis & Strategic Management,* Vol. 16, No. 2, 241-259, June.
Klotzle, M. C. (2002). Alianças Estratégicas: Conceito e Teoria. S. Paulo: RAC.
Koenig, C., Thietart, R. (1988). Technology and Organization: The Mutual Organization in The Euopean Aerospace Industry. *Int. Studies of Mgt. & Org.,* Vol XVII, No 4, pp. 6-30, M.E. Sharpe Inc.
Kogut, B.(1988). Joint Ventures: Theoretical and Empirical Perspectives. *Strategic Management Journal,* 9: 319:332.
Krishnan, R. T. (2003). Where core competence soars. Hindu Business Line, http://www.blonnet.com/2003/10/01/stories/2003100100020800.htm.
Liboni, L. B. (2004). TAKAHASHI. Sérgio. MAUAD. Talita Marum. Alianças Estratégicas para o Desenvolvimento de Novos Produtos. Curitiba: Enanpad. 2004
Mendonca, M. (1997). Incentives to Embraer's Productive Chain Densification. Final Report, mimeo, December.
Nyiri, L. (2003). Foresight as a Policy-making Tool. Turkey: Technology ForeSight Initiative.
Oliveira, C. A., Oroslinda, M. T. (2003). Alianças como Instrumento Eficaz de Inovação. Atibaia: Enanpad.

Oliveira, Bernardes, R. (2002). O desenvolvimento do design em sistemas complexos na indústria aeronáutica: o caso de gestão integrada de projetos aplicada ao programa ERJ-170/190. Salvador: Enanpad.

Powell, W.W. (1987). Hybrid Organizational Arrangements: new form or transitional development? *California Management Review,* 30 (1): 67-87.

Pritchard, D. (2002). The global decentralization of commercial aircraft production: implications for U.S. based manufacturing activity. Doctoral Thesis, University of New York.

Seixa, C. M., Grave, P. S., Gimenez, F. A. P. (2001). Globalização, Aliança Estratégica e Desenvolvimento Tecnológico: estudo do caso de uma empresa de alta tecnologia. Campinas: Enanpad.

Stalk, G. Jr., Hout, T.M. (1993). Competindo contra o tempo. RJ: Ed Campus.

Vedovello, Conceição, Melo. (2004). Marne Santos de. MARINS, Luciana Manhães. Globalização de Competências Inovadoras e o Papel de Infra-estruturas Tecnológicas: Evidências de Institutos de Pesquisa e Desenvolvimento (P&D) em Telecomunicações no Brasil. Curitiba: Enanpad.

White Blake, L. (1986). Key Considerations in the Technology Assessment Process. USA: Strategic Technology Institute - 58th Annual Conference of the National Technical Association Washington, D.C.

Williamson, O. (1975). Markets and Hierarchies: Analysis and antitrust implications. New York: Free Press.

Williamson, O. (1981). The Modern Corporation: Origins, Evolution, Attributes. *Journal of Economic Literature,* XIX (4), 1537-68.

Yin, R.K. E., Campbell D.T. (1994). Case study research: design and methods. Sage publications.

Zirger, B.J. E, Hartley, J. L. (1996). The effect of acceleration techniques on product development time. *ieee Transactions on Engineering Management,* Vol. 43, n. 2, may.

CHAPTER 18

THE LIMITS OF BUSINESS DEVELOPMENT

Mats Larsson

School of Economics and Management, Lund University, Sweden

In this article two key propositions are made: (1) There is a definite limit to the level of efficiency that can be achieved in any part of a business; and (2) The number of processes in a business available for companies to excel in is either finite or it increases very slowly.

1. Introduction

It has been assumed that the opportunities for business development are infinite and thus the potential for economic growth is also infinite. This is a key assumption in economics (Larsson 2004). In this article it is argued that from some important aspects business development has very definite limits. Business Process Reengineering, Supply Chain Management and electronic business are all wide-spread examples of the cost reduction trend in various areas of business and they all drive development in the direction of zero time and cost.

All these face the limit of attainable cost reduction and time compression that is built into the very fabric of reality: Nothing can be done at a cost that is lower than zero or in a time that is less than no time. As we reduce cost and time we come closer to zero and the value of further reductions are likely to be lower than the large reductions that were initially possible.

Historically, philosophy about quantitative limits to growth may not have been relevant, because the limits have been distant and, even if they existed, they may not have had any practical relevance for economists, strategists or operations analysts. With modern tools for operational improvement and electronic business, we are now able to see that many administrative processes of considerable complexity can now be performed automatically in almost no time at all. Through process reengineering and supply chain management companies have also been able to reduce non-value added time and cost to a fraction of the previous amount, indicating that there may not be much more time and cost left to eliminate in the future. Porter (1996) argues that productivity can be improved infinitely, but, according to the argument in this article and in Larsson (2004) this may not be the case.

Further, in the field of business strategy, it has been assumed that new value can always be added to products and services and that this is the primary task of strategy to find ways of adding new value (Porter 1996), and that good strategies create sustainable competitive advantages, while operational improvements can be copied. Yet, even the companies given as examples by Porter (1996) of good and sustainable strategies sometimes build their strategies on factors that clearly resemble operational efficiency. Southwest Airlines is one case in point. Since 1996 the low price travel concept of this company has been extensively copied by other airlines in the US and in Europe. In this and many other industries demand growth is based on the ability to reduce the unit production cost of the product. Traditional full service carriers have been forced to increasingly compete on price and reduce service levels in the direction of offering fewer free meals and other complimentary items. We may assume that the number of complimentary items and other service features can not be reduced lower than zero and that one very important competitive advantage of Southwest Airlines is that it has reduced such extra cost to a minimum, resulting in increased demand.

If we look at other industries, we find that one of the main aspects of development has been the reduction of the unit production cost in the direction of zero (Freeman & Louca 2001). Many times the market leaders in an industry have been the leaders in the development towards zero time and cost. In the printing machine industry Heidelberg has been the leader. This company has offered the most expensive, but also the fastest and most reliable printing machines, offering the lowest per unit cost per print, if maintenance, manual operating cost, set-up time, make-ready and other cost is included. In the case of digital printing, where Xerox is the leader, this technology offers a number of new values, but it also has the consequence of reducing the amount of manual labour in the printing process to a minimum, which has a significant impact on the cost per printed item (Larsson 2004).

2. Examples of Situations Where Time and Cost Approach Zero

In order to illustrate how different aspects of time and cost in business approach zero, we need to look at a number of examples:

2.1. *Information flows*

In the case of information flows these can be brought very close to zero using modern information technology. In the automotive industry one auto maker fifteen years ago sent monthly purchasing forecasts for the next six months to twelve thousand different suppliers. These forecasts took two weeks for someone at each of these suppliers to manually enter into the supplier's order system (Larsson 2004).

Today, every supplier receives one six month forecast every day via EDI (Electronic Data Interchange). Each day the forecast of the day before is updated with the new information. This has increased the information flow, but the cost of sending forecasts and entering them into the suppliers' systems has decreased to a few cents per day for the electronic exchange, which means that it is now close to zero (Larsson 2004).

2.2. *Material flows*

Mason-Jones & Towill (1998) make the distinction between information flows and material flows. Material flows are flows in production and logistics where components, products and materials are handled. In these cases we will not arrive at zero time and cost. Instead, many companies have focused on the reduction of non-value added time and cost. This means reducing waiting times for products between process steps, intermediary storage and other activities that don't add value to products.

Stalk (1988) mentions a company where delivery used to take nineteen weeks and this was reduced to less than two weeks. Everybody who has ordered a customized computer from Dell and received the product in less than ten working days has experienced an example of this change. In 1988 it took four months from order to delivery to receive a new computer.

In some industries companies also work hard to reduce time and cost in value added processes. In the auto industry companies consider changing over to plastic car bodies. This reduces the time and cost needed to produce a car body. When a steel body is produced the steel first needs to be produced, then it has to be rolled and delivered in rolls to the car manufacturer. The auto maker then cuts it into pieces and dies each piece to the form of a car body. After the body has been formed it needs to be painted, which means a number of steps in the painting process. In the case of plastic bodies the plastic is made in a chemicals plant and granulated. The granules are delivered to the car plant, where a car body is molded. The plastic is already colored, so the body doesn't need to be painted. This reduces the number of steps, the time taken to produce a car body, and consequentially the cost is reduced (Larsson 2004).

2.3. Products

The cost of many products is reduced and sometimes the capacity is improved. In the case of a mature product, such as a book, the changes to the product itself over time have been minimal, even though the quality of printing, especially the printing of color pictures, has been improved. In addition to this, the production cost has gradually been reduced through new and improved production technologies. The fastest analogue presses from Heidelberg today produce 15.000 identical prints per hour. This is a very high speed, but books need to be finished in a process where pages are sorted and bound in separate process stages afterwards, and the manual tasks of emptying finished prints from the printing machine, carrying them to the storage facility awaiting the rest of the pages to get ready and carrying the prints to the sorting machine for sorting and binding.

Digital printing machines have not yet arrived at the same capacity as an analogue machine and printing experts agree that the quality of print of color pictures is not as high. Yet, for the layman, the differences in the quality of print between an analogue press and a digital machine are hardly discernible to the eye. The speed of the most modern digital printing machines is 12.000 copies per hour, which is not directly comparable to the offset machine above, because offset machines print large sheets and digital machines print double pages, but it is indicative of the fact that digital printing is catching up on offset printing.

The actual printing of a book or a magazine becomes slightly more expensive using a digital printing press, but in the digital machine pages don't need to be sorted and the manual work that precedes sorting is taken away, as all pages are printed in the sequence that they are going to appear in in the finished book so that the book can be bound or the magazine can be stapled immediately after printing is ready. This reduces the total cost of the printing process for smaller runs and it "flattens out" the cost function, so that there may no longer be the same need in the future as there is today to print long runs to reduce cost. Instead, the reduction of the length of print runs reduces cost of storage and obsolescence (Larsson 2004).

In the case of many other mature products, the production cost is gradually reduced and at the same time the product is improved slightly for each new generation. Refrigerators, cars and vacuum cleaners are cases in point. Customers find that the product is improved, but due to modern construction and production philosophies, the average cost of production is reduced in the same way that the production of books in the case above indicates and one of the main features of product development is this reduction of the production cost. With mature products, however, the largest possible cost reductions have many times already been made and improvements are increasingly difficult to achieve.

2.4. *Technologies*

Digital printing is an example of a new technology that reduces the cost of printing a book. In the case of a book, which is an information product, the content can also be downloaded over the Internet or over a broadband network. The digital printing technology brings printing and binding costs a step in the direction of zero for smaller runs, while electronic distribution of information products brings the cost of reproducing and distributing a book or some other piece of information very close to zero altogether. At the same time the time it takes to access a book and receive it to my computer is decreased in the same direction.

Similar developments are going on in other industries, such as the knitting industry, where modern machines knit a garment in one piece. Jumpers no longer need to be sewn together from five different pieces, which saves time and cost in the direction of zero. In the automotive industry, as discussed above, the use of plastic bodies instead of steel ones will have the same effect. The cost of car manufacture will drop (Larsson 2004).

2.5. *Strategies*

It is the purpose of strategy development to find "uncontested market space" (Kim & Mauborgne 2005) or "to achieve sustainable competitive advantage by preserving what is distinctive about a company. It means performing *different* activities from rivals, or performing *similar* activities in different ways" (italics in original) (Porter 1996).

The strategies of successful companies have often been based on offering better products or services than the competition. Many times, at least in the case of production equipment, the purchasing price of a leading brand has been higher than the price of a low price alternative, but the leading brand has often been sold on the argument that the unit cost of the product that is produced is still lower, because of lower costs of maintenance, improved reliability and less scrap in the production process. Today, the leading brands find it increasingly difficult to defend such arguments. In many cases low price alternatives offer similar quality and performance as the leading brands and increasingly loyalty to the historically leading brands is weakening. The reason why followers are catching up with the leaders may be that technologies that were previously invented by the leading company is maturing, they are becoming generic and the knowledge about these technologies, that was previously concentrated to a few companies, may diffuse and become generally accessible and built into machinery from other producers as well (Larsson 2004).

In the case of consumer products a similar development is underway. The quality of a product is increasingly built into the production equipment and less than before a product of the craftsmanship or skills of individuals at the leading companies. Therefore, it is now possible for supermarket chains to offer private label alternatives of a quality that is similar to that of the leading brands. One or two decades ago, production companies in the food and beverage sectors bought machinery from different manufacturers and the competence to integrate this machinery into a production line and operate this line to produce a superior product rested with the production company itself. Today, Tetra Pak and other companies increasingly offer turn-key solutions, making it possible for any would-be producer to start high quality production (Larsson 2004).

Many people see new innovations, such as digital printing technology, computers and mobile phones, as completely new inventions that primarily add new value to the economy. If we analyze these developments in detail, the primary advantage that these technologies offer is that they reduce the unit cost of doing things that we have already done for many years, further in the direction of zero. Digital printing reduces the per unit cost of printing, computers reduce the unit cost of doing administrative tasks, communicating and handling information and mobile telephony reduces the per unit cost of telephoning, if the calculation takes into account the time it previously took for a businessman to be at a fixed line telephone. Now, the businessman can use time between meetings, time at the airport or in the car to do business and the cost reduction of the mobile phone could be seen as reducing the waste of time in the direction of zero. With a cost per hour of a manager at, perhaps, 100 USD, 20-30 cents per minute for making a call is outweighed by the saving in terms of improved efficiency.

3. Proposition

Based on the above reasoning, two key propositions are made:

1. There is a definite limit to the level of efficiency that can be achieved in any part of a business and the opportunity for companies to invent new products and services is also limited, in that new products, technologies and strategies are often based on time and cost reduction in one way or another.
2. The number of processes in a business available for companies to excel in is either finite or it increases very slowly.

I also propose that many aspects of current business development make companies increasingly similar to one another and that this development, will decrease overall profitability in companies and industries. The reduction of time and cost in the direction of zero through the use of generic management practices, technologies and increasingly also strategies makes it increasingly difficult for companies to differentiate from the competition and also increasingly difficult to stay profitable.

4. Theoretical Foundation

What are the implications of the development towards time compression and cost reduction, described above, for strategy and operational improvement? Nattermann (2000) argues that downright benchmarking and copying of competitors' strategies results in an erosion of profitability in an industry. Instead of copying the profitability of the most successful company, everyone loses, because they compete with the same strategic formula. Hamel & Välikangas (2003) argue that it leads to "strategy decay" when it is possible for competitors to copy a strategy.

Larsson (2004) argues that companies don't need to copy each other's benchmarks exactly. When a number of companies use the same models for operational improvement, such as Business Process Reengineering (Hammer & Champy 1995), Supply Chain Management (Towill 1996), modularization (Sanchez 2004) or eBusiness (Beynon-Davies 2004) that have the same goals (time and cost reduction) they all move their time and cost positions along the same axis in the direction of zero. In the line of the Resource Based View (Collis & Montgomery

1995), they all acquire the same or similar resources in the form of fast processes, integrated ERP systems and management mind-sets. They gradually move time and cost in the direction of zero and nobody benefits from this, exactly as Porter (1996) points out, because, in the words of Nattermann (2000) you need a unique strategic position in the matrix presented by Nattermann in order to distinguish yourself and make a profit. Along the lines of the Resource Based View of strategy (Collis & Montgomery 1995) you need unique resources that are not available on a ubiquitous factor market. ERP systems, processes that are completely or semi-completely automated and management mind-sets that are focused along the lines of BPR, SCM and eBusiness are all available on factor markets along with a host of other factors that contribute to low cost and time compression.

It may be, however, that there is not so much substance in the argument of Porter (1996), that there is a clear difference between operational efficiency and strategy. If strategy rests on performing different activities or performing similar activities in different ways, then these activities, that ought to form part of operations, ought to be possible to copy as well. How does Porter identify the operational activities that are possible to copy and distinguish them from the *different* activities and the *similar* activities that are performed in different ways? There is no attempt in the article to categorize activities into those that are copyable and those that are not.

The argument of this article thus extends the argument of Nattermann (2000) so that the problematic situation does not arise from benchmarking only. It arises in all situations where companies strive to dramatically reduce time and cost towards zero, which is a fairly widespread practice. I argue that there are at least four different distinct situations where we now see that companies seem to converge in their pursuit of these factors:

- Reduction of time and cost in information flows to a situation very close to zero
- Reduction of time and cost in material flows to a situation where non-value added time approaches zero

- Improvement of the capacity and cost efficiency of products so that time and cost approaches zero, such as in the case of digital cameras, mobile phones, pharmaceuticals etc
- Development of new technologies that aim at reducing time and cost towards zero, such as in the development of internet technologies, digital printing, digital production technologies in a number of other areas, plastics technologies, nano-technologies, bio-tech etc.

This development compounded may lead to "strategy decay" (Hamel & Välikangas 2003), because many companies gradually become increasingly similar to one another.

5. Discussion

Companies gradually are bound to become more similar in several ways:

1. If they apply "best practice" they copy their competitors and become gradually more and more similar. This results in increasing competition and overall lower profits (Nattermann 1999).
2. If they don't apply "best practice" and pursue improvements in their own way, they are bound to use tools and practices that are similar to the ones used by their competitors, such as "Time Compression" (Towill 1996), "Business Process Reengineering" (Hammer & Champy 1995) or "Supply Chain Management" (Towill 1996) or "Time Pacing" (Brown & Eisenhardt 1998). These all have the overall goal of reducing time and cost in various areas of the business. As this is done increasingly expertly by a number of competitors companies in an industry are bound to move towards the best achievable position in their industry, namely as close as possible to zero time and zero cost as it is possible to get.

The Resource Based View of Strategy (Collis & Montgomery 1995) tells us that only resources (processes, technologies, people etc) that are in some sense unique to a company can create profit for a company. As companies increasingly utilize the same methods and tools to achieve competitive advantage, the number of areas available for uniqueness decreases as companies move towards zero time and cost in area upon area.

Some readers may at this point argue that it may be true that there is a limit to excellence in each area of current expertise, but the number of areas to excel in is increasing at a relatively high pace. The emergence of new management disciplines like e-business, Supply Chain Management and Human Resource Management is testimony to this fact.

The answer to this argument is that neither of these three management disciplines is new, they just represent new ways to perform existing tasks. E-business is a set of new tools for communication with customers and suppliers that reduce cost and increase reach compared to old tools like the telephone and the fax machine. Supply Chain Management was previously called "Purchasing" and is only a more potent tool to organize supply chains and set requirements for suppliers, so that the unit cost of each item that is purchased is reduced. Human Resource Management is a development of the personnel department and it represents a way to manage this department in a more powerful way.

Sanchez and Heene (1997) refer to the early French management thinker Henri Fayol, who identified six areas in companies that need to be administrated. At the beginning of the twentieth century these were technical operations, commercial operations, financial operations, security operations, accountancy operations and administrative operations. These are still largely the areas that companies need to administrate today.

One reason why we do not see directly how companies become increasingly similar is that they go about the change process in different ways. Larsson (2004) compares the change process to a situation where, initially, we have two villages in a mountain landscape with a number of different routes to go from one to the other. Each of the routes is unique and it symbolizes the paths chosen by a number of different companies to satisfy a particular customer demand. One route winds a long way into a valley and up on the other side of the mountain. Another is shorter, but it is sometimes steep and dangerous and crosses a number of mountain passes.

At one point it is decided that there is to be a competition between road planners. Three such planners are given the task of drawing the optimal route between the two villages, using bridges, tunnels and other modernizations to arrive at the best possible route. Each of the planners starts with a ruler and draws a straight line between the villages, then he plans a route that is as close to the straight line as possible, allowing for deviations in order to make a realistic suggestion from a cost perspective. The results, drawn by the three planners, are remarkably similar, but they vary in some details. The total driving time for the three alternatives would differ only by a few seconds.

Now, three different construction companies get the assignment to make project plans for the three alternatives. The three plans, again, are widely different. One company suggests that the construction should start at village A and finish at village B four years later. One other company suggests that the project should start at B and finish at A. The third company suggests that four parallel projects should be run at different parts of the route, which would bring the total construction time down to eighteen months.

Now, it is obvious in reality that there would not be three different roads built between the villages. As mentioned above, the different routes represent different strategies and operational solutions developed by a number of companies initially. The solutions developed by the road planners symbolize the "best practice" solutions that each company could arrive at for the type of production and administration in question. The three different projects represent the change programs undertaken by each of the competitors. Even if the end results are similar, the projects that lead up to these similar results may choose different starting points, different paths and it may be difficult, if you have not seen the best practice routes, to see that all projects will arrive at similar end results.

In business change, the time span for change may be more than one decade. The changes undertaken in the automotive industries of the US and western Europe in order to apply time compression, quality management and other management principles emulated primarily from Toyota (Womack *et al.* 1991), has been going on for almost twenty years. Through this development, IT systems have been implemented, just-in-time deliveries have become standard and supply chain management has been implemented in steps both at automotive companies and their suppliers. In order to facilitate production, constructions have become modularized, the number of parts has been reduced and functions in the cars have become computerized instead of mechanical. All these efforts may have been gone through in different sequences in different companies, but the end results are likely be relatively similar. Womack *et al.* (1991) attest to the fact that the original situations were highly different at different plants, maybe because there were no universal management principles that were applied by all competitors, prior to 1990.

In short, improvement is not done sequentially in the sense that a company takes one area first and continues with the next. Instead, managers of different areas work at the same time to improve their areas. In addition to this, new tools, such as Business Process Reengineering, ERP systems, operations management and tools for strategic management facilitate improvements across the whole organization, but the sequence of projects may at times seem haphazard and it may be difficult to see that each company is aiming at the same end result as the others.

6. Rebuts

Several rebutting evidence are offered in the literature to the proposition that profitability will decrease as companies become more similar to one another.

> *Companies have always made profits and they will find ways of making profits in the future as well.*

It is true that there have always been profitable companies. However, not all companies have been profitable at all times. During the last decades companies have used methods such as the ones mentioned above to improve profitability. These methods will not necessarily ascertain that companies will be profitable in the future. Methods that have been used in the past have been given up in favor of new methods, even though they for a number of years showed promise. The BCG (Boston Consulting Group) matrix that was widely used in the nineteen-seventies and nineteen-eighties provides one such example.

Companies have always found new ways of improving profitability in the past and they will do this in the future as well.

This is true, but companies and authorities have also over many years pursued paths that eventually were abandoned, because they didn't pay off. After decades of success in the tanker industry Swedish shipyards came into a crisis in the nineteen-eighties. For a number of years, the strategy among the companies and Swedish authorities was to wait until the market would improve. Too late companies and authorities realized that this market was lost to Swedish shipyards and structural change could begin (Carlsson, 2001).

The aim of the article is to argue that development towards reduced cost and time will not continue to increase profitability indefinitely, and it may even become increasingly difficult to develop successful strategies, even for the companies that today are the leaders in these fields. Companies will need to find new directions of development and they should start to do this now and not wait for the situation that this article warns against to emerge.

7. Conclusion

The reasoning above means that companies pursue goals that will inevitably reduce their uniqueness and increase competition. This will, in its turn, reduce profitability in all industries where this is true.

We are not there yet. In some areas companies are close to zero in terms of cost and time and we can't expect development to go any further in the same direction in the future. This will gradually become true for information processes. In other areas companies have been highly successful in the last decades to reduce non-value added time and cost in their materials processes and the opportunity to continue to take away non-value added time is limited. This is true for many production and logistics processes, where total cycle time has been reduced from three to four months (as in the case used by Stalk 1988) to less than two weeks.

In other situations new technologies have recently been invented that promises substantial time and cost reduction, but these technologies have yet to be perfected and spread widely. Among such technologies are various Internet technologies and other computer technologies, digital production technologies (such as digital printing), nano technologies and bio technologies (Ratner & Ratner 2003; Robbins-Roth 2000) offering such promises.

During the next two decades companies will arrive close to zero time and zero cost in different areas at different times. In some areas the limits have already been reached, in others the limits will be reached within the next few years and in yet others it may take ten or twenty years before the limits are reached.

In order to avoid a situation where it becomes increasingly difficult to make a profit, companies need to find areas outside of the ones presented above in which they can develop new competitive advantages and find new profitability.

Acknowledgments

The research behind this article has generously been financed by Sparbanksstiftelsen Skåne.

References

Beynon-Davies, P. (2004) E-Business. Palgrave Macmillan, Basingstoke.
Brown, S. L. & Eisenhardt, K. M. (1998) Competing on the Edge, Harvard Business School Press, Boston, Mass.
Carlsson, R. (2001) Konsten att tjäna pengar, Ekerlids Förlag, Stockholm. (in Swedish)
Collis, D. J. & Montgomery, Cynthia, (1995) Competing on Resources: Strategy in the 1990s Harvard Business Review, July.
Freeman, C. & Louca, F. (2001). As Time Goes By. Oxford University Press, Oxford.
Hamel, G. & Välikangas, L. (2003) The Quest for Resilience. Harvard Business Review, sept.
Hammer, M. & Champy, J. (1995) Reengineering the Corporation, Nicholas Brealey Publishing, London.
Kim, W. C. & Mauborgne, R. (2005) Blue Ocean Strategy. Harvard Business School Press, Cambridge.
Larsson, M. (2004) The Limits of Business Development and Economic Growth, Palgrave Macmillan, Basingstoke.

Mason-Jones, R. & Towill, D. R. (1998) Time compression in the supply chain: information management is the vital ingredient, Logistics Information Management, vol 11, iss 2.

Mason-Jones, R. & Towill, D. R. (1999) Total cycle time compression and the agile supply chain, International Journal of Production Economics, 62 (1999) 61-73.

Nattermann, P. M. (2000) Best practice (not equal to) Best Strategy, McKinsey Quarterly.

Porter, M. E. (1996) What Is Strategy? Harvard Business Review, Nov-Dec 1996.

Ratner, M. & Ratner, D. (2003) Nanotechnology: A Gentle Introduction to the Next Big Idea, Prentice Hall, Upper Saddle River.

Robbins-Roth, C. (2000) From Alchemy to IPO, Basic Books, New York.

Sanchez, R. (2004) Creating Modular Platforms for Strategic Flexibility, Design Management Review, Winter.

Sanchez, R. and A. Heene. (1997). Competence Based Strategic Management. John Wiley & Sons, New York.

Stalk, G. (1988) Time: The Next Source of Competitive Advantage. Harvard Business Review, July-August.

Towill, D. R. (1996) Time compression and supply chain management – a guided tour. Logistics Information Management, vol 9, iss 6.

Womack, J. P.; Jones, D. T. & Roos, D. (1991) The Machine That Changed the World, Harper Collins, New York.

CHAPTER 19

KNOWLEDGE MANAGEMENT ACROSS BORDERS: EMPIRICAL EVIDENCE OF THE CURRENT STATUS AND PRACTICES OF KNOWLEDGE MANAGEMENT IN MULTINATIONAL CORPORATIONS

Helmut Kasper and Florian Kohlbacher

Vienna University of Economics and Business Administration, Vienna, Austria

Knowledge management seems to have become a ubiquitous phenomenon both in the academic as well as in the corporate world. However, according to the authors, there has not been any comprehensive and holistic empirical study of the current status and practices of knowledge management in corporations. Therefore, based on our recent global study on knowledge management and organizational learning in multinational companies (MNCs), this paper wants to make up for this shortcoming in knowledge management research. In nine renowned MNCs, three interviews with respondents from the top and upper management level were conducted in the headquarters and in two different subsidiaries respectively. Thus, both quantitative and qualitative data from 81 interviews in total were earned. This paper focuses on the use of knowledge management tools and shows that not all tools have the same impact, and some of them even influenced knowledge management processes negatively.

1. Introduction

There has not been any comprehensive and holistic empirical study of the current status and practices of knowledge management (KM) in corporations. Looking at the current status and practices of KM in the corporate world might help to find an answer to the question of whether this passion for KM is a passing fad. Therefore, based on our recent global study on KM and organizational learning in multinational companies (MNCs), this paper wants to make up for this shortcoming in KM research. The focus lies on the use of different KM tools and their impact on knowledge transfer within MNCs.

2. Research Methodology

The insights offered in this paper are based on a recent global study (2001-2005) on KM and organizational learning in MNCs conducted by the authors. Nine renowned MNCs were selected to serve as our sample. In each MNC, three interviews with respondents from the upper management level (mainly CEOs, HR-managers, CFOs) were conducted in the headquarters and in two different subsidiaries respectively. Thus, we earned both quantitative and qualitative data from 81 interviews in total.

For our theoretical sample we attempted to select companies that would provide us with an opportunity to collect rich data and to compare different approaches on KM and the way knowledge is handled in a variety of different contexts. Specifically the research sample consists of 27 units of 9 MNCs from different branches. The headquarters and two subsidiaries are each chosen to reflect as many regional and cultural differences as possible. Consequently, it was our aim to gain the support of units located in very different regions.

In accordance with our qualitative research design for our explorative study, in-depth interviews (qualitative, semi-structured) based on an interview guideline were conducted with respondents in the respective countries. The interviews were transcribed and encoded based on our system of categories so that they could be used not only for qualitative word context analysis supported by NVivo but also for quantitative analysis using logistic data regression, MANOVA and ANOVA.

To analyze and interpret the data, we used qualitative content analysis according to Mayring, which is "an approach of empirical, methodological [sic] controlled analysis of texts within their context of communication, following content analytical rules and step by step models, without rush quantification" (Mayring 2000). Following our research questions, the aspects of text interpretation are put into categories which are formed inductively and/or deductively and revised within the process analysis (feedback loops).

To lend quantitative support to the observations that emerged from the interviews, we conducted several additional surveys. Central to the findings presented in this paper were two different collection instruments. First, a questionnaire on KM tools and processes used in the organization was employed. On a seven-point scale the usage frequency of 19 common KM tools was surveyed. The influence of these KM tools on the inter-organizational knowledge transfer was analyzed in a multivariate fashion using a logistic regression model. The antilogs of the model-coefficients were interpreted as the corrected odds ratio.

Second, an illustration prepared in accordance with structure formation technique was used to visualize the knowledge flows on both the personal and the technical level between the different units as perceived by the interviewee.

3. The Results: Current Status of KM in MNCs

3.1. *Use of KM tools in MNCs*

Nonaka and Takeuchi (1995) distinguish between explicit and tacit knowledge (cf. also Inkpen and Ramaswamy 2006 for a discussion in a global KM context). Their well-known spiral of knowledge illustrates the process of creating knowledge in an organization through the interaction between tacit and explicit knowledge. Although many studies apply and extend the Nonaka and Takeuchi (1995) model, the lack of classification and categorization of existing KM systems (KMS) is especially problematic. In fact, a survey of the KMS literature indicates that there appears to be no generally accepted systematic framework guiding KMS research (Gallupe 2001). However, there are some contributions that attempt to provide insight into the entire process of KM. These processes, each including various forms of knowledge, are very complex. Consequently, to capture an organization's entire KMS seems to be very difficult. Notwithstanding the problems arising from this complexity, some authors have made attempts to categorize KMS (e.g. Birkinshaw 1999; Davenport 2005; Davenport and Harris 2005; Hansen, Nohria et al. 1999; Hong 1999; McAdam and McCreedy 1999; Zack 1999; Bloodgood and Salisbury 2001; Earl 2001; Gallupe 2001; Holsapple and Joshi 2001).

For our purposes, the approach suggested by Hansen et al. (1999) renders itself particularly useful. The authors found that in some companies, KMS center around the technological infrastructure, while other companies primarily foster personal communication and contact. Technologically focused companies, it is argued, attempt to codify and store knowledge in databases to make it easily accessible to anyone in the company. The authors call this a codification strategy. A personalization strategy, in contrast, implies that knowledge is closely tied to the individuals who develop it. In these companies, information technology primarily serves to enable communication among the members.

The Hansen et al. (1999) approach does not only have a high face validity, but a suitable scale, based on Nonaka and Takeuchi's knowledge spiral, has also been developed by Becerra-Fernandez and Sabherwal (2001). We slightly modified this scale and aggregated the different KM tools as personalization and codification instruments. To this end, we identified the personalization and codification tools and created two dimensions by splitting the scale. In the following exhibits, we list the items associated with codification (Table 1) or personalization (Table 2).

The following two exhibits show the average use of the codification (Figure 1) and the personalization tools (Figure 2) in the nine MNCs surveyed in our study. The average use was surveyed by using a seven-point scale going from "very infrequently" (value 1) to "very frequently" (value 7) and "not applicable" coded by value 0. With a total mean of 4.56, the personalization tools are used a little bit more frequently than the codification tools (total mean of 4.33).

Table 1. Codification KM tools.

Capture and transfer of experts' knowledge
Decision support systems
Modeling based on analogies and metaphors
Groupware and other team collaboration tasks (e.g. document sharing)
Databases
Web-based access to data
Pointers to expertise (skills "yellow pages" within the company)
Repositories of information, best practices, and lessons learned
Web pages (Intranet and Internet)
A problem-solving system based on a technology like case-based reasoning

Table 2. Personalization KM tools.

Learning by observation
Chat groups/Web-based discussion groups
Employee rotation across areas
Cooperative projects across subsidiaries
The use of apprentices and mentors to transfer knowledge
Brainstorming retreats or camps
Learning by doing
On-the-job-training
Face-to face meetings

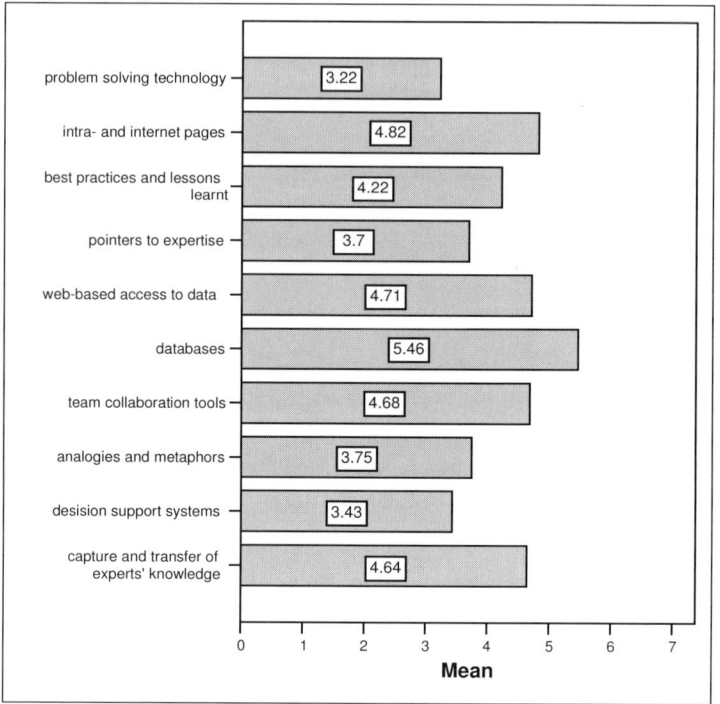

Figure 1. Average use of codification tools.

3.2. *Impact of tool use on the inter-organizational knowledge transfer*

Knowledge is transferred in organizations whether or not the process is managed at all and the everyday knowledge transfers are part of organizational life (Davenport and Prusak 2000). However, there are certain factors that influence the transfer of knowledge and different strategies to manage knowledge sharing in firms can be applied.

The authors have developed a comprehensive model of knowledge sharing in MNCs, which can basically be divided into three sub-models (Kasper and Haltmeyer 2002; Kasper and Mühlbacher 2004): A model describing the process of inter-organizational knowledge sharing, a model of the organizational context factors and a model of international/inter-organizational context factors influencing the process. Since the context factors have a strong impact on the process of KM, the process and context of KM are highly intertwined. For a successful management

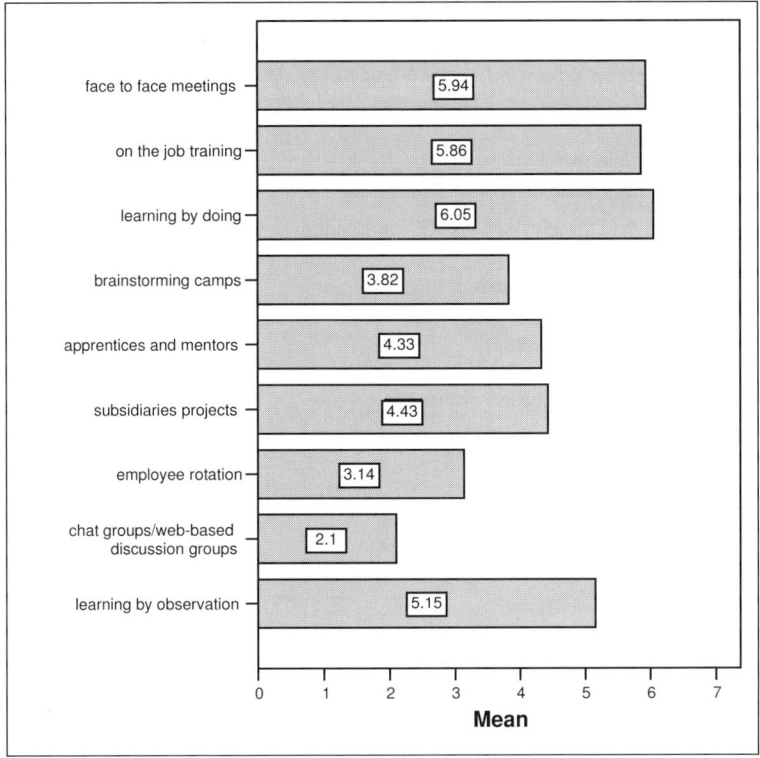

Figure 2. Average use of personalization tools.

of knowledge sharing between these organizations it is absolutely necessary to be aware of these different context factors, to know about their influence on the knowledge sharing process, and to adapt the KMS to these conditions. (Kasper and Haltmeyer 2002).

Among the different context factors we identified, it is the factor of appropriate structures/infrastructure that comprises the KM tools. As a matter of fact, learning in organizations requires an adaptation of the organizational structures. In order to enable learning and knowledge sharing, a flexible structure that encourages growth and experimentation, creative problem solving and flexibility must be in place (Kasper and Haltmeyer 2002). Besides, fostering both formal and informal communication and providing sophisticated KM tools to enable the capture, storage and dissemination of knowledge are also a conditio sine qua non. Davenport *et al.* (1998) in their study of successful KM projects put it like this: "Knowledge projects are more likely to succeed when they use the broader infrastructure of both technology and organization" (p. 51).

Table 3. KM tools.

Tool 1 Capture and transfer of experts' knowledge
Tool 2 Decision support systems
Tool 3 Learning by observation
Tool 4 Chat groups/Web-based discussion groups
Tool 5 Employee rotation across areas
Tool 6 Cooperative projects across subsidiaries
Tool 7 Modeling based on analogies and metaphors
Tool 8 Groupware and other team collaboration tasks (e.g. document sharing)
Tool 9 Databases
Tool 10 Web-based access to data
Tool 11 Pointers to expertise (skills "yellow pages" within the company)
Tool 12 The use of apprentices and mentors to transfer knowledge
Tool 13 Brainstorming retreats or camps
Tool 14 Repositories of information, best practices, and lessons learned
Tool 15 Web pages (Intranet and Internet)
Tool 16 Learning by doing
Tool 17 On-the-job-training
Tool 18 A problem-solving system based on a technology like case-based reasoning
Tool 19 Face-to face meetings

Using a logistic regression model, we looked at the impact of the average use of the 19 different KM tools (see Table 1 and Table 2) on the inter-organizational knowledge transfer within each of our 9 target companies. The inter-organizational knowledge transfer in MNCs is represented by the knowledge flows between headquarter and subsidiary and between subsidiaries respectively. Besides, personal knowledge flow and technical knowledge flow can be distinguished. We defined personal knowledge flow as a more or less direct exchange of knowledge on a person-to-person basis. This includes face-to-face meetings, telephone, e-mail, videoconferences etc (cf. also Hansen, Nohria *et al.* 1999). Technical knowledge flow, in contrast, means the sharing of knowledge via a technical intermediary with the collectivity. Here, knowledge needs to be codified and transmitted to the intermediary first, before it is transferred further to or 'picked up' by the final recipients. Table 3 gives an overview of the KM tools as they were surveyed by our questionnaire.

Table 4 shows the impact of the KM tools on the personal knowledge flow. The significant values are highlighted in bold: Decision support systems (Tool 2), Databases (Tool 9), Web-based access to data (Tool 10), Pointers to expertise (skills "yellow pages" within the company) (Tool 11), and Web pages (Intranet and Internet) (Tool 15).

While Tool 2, Tool 10 and Tool 15 reduce the chance for a high personal knowledge flow, Tool 9 and Tool 11 increase it. In fact, Tool 2 reduces the chance for a high knowledge flow per unit by 32%, Tool 10 by 28% and Tool 15 by 36%. Tool 9 increases it by 70% and Tool 11 by 38%. In other words, an increased use of Decision support systems (Tool 2), Web-based access to data (Tool 10), and Web pages (Intranet and Internet) (Tool 15) reduces the personal knowledge flow, while an increased use of Databases (Tool 9) and Pointers to expertise (skills "yellow pages" within the company) (Tool 11) increases it.

How come that tools which are supposed to support KM (processes) in firms, have a reverse effect? As matter of interest, all of the KM tools that had a significant impact on the personal knowledge flows belong to the KM tools used for codification according to our attribution. It should be plausible to assume that codification tools have influence on the technical flow of knowledge while personalization tools have impact on the personal knowledge flow. However, this reasoning takes only the

Table 4. Impact of KM tools on personal knowledge flow.

	Beta	SE Beta	OR	-95%CL	+95%CL	p value
Const.B0	-4.067884	2.587898	0.01711356	9.55073E-05	3.066508	0.1159866
TOOL_1	0.003807144	0.2103082	0.9962001	0.6534758	1.518671	0.985557
TOOL_2	**-0.3869662**	**0.1869819**	**0.679114**	**0.4668055**	**0.9879828**	**0.03850364**
TOOL_3	-0.06062911	0.1927317	0.9411722	0.6395229	1.385103	0.7530844
TOOL_4	0.2383463	0.2163488	1.269149	0.8224999	1.958345	0.2706108
TOOL_5	0.1659317	0.1927571	1.180493	0.8020992	1.737394	0.3893364
TOOL_6	0.1566458	0.2012908	1.169581	0.7812048	1.751039	0.436453
TOOL_7	0.1225021	0.1713576	1.130322	0.8016772	1.593692	0.4746814
TOOL_8	0.09255841	0.217329	1.096977	0.7095248	1.696007	0.6701903
TOOL_9	**0.5313026**	**0.2571841**	**1.701147**	**1.015803**	**2.848879**	**0.03885108**
TOOL_10	**-0.329154**	**0.1963935**	**0.7195322**	**0.4853431**	**1.066723**	**0.09374947**
TOOL_11	**0.3223179**	**0.1869587**	**1.380324**	**0.9488431**	**2.008017**	**0.08471654**
TOOL_12	0.133987	0.2236924	1.143378	0.7301617	1.790444	0.5491912
TOOL_13	0.1829853	0.2304218	1.200797	0.756553	1.905898	0.4271247
TOOL_14	0.08610144	0.1928761	1.089917	0.7403799	1.604472	0.6553056
TOOL_15	**-0.4461421**	**0.2112163**	**0.6400928**	**0.4191169**	**0.9775763**	**0.03467255**
TOOL_16	-0.1232741	0.3266165	0.8840213	0.4592775	1.701572	0.7058585
TOOL_17	0.3885334	0.3177485	1.474816	0.779959	2.788714	0.2214255
TOOL_18	0.2077617	0.1717025	1.23092	0.8724228	1.736731	0.2262842
TOOL_19	-0.1068618	0.3180194	0.8986498	0.4749944	1.700171	0.7368557

positive effect into account and neglects a possible negative impact. In fact, the increased use of codification tools which again increase the availability of information and knowledge might render personal contact unnecessary up to some extent and thus decrease the personal knowledge flow between different units within MNCs. If the desired information or knowledge can easily be accessed from a repository such as decision support systems or the intranet, knowledge exchange on a personal basis might become superfluous. This explains reasonably well why Tools 2, 10 and 15 have a negative impact on the personal knowledge flow. Moreover, it is not surprising that an increased use of pointers to expertise (Tool 11) leads to a higher personal knowledge flow. Pointers to expertise like yellow pages for example, do not carry or contain the knowledge itself, but – as the term already suggests – point to the place or person where the knowledge is located. Thus, having identified the knowledge source, it needs to be tapped to initialize the knowledge exchange. Usually this implies direct contact with the person that possesses the relevant knowledge. This again, increases the personal knowledge flow as a consequence.

But why is it that a higher use of databases (Tool 9) has a positive impact on the personal knowledge flow? A feasible explanation of this phenomenon is that in contrast to web-based access to data and intranets for instance, databases might only be accessible locally but not inter-organizationally (i.e. across all subsidiaries). Or, even if the databases are available globally they might not also be completely up-to-date or it might be more difficult to make them become available across all subsidiaries. As a matter of fact, this is the case with one of the most widely used groupware Lotus Notes, where databases first need to be replicated to another location and are usually only updated once per day. Therefore, if a lot of knowledge is stored in databases locally, this knowledge still needs to be shared and transferred inter-organizationally via personal contact, thus leading to a higher personal flow of knowledge. Moreover, even though explicit knowledge can be shared through contributing to and referring to databases and other documents that can be placed in various searchable forms, the knowledge encoded in databases is never complete (Mohrman, Finegold *et al.* 2002). In fact, the embedded assumptions and tacit understanding behind it must be shared in person-to-person interactions (ibid.; cf. also Leonard and Swap 2005).

Table 5 shows the impact of the KM tools on the technical knowledge flow. The significant values are highlighted in bold: Capture and transfer of experts' knowledge (Tool 1), Decision support systems (Tool 2), Learning by observation (Tool 3), Employee rotation across areas (Tool 5), and Modeling based on analogies and metaphors (Tool 7).

While Tool 2, Tool 3 and Tool 7 reduce the chance for a high technical knowledge flow, Tool 1 and Tool 5 increase it. In fact, Tool 2 reduces the chance for a high knowledge flow per unit by 27%, Tool 3 by 27% and Tool 7 by around 30%. Tool 1 increases it by 54% and Tool 5 by 51%. In other words, an increased use of Decision support systems (Tool 2), Learning by observation (Tool 3), and Modeling based on

Table 5. Impact of KM tools on technical knowledge flow.

	Beta	SE Beta	OR	-95%CL	+95%CL	p value
Const.B0	-1.026577	2.323514	0.3582312	0.00340332	37.70716	0.658622
TOOL_1	**0.4311437**	**0.2242218**	**1.539017**	**0.9819578**	**2.412092**	**0.05450848**
TOOL_2	**-0.3142127**	**0.1564019**	**0.7303637**	**0.5338454**	**0.9992241**	**0.04454466**
TOOL_3	**-0.3180095**	**0.1869631**	**0.7275959**	**0.5002277**	**1.05831**	**0.08896745**
TOOL_4	0.1341113	0.1780405	1.14352	0.8003629	1.633806	0.4512978
TOOL_5	**0.4102652**	**0.1818746**	**1.507217**	**1.046844**	**2.17005**	**0.02409245**
TOOL_6	-0.1974307	0.1950358	0.820837	0.5552754	1.213404	0.3114122
TOOL_7	**-0.3517492**	**0.1644112**	**0.7034565**	**0.5059908**	**0.9779842**	**0.03240697**
TOOL_8	0.1752157	0.1861984	1.191503	0.8204239	1.730422	0.3467036
TOOL_9	0.1880341	0.2102614	1.206875	0.7918849	1.839341	0.3711749
TOOL_10	-0.04808488	0.1883939	0.9530529	0.653355	1.390224	0.7985425
TOOL_11	-0.1640579	0.1597696	0.8486929	0.6161633	1.168975	0.3045035
TOOL_12	-0.143622	0.1940553	0.8662151	0.587125	1.277971	0.4592396
TOOL_13	0.2091027	0.2171909	1.232571	0.7975923	1.904773	0.3356754
TOOL_14	0.00316004	0.1590544	1.003165	0.7293569	1.379763	0.984149
TOOL_15	-0.1609621	0.1865019	0.8513243	0.5858332	1.237132	0.3881115
TOOL_16	0.3844723	0.3262531	1.468839	0.7638733	2.824406	0.2386267
TOOL_17	-0.227724	0.2915341	0.796344	0.4439819	1.428355	0.4347359
TOOL_18	-0.08180435	0.1577127	0.9214522	0.671751	1.263972	0.6039789
TOOL_19	0.1914657	0.3014144	1.211023	0.6619388	2.215579	0.5252867

analogies and metaphors (Tool 7) reduces the technical knowledge flow, while an increased use of Capture and transfer of experts' knowledge (Tool 1) and Employee rotation across areas (Tool 5) increases it.

Here, Tools 1, 2 and 7 are codification KM tools while Tools 3 and 5 are personalization KM tools. Of the former, only Tool 1 has a positive impact on the technical knowledge flow, while Tool 2 and 7 correlate negatively. The finding that Capture and transfer of experts' knowledge (Tool 1) increases the chance for a high technical knowledge flow is hardly surprising. However, the fact that Decision support systems (Tool 2) influence the technical knowledge flow negatively is a rather puzzling result. At this point of the study, the authors see the need for further research and investigation. The negative impact of Modeling based on analogies and metaphors (Tool 7) might be due to the fact that even though it is a codification tool, steps of personal rather than technical knowledge transfer and communication are involved. Indeed, analogies and metaphors represent a way to codify tacit knowledge. This is a difficult and time-consuming task, which might prevent employees from making use of this codification tool. Hence, it rather hinders than fosters the technical knowledge flow.

Tools 3 and 5 belong to the personalization KM tools. The outcomes of Learning by observation (Tool 3) are probably very hard to codify and need to be shared personally, thus reducing the chances for a high technical knowledge flow. Interestingly, Employee rotation across areas (Tool 5) increases the chance for a high technical knowledge flow. One reason for this might be that the changing of locations and positions makes it necessary to codify and store relevant knowledge to make it become available to successors and other (i.e. former) colleagues. Of course, this can also be done personally, but the company might request people who are to be moved to codify and store as much knowledge as possible and then transfer it to their colleagues. Additionally, employee rotation strengthens the knowledge connections between the different locations, resulting in better interpersonal relationships. This might lead to a higher level of trust, not only in the personal knowledge of the colleagues but also in the external knowledge stored in different codification tools.

4. Conclusions

This study provides empirical evidence for the impact of commonly used KM tools on the personal and technical knowledge flow within MNCs. It yields two interesting findings. First, not all of the prevalent KM tools show a significant impact on either the personal or the technical inter-organizational knowledge flow. In fact, only 9 out of 19 tools (Tool 2 was significant both for the personal and the technical flow) proved to have a significant effect. Second, not all of the significant tools displayed a positive impact. Indeed, 5 of the 9 significant tools have a negative influence on the knowledge flow (Tool 2 shows a negative impact in both cases of personal and technical knowledge flow).

Moreover, of the 2 personalization tools one has a negative and the other one a positive impact. Interestingly, there are by far more significant codification tools than personalization tools, even though we found that personalization tools are used slightly more often than the codification tools (see above). Of the 7 codification tools, 4 reduce the chance for a high knowledge flow, and only 3 increase it. This seems to be rather astonishing given the fact that technology has been frequently viewed as both a key contributor to and enabler of KM (cf. e.g. Davenport and Prusak 2000). Indeed, expectations for knowledge technologies were or still are quite high as the following statement shows:

"Knowledge technologies attempt to push users to think beyond their current boundaries, thus facilitating organizational activity, promoting continuous improvement and growth through innovation" (Moffett, McAdam *et al.* 2004, p. 176).

Does this mean that KM tools – especially the codification ones – should be discarded? According to the authors, this is not necessarily the case. However, arbitrary and incautious use of KM tools might be a waste of money and even lead to counter-productive effects. For tools to be effective they have to be widely accepted and perceived as useful by those who are supposed to employ them, i.e. the employees. Besides, as our results suggest, one has to be aware of the fact that different tools have different effects on the way knowledge is shared. Hence, depending on what kind of knowledge flow is to be encouraged the appropriate tools have to be applied. Above all, KM is more than simply implementing KM tools and a lot of different factors have to be taken into account. Davenport *et al.* (1998) put it like this: "Effective KM is neither panacea nor bromide; it is one of many components of good management" (p. 56).

One of these factors that are necessary for KM to become effective is a "knowledge-friendly" culture (Davenport, De Long *et al.* 1998). Zack (1999) puts it like this: "Effective use of information technology to communicate knowledge requires that an organization share [sic] an interpretive context" (p. 50). In fact, "if the cultural soil isn't fertile for a knowledge project, no amount of technology, knowledge content, or good project management practices will make the effort successful" (Davenport, De Long *et al.* 1998, p. 53). Nevertheless, even though technology alone does not make a firm become a knowledge-creating company, at least the presence of KM technologies may even have a positive effect on the knowledge culture of the organization (Davenport and Prusak 2000).

References

Akhavan, P., M. Jafari et al. (2006). Critical success factors of knowledge management systems: A multi-case analysis. *European Business Review*, 18,(2), 97-113.

Becerra-Fernandez, I. and R. Sabherwal (2001). Organizational knowledge management: A contingency perspective. *Journal of Management Information Systems*, 18(1), 23-55.

Birkinshaw, J. (1999). Acquiring intellect: Managing the integration of knowledge-intensive acquisitions. Business Horizons May-June, 33-40.

Bloodgood, J. M. and W. D. Salisbury (2001). Understanding the influence of organizational change strategies on information technology and knowledge management strategies. *Decision Support Systems,* 31, 55-69.

Davenport, T. H. (2005). Thinking for a living: How to get better performance and results from knowledge workers. Harvard Business School Press, Boston.

Davenport, T. H., D. W. De Long, et al. (1998). Successful knowledge management Projects. *Sloan Management Review,* 39, 43-57.

Davenport, T. H. and J. G. Harris (2005). Automated decision making comes of age. MIT *Sloan Management Review,* 46(4), 83-89.

Davenport, T. H. and L. Prusak (2000). Working knowledge: How organizations manage what they know. Harvard Business School Press, Boston.

Earl, M. (2001). Knowledge management strategies: Towards a taxonomy. *Journal of Management Information Systems,* 18(1), 215-233.

Gallupe, B. (2001). Knowledge management systems: Surveying the landscape. *International Journal of Management Reviews,* 3(1), 61-77.

Hansen, M. T., N. Nohria, et al. (1999). What's your strategy for managing knowledge? *Harvard Business Review,* 77(2), 106-116.

Holsapple, C. W. and K. D. Joshi (2001). Organizational knowledge resources. *Decision Support Systems*, 31, 39-54.

Hong, J. (1999). Structuring for organizational learning. *The Learning Organization*, 6(4), 173-185.

Inkpen, A. C. and K. Ramaswamy (2006). Global strategy: creating and sustaining advantage across borders. Oxford University Press, New York.

Kasper, H. and B. Haltmeyer (2002). Knowledge sharing in multinational organizations. *Journal of Cross-Cultural Competence & Management*, 3, 279-313.

Kasper, H. and J. Mühlbacher (2004). Entwicklung des organisationalen Wissens in lernenden Organisationen, Zur Differenz zwischen theoretischem Anspruch und Unternehmenswirklichkeiten. Strategien realisieren - Organisationen mobilisieren, Das neueste Managementwissen aus dem PGM MBA. H. Kasper. Linde International. Vienna: 241-261.

Leonard, D. A. and W. C. Swap (2005). Deep smarts: How to cultivate and transfer enduring business wisdom. Harvard Business School Press, Boston.

Mayring, P. (2000). Qualitative content analysis [28 paragraphs]. Forum Qualitative Sozialforschung / Forum: *Qualitative Social Research* [On-line Journal], 1(2).

McAdam, R. and S. McCreedy (1999). A critical review of knowledge management models. *The Learning Organization*, 6(3), 91-100.

Moffett, S., R. McAdam, et al. (2004). Technological utilization for knowledge management. Knowledge and Process Management 11(3), 175-184.

Mohrman, S. A., D. Finegold, et al. (2002). Designing the knowledge enterprise: Beyond programs and tools. *Organizational Dynamics*, 31(2), 134-150.

Nonaka, I. and H. Takeuchi (1995). The knowledge-creating company: How Japanese companies create the dynamics of innovation. Oxford University Press, New York.

Zack, M. H. (1999). Managing codified knowledge. *Sloan Management Review*, 40(4), 45-58.

SECTION III

TECHNOLOGY AND INNOVATION MANAGEMENT

CHAPTER 20

THE INFLUENCE OF ORGANIZATIONAL MATURITY IN THE PROJECT PERFORMANCE: A RESEARCH IN THE INFORMATION TECHNOLOGY SECTOR

Renato de Oliveira Moraes* and Isak Kruglianskas**

*Departamento de Ciências Exatas e Aplicadas da Universidade Federal de Ouro Preto, Brazil; **School of Economics, Business Management and Accountancy of the University of São Paulo, Brazil

This work is the result of a field research on software development projects organizations in order to see if there is a relationship between project performance and project management maturity of the organization. The concept of maturity in project management, which arises with proposals for maturity models in project management (Goldsmith, 1997; Ibbs and Kwak, 2000; Fincher and Levin, 1997; Schlichter, 2001), is very closely linked to two other models: the Capability Maturity Model (CMM), developed by Carnegie Mellon University, which deals with software development processes, and PMBoK (*Project Management Body of Knowledge*). In this work maturity is understood as being the degree to which the project management processes are formalized. Project performance was assessed using the project performance dimensions suggested by Shenhar: project efficiency (the observance of costs and timescales) and customer satisfaction (which refers to the technical quality and use of the product that is developed). The organizations that composed the sample was divided in two groups. One of the groups was composed of the less mature organizations and the other group was composed of mature organizations. After comparing the performance of the projects executed in these two groups it was found a significant difference regarding the performance dimension "client satisfaction".

1. Introduction

The appeal that maturity models have to project management comes from the expectation that the organization may manage to achieve a general improvement in the performance of the projects it develops. In identifying the different visions of quality, Garvin (1988) highlights that the process approach is based on the principle that the product cannot have superior quality (product quality) than the quality of the process that produced it. This justifies the existence and the adoption of various systems for guaranteeing quality, such as the ISO 9000 series. The CMM model – *Capability Maturity Model* (Paulk, 1994), which is a system for guaranteeing the quality of software development processes, also starts from this assumption: the quality of the product is subordinated to the quality of the process that has created it. In maturity models for project management it is easy to perceive the influence of CMM, because several of these models use the same five maturity levels.

Therefore it seems natural that the maturity models in project management should create the expectation of some type of improvement in project performance. There is still no maturity model that has been generally accepted. Of all the current models, it is possible that OPM3, which is sponsored by the PMI, will be the one with the greatest chance of becoming a world standard, as happened with PMBoK (*Project Management Body of Knowledge*). This lack of a standard leads to a more conceptual treatment that is independent of these models. Instead of working with a specific model we chose to adopt a definition of maturity in project management that is above these models.

We chose to define maturity in project management as the degree to which the processes of project management described in PMBoK are formalized. Several maturity models are strongly influenced by PMBoK and this definition is consistent with quality management systems, such as CMM and ISO 9000 that try to guarantee stability in development processes as a way of achieving process quality.

All of this suggests that maturity in project management, regardless the model adopted by the organization, leads to projects with superior performance. Unfortunately there is still a lack of empirical evidence to confirm this line of reasoning.

This research was performed through a survey of 131 software projects, in which the maturity in project management of the organizations carrying them out and the performance of the projects, were evaluated in order to find statistically acceptable evidence that superior maturity is linked to superior performance.

2. Project Performance

There is a striking difference between works dealing with project performance that has to do with the discussion about the issue of the number of concepts related to performance. While some authors (Lim and Mohamed, 1999; Cook-Davies, 2000; Baccarini, 1999; Munns, 1997) refer to two different concepts – success in managing the project (focus on the development process) and the success of the project (focus on the product resulting from the project) – others (Shenhar et al., 2001; Baker et al., 1983; Pinto and Slevin, 1988) understand that there is a single element under discussion that has multi-dimensional characteristics, and in which the relevance of each dimension fluctuates over time.

Shenhar et al. (2001) do not recognize the existence of two different concepts of success – project success and product success – and defend the idea that the relative importance of the success dimensions of the project change over time. These authors identified the following dimensions of success:

- Project efficiency (complying with timetables and budgets);
- Impact on the client (customer satisfaction and product quality);
- Business success (generating revenue, profit and market share and other benefits for the mother organization); and
- Preparing for the future (the development of an organizational and/or technological infrastructure for the future).

However the proposal of these authors also recognizes that the assessment of each dimension cannot be carried out entirely at the same point in time. The relative importance of each dimension varies over time and with technological uncertainty. In the very short term the efficiency of the project is more important and is also the only one that can be measured with any reliable degree of precision. With the use of the product that was developed, it becomes possible and relevant to assess the other dimensions.

Table 1. Project success dimensions.

Project Success Dimensions	Metrics/Variables Used
Project Efficiency	Time Goal
	Budget Goal
Impact on the Client	Functional Performance
	Conformity to Technical Specifications
	Attendance of Client Needs
	Client's Problem Solution
	Utilization of Project Products by the Client
	Client Satisfaction
Business Success	Commercial Success
	Market Share Increase
Preparation for the Future	Creation of a New Market
	Creation of a New Product Line
	Development of a New Technology

Source: Shenhar et al. (2001).

In projects with low technological uncertainty the expectations in relation to the Project are much more linked to marginal contributions in which the efficiency of the development is the determining factor. For example, on product update the interest is on maintaining the product in accordance with market specifications while dramatically changes in the life cycle of the product. are not expected When working with large innovations and great technological uncertainty, organizations become more tolerant of low project efficiency. This is because there is the expectation that the project might possibly generate internal competence in a new and emerging technology.

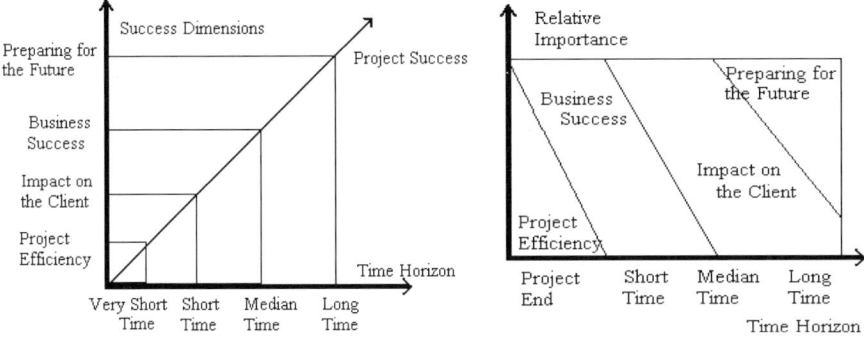

Figure 1. Project success dimensions × time.

Figure 2. Relative importance of success dimensions × time (Source: Shenhar et al., 2001).

3. Methodology

The population considered in this work comprises software development projects that have the following characteristics:

- Initial cost of no less than U$7,000.00
- Completed no more than five years and no less than two years ago

The sample used was made up from mailing lists with professionals from the Information Technology (IT) area. The choice of an intentional sample limited the possibilities of generalizing the statistically relevant findings of the sample. However this option enabled us to obtain a sufficient number of replies for applying the statistical techniques we used.

The elements of the sample were contacted via e-mail and invited to take part in the research. They could reply to the questionnaire by e-mail or directly through the site especially constructed for this purpose. The response rate was between 2% and 3%, which can be considered reasonable for this type of research.

The original questionnaire contained 51 questions, distributed in three parts:

(i) Interviewee: the interviewee is identified and screened.
(ii) Company: this part contains questions on the "umbrella" organization of the Project. This includes information relating to the project management maturity of the organization.
(iii) Project: questions divided into two groups:
   a) Project development environment: questions about the Project and the conditions under which it was developed.
   b) Project performance: questions relating to project performance and the relative importance of different performance criteria.

In order to achieve the objective of this work and check the conceived hypothesis the data collected in the sample were analyzed in accordance with the following:

a) An initial analysis was made of each of the completed questionnaires in order to check for errors that occurred when filling them in and that could be identified and corrected even before the replies were tabulated.
b) Each of the quantitative variables (ordinal variables considered as interval ones) was evaluated in isolation to check if the requirements for the application of the multivariate methods were satisfactorily met.
c) A series of factor analyses was made with regard to the following groups of variables:
   I. The performance variables used in evaluating the project;
   II. The organizational level of maturity of the project management.
   Because the organizational maturity level in project management and the project performance models used in the research did not allow a direct measurement, a factor analysis approach was applied to permit the reduction of the number of dimensions to be analyzed.
d) Bivariate Correlation Analysis - The objective of this procedure was to verify the degree of association among the factors extracted in two factorial analyses – factors related to the maturity level and factors related to the project performance.
e) Cluster analysis - The formation of the groups was based on the dimensions of organization maturity levels obtained from the factor

analysis. With this approach it was possible to create groups of organizations with homogeneous project management maturity levels. According to Hair *et al.* (1998) an important aspect in this process is the use of manifest variables or extracted factors (from previous factor analyses).

f) Variance analysis - here we tried to see whether any relationship exists between maturity in project management - as represented by the maturity groups formed in the cluster analysis - and project performance - as represented by the factors we extracted during the factor analysis of the variables linked to project performance.

4. Results

The data collected were submitted to factor analysis – which allowed for a significant reduction in the size of the problem – and to cluster analysis, which allowed for a grouping of the sample elements into sets that had internal homogeneity and large external heterogeneity.

In order to study the variables related to the organizational maturity on project management a factor analysis was carried out to reduce the elements and thereby simplify our understanding and interpretation of the results. The results obtained were satisfactory. As the KMO coefficient and the significance of the Bartlett Test (Table 2) show, factor analysis proved to be appropriate for the data we collected.

Table 2. KMO and Bartlett's test for the maturity in project management variables.

Kaiser-Meyer-Olkin Measure of Sampling Adequacy		0.898
Bartlett's Test of Sphericity	Approx. Chi-Square	1479.382
	Df	91
	Sig.	0.000

We extracted all the eigenvalue factors greater than 1 (one). Table 3 shows the factors extracted and Table 4 shows the factor loading after rotation. As a result of the factor loadings the factors were called:

- Factor 1 – Internal development management. This refers to the maturity of the processes for managing those development activities that are carried out internally.
- Factor 2 – Third party management. Only the processes for managing contracts were "loaded" into this factor.

An explanation for the formation of these factors may be in how little tradition there is in sub-contracting out software projects when compared with other types of projects. Because subcontracting is a recent practice in the studied sector perhaps it is still not completely incorporated and integrated into other project management processes.

Table 3. Extracted factors and explained variance.

Component	Initial Eigenvalues			Extraction Sums of Squared Loadings			Rotation Sums of Squared Loadings
	Total	% of Variance	Cumulative %	Total	% of Variance	Cumulative %	Total
1	8.043	57.449	57.449	8.043	57.449	57.449	7.723
2	1.214	8.674	66.123	1.214	8.674	66.123	3.830
3	.890	6.356	72.479				
4	.818	5.845	78.323				
5	.668	4.775	83.098				
6	.471	3.362	86.460				
7	.418	2.984	89.444				
8	.373	2.667	92.111				
9	.345	2.461	94.572				
10	.235	1.679	96.251				
11	.179	1.277	97.528				
12	.149	1.067	98.595				
13	.121	.865	99.460				
14	.076	.540	100.000				

Extraction Method: Principal Component Analysis.

Factor analysis extracted only 65.6% of the common behavior from the original variables. This apparently low figure becomes acceptable when we consider that the concept of maturity is relatively new; this means that this research takes on the aspect of exploratory research.

Table 4. Factorial loads after rotation.

Pattern Matrix	Factor	
	1	2
Project Duration Estimates	.934	-.209
Chronogram Control	.910	-.102
Cost Estimates	.905	-.215
Cost Control	.844	-.147
Quality Assurance	.750	.086
Scope Change Control	.699	.267
Risk Planning	.654	.264
Team Development	.653	.188
Risk Monitoring and Control	.639	.312
Information Distribution	.639	.278
Integrated Change Control	.606	.277
Comunication Planning	.581	.314
Suppliers Selection	-.004	.858
Contract Management	.168	.706

Extraction Method: Principal Component Analysis; Rotation Method: Oblimin with Kaiser Normalization.

To evaluate the confidence of the results the factorial analysis were calculated the Alpha of Cronbach of the constructed scales, respectively "Internal Project Management" and "Management of Project Outsourcing". The first factor presented the value 0.944 and the second the value 0.766 what is very satisfactory.

The cluster analysis that followed led to the identification of two groups as far as the level of maturity is concerned:

- Group 1: Projects of organizations with inferior maturity. In this group, with 96 elements, was found 74% of the valid cases.
- Group 2: Projects of organizations with superior maturity. This group had 35 elements, which correspond to 26% of the valid cases.

Figure 3 gives a spatial representation of the projects of these two groups as a function of these two dimensions of maturity in project management.

As expected, the group of organizations with greater maturity in project management is smaller and showed greater formalization and stability in both dimensions.

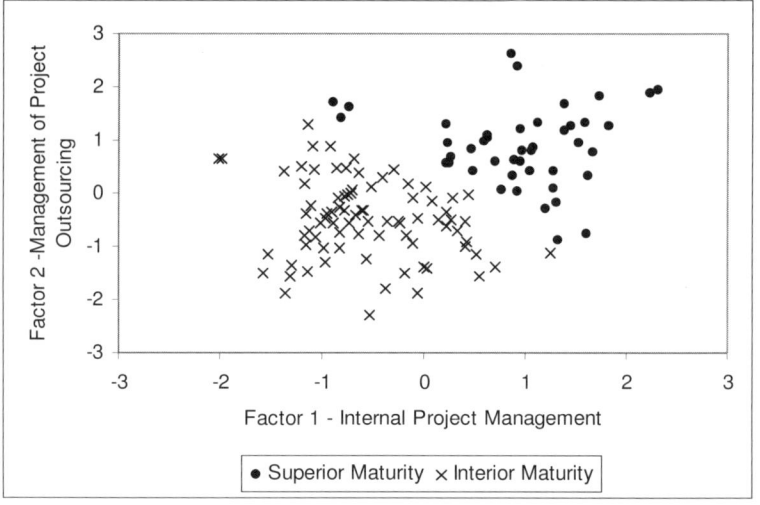

Figure 3. Formed groups: Project of organizations of superior and of inferior maturity levels.

For studying the variables related to performance, a factor analysis was carried out to reduce the elements and simplify our understanding and interpretation of the results. The results obtained were quite satisfactory. As the KMO coefficient and the significance of the Bartlett Test (Table 5), show, factor analysis proved to be appropriate for the data collected.

Table 5. KMO and Bartlett's test for the project performance variables.

Kaiser-Meyer-Olkin Measure of Sampling Adequacy		0.851
Bartlett's Test of Sphericity	Approx. Chi-Square	521.55
	Df	28
	Sig.	0.000

We extracted all the eigenvalue factors greater than 1 (one). The final result, composition of factors, represented 72.2% of the variance of the original variables and coincided with the theoretical model. Table 6 shows the factor loading after rotation.

The first factor corresponds to the Performance Dimension "Client Satisfaction" and the second to "Project Efficiency" of (Shenhar et al., 2001) model. The confidence of the results was evaluated by calculating the Alpha of Cronbach of the constructed scales, presenting respectively the values 0.904 and 0.761, what is again very satisfactory. To verify the existence of correlation between the maturity level and the Project performance was calculated the bivariate correlation that is shown in Table 7.

Table 6. Factorial loads after rotation.

Pattern Matrix	Factor	
	1	2
Client Problems Resolution	.949	-.166
Atendance of Client Needs	.921	-.075
Client Satisfaction	.879	-.039
Utilization of the Product by the Client	.722	.105
Functional Performance	.710	.171
Atendance of Technical Requirements	.695	.178
Budget Goal	-.034	.920
Time Goal	.088	.839

Extraction Method: Principal Component Analysis; Rotation Method: Oblimin with Kaiser Normalization.

Table 7. Pearson bivariate correlation between maturity in project management and project performance.

		Impact on the Client	Project Efficiency
Maturity of the Internal Project Management	Pearson Correlation	.195(*)	.329(**)
	Sig. (2-tailed)	.039	.000
	N	112	112
Maturity of the Management of Project Outsourcing	Pearson Correlation	.121	.046
	Sig. (2-tailed)	.204	.633
	N	112	112

** Correlation is significant at the 0.01 level (2-tailed).
* Correlation is significant at the 0.05 level (2-tailed).

Even though not very high, there is a significant correlation between the Project Performance and the Maturity Level of Internal Project Management (one of the dimensions of the maturity level of the executing organization).

To check, in a different way, the relationship between the maturity in project management of the organizations carrying out the project and the performance of the software projects, a variance analysis was carried out, comparing the groups of different levels of maturity: Group 1 – organizations with inferior maturity – and Group 2 – organizations with superior maturity in respect of their project performances: project efficiency and impact on the consumer (Table 8).

Table 8. Variance analysis – Comparisons of project performance inside and among organizations of different levels of maturity.

		Sum of Squares	Freedom Degrees	Mean Square	F	Sig.
Impact on the Client	Between Groups	4.713	1	4.713	4.924	.029
	Within Groups	105.293	110	.957		
	Total	110.006	111			
Project Efficiency	Between Groups	5.242	1	5.242	5.479	.021
	Within Groups	105.236	110	.957		
	Total	110.478	111			

As shown in Table 8, the hypothesis that there is relationship between maturity and performance within the sample studied was accepted.

At the 5% significance level there is a relationship between maturity and performance. The Tables 7 and 8 show similar results. Pointing to the same direction the Internal Project Management is more relevant in representing the maturity level because it is more strongly related to project performance and this can be explained, probably, because of the low utilization of external suppliers (outsourcing) for the development of software projects in most of the organizations that participated in the research.

5. Final Considerations

Despite the characteristics of the sample which didn't allow to make generalizations about the results, it can be concluded that the results obtained here strongly suggest the need for greater attention when considering the benefits of the organizational maturity for the performance of information technology development projects.

Also, as a conclusion of the research it can be stated that an increase in the level of organizational maturity in project management in an organization not necessarily conducts to a higher project performance, but it can conduct to some specific benefits such as higher client satisfaction.

However, considering the limitations and characteristics of the present research that used a partial vision of the concept of project performance as well as of the organization maturity level in project management it is highly recommendable to perform more studies in this field to go deeper in the understanding of the processes involved in order to have more confident conclusions regarding the advantages of looking for higher maturity in organizations developing software projects.

References

Baccarini, D. (1999). The Logical Framework Method for Defining Project Success. *International Journal of Project Management,* Vol. 30, no. 4, pp 25-32.

Baker, Bruce N., Murphy, David C. & Fisher, Dalmar (1983). "Factors Affecting Project Success" in Cleland, D. I. & King, W. R. Systems analysis and Project management. New York: McGraw Hill.

Cook-Davies, T. (2000). The real success factors on projects. *International Journal of Project Management*, Vol. 20, pp. 185-190.

Fincher, A. & Levin, G. (1997). Project Management Maturity Model. Project Management Institute 28th Annual Seminar/Symposium, Chicago, Ill., pp. 48-55.

Garvin, D. (1998). Managing Quality. Free.

Goldsmith, L. (1997). Approaches Towards Effective Project Management, Project Management Maturity Model. Project Management Institute 28th Annual Seminar/Symposium, Chicago, Ill., pp. 49-54.

Hair, J.R., J. F. et al. (1998). Multivariate Data Analysis - New York, Macmilan Publishing Company.

Hartman, F. T. & Skulmoski, G. (1998). Project Management Maturity. *Project Management Journal*, pp. 74-78.

Ibbs, W. & Kwak, Y.H. (2000). Assessing Project Management Maturity. *Project Management Journal*, Vol. 31, no. 1, pp. 32-43, March.

Levantamento do Universo de Empresas Associadas SOFTEX – Pesquisa Censo SW – Agosto de 2001. Ministério da Ciência e Tecnologia

Lim, C. S. & Mohamed, M. Z. (1999). Criteria of project success: an exploratory re-examination. *International Journal of Project Management*, Vol. 17, no. 4, pp. 243-248.

Munss, A. K. & Bjeirmi, B. F. (1997). The role of project management in achieving project success. *International Journal of Project Management*, Vol 14 no. 2 pp. 81-87.

Paulk, Marc C. et al. (1994). The Capability Maturity Model: Guidelines for Improving the Software Process Addison-Wesley.

Pinto, J. K. & Slevin, D. P. (1988). Project Success: Definitions and Measurement Techniques. *International Journal of Project Management*.

Pmbok (2000). A guide to the project management body of knowledge PMI - Project Management Institute.

Schlichter, J. (2001). PMI's Organizational Project Management Maturity Model: Emerging Standards. Proceedings of the Project Management Institute Annual Seminars & Symposium, USA: Nashville, Tennessee, Nov.

Shenhar, A. et al. (2001). Project success: a multidimensional strategic concept. *Long Range Planning*, No. 34, pp. 699-725.

CHAPTER 21

ORGANIZATION FOR PRODUCT DEVELOPMENT: THE SUCCESSFUL CASE OF EMBRAER

Paulo Tromboni do Nascimento*,
Eduardo Vasconcellos*, and Paulo César Lucas**
* School of Economics, Business Administration & Accounting,
University of Sao Paulo, Brazil; **Embraer, Brazil

Embraer is a Brazilian aircraft manufacturer and a global leader in the market for regional commuter aircraft. Among the factors that explain its success is the organizational structure used to coordinate large volumes of material and human resources located in several countries and various different firms for the conception, design and construction of new aircraft to meet market needs competitively in terms of lead times, pricing and performance. The article analyzes this structure, which consists of a hierarchy of cells operating through an international matrix with multiple axes. The use of cells represents an important addition to the virtues of the matrix structure. The matrix promotes an adequate division of labor between routine and project requirements, while cells assure the integration of the sophisticated competencies necessary to achieve high performance in large-scale projects, in terms of quality and time at all levels of the product tree for a complex product requiring the participation of many individuals with different specialties and belonging to different organizations.

1. Introduction

Embraer sells aircraft to regional airlines around the world. It also has military programs such as the AMX attack jet developed in partnership with Italian manufacturers, and the Tucano trainer. In the regional segment, operating cost and customer service are of paramount importance in choosing a supplier, alongside financing, which is essential for a product that costs around 20 million US dollars. No new aircraft can be marketed without certification from the national agencies responsible for this sector. Brazil's agency is the IFI, subordinated to the CTA. In the U.S. it is the Federal Aviation Administration (FAA). FAA certification is of course indispensable for any manufacturer selling to international markets.

The complexity of regional jets requires numerous suppliers. In Embraer's case many suppliers are large global corporations located outside Brazil. For example, there are only three major global suppliers of jet engines, which account for about 30% of an aircraft's value: GE, Pratt & Whitney, and Rolls-Royce. In sum, Embraer deals with an international network of suppliers that are themselves global in scope. It has chosen to integrate these solutions and focus on a product designed for the global market – regional commuter jets.

To develop new products that meet market needs in terms of leading times, performance and pricing, and shareholder profit expectations is a major challenge. Especially in the case of complex products that involve many people in different specialties and organizations. This requires a high level of integration among specialist areas. Integrated Product Development has been increasingly used as the approach to tackling this challenge over the last decade and is now almost a paradigm (Gerwin & Barrowman, 2002).

This chapter describes and discusses the organizational structure used by Embraer to develop new aircraft, which consists of a hierarchy of cells operating through an international matrix with multiple axes. The use of cells represents an important addition to the virtues of the matrix structure. This chapter is partly based on 11 semi-structured interviews with several managers responsible for the EMB 170 program. The final text has been validated for Embraer by the third author.

2. Organizational Structure for New Product Development

Conceiving a new aircraft is a multidisciplinary activity restricted to a small group of highly skilled specialists using a small volume of physical resources.

The next step - product definition - is characterized by fluidity with fuzzy borders in terms of scope and interfaces. Because aerospace projects are complex the product tree has many levels, descending to subsystems, equipment items, and components. Each level has its own specifications and interface documents, which are absolutely essential for the intellectual division of labor that characterizes the effort to develop such products.

Once the product tree and interfaces are defined, a natural structure emerges for the program. The natural approach is to attribute development of each subsystem to a group with technical targets defined in interface documents, and costs and deadlines defined in management documents. As highlighted by Freeman (1982) in the economic literature on innovation and by Clark & Wheelwright (1993) in the literature on product development, integration is the key issue throughout the process.

Shtub *et al.* (1994) argue that the matrix is an effective tool for integrating several functional areas in multidisciplinary projects but stress the following as the main risks of this approach: conflicts of interest, double subordination, total responsibility of the project manager for the end-product without having competencies in all the technical areas involved, conflicts created by horizontal/diagonal communication, and differing objectives for the project manager and functional managers.

This is not a recent problem. In the 1960s, Lockheed implemented a matrix structure to develop the C-141 program, creating a program management unit to integrate the various functional areas involved (Corey & Star, 1971).

The matrix structure is most effective when the project manager's power is balanced with that of functional managers (Bernasco *et al.*, 1999). The project manager is strongest in the project matrix and functional managers hold sway in the functional matrix.

A survey of 500 managers in companies that used various types of matrix demonstrated the predominance of this structure in new product development. All three types of matrix were used in the companies surveyed but the project matrix proved most effective (Larson & Gobeli, 1987).

Texas Instruments succeeded in shortening the time to market for new products by 50% using a matrix structure (Bernasco *et al.*, 1999).

Filkelstein (1991) proposes the "flattest" possible matrix for the 21st century, with only a few hierarchical levels and as much adaptability to change as can be engineered. Frank (1992) goes further, arguing that the matrix will gradually be replaced by a "flatter networked organization". It is a trend that relates directly to new information technologies and their role in facilitating the integration of large organizations via networks.

3. Organizing for New Products at Embraer

By the second half of the nineties, the 170/190 program was conceived to develop a new, technically more complex family of aircraft seating 70-116 passengers. It was an investment of 850 million US dollars and a new level of engagement with partners and customers in a more aggressive and competitive environment.

In 1998 Embraer's new Department of Programs took over the coordination of all projects and programs subordinated to the Industrial vice Presidency, covering Engineering, Procurement and Production. When it was decided to develop the EMB 170/190 series, a second stage in the use of matrix structures began with the creation of a dedicated department. Thus there was a time when there was a specific department responsible for the EMB 170/190 program and a separate department for all other programs. The 170/190 department, like the rest, relates via a matrix structure to the various units of the Industrial, Commercial and Customer Service vice Presidencies.

3.1. *Integrated product development*

Don Clausing's book *Total Quality Development* (American Society of Mechanical Engineers, April 1994) represented a landmark in Integrated Product Development. It deals with issues such as robust design, incremental product definition, and multifunctional teams. In particular, it takes a global view of the development process from the customer to the concept and back to the customer.

In an interesting review of recent empirically based studies (Gerwin & Barrowman, 2002), the authors set out to refine the definition of IPD, identifying its distinctive characteristics and relating them to the performance of products during the development process.

While recalling Griffin's finding (1997) that 64% of projects had multifunctional teams as an indication of the growing use of IPD in the United States, Gerwin & Barrowman (2002) found few studies that related organizational characteristics to performance in IPD. Furthermore, the only characteristic examined in the empirical studies they reviewed – enrichment of development tasks – proved not to correlate significantly with performance in the overall set of studies analyzed. Griffin (1997) found that the use of multifunctional teams was not a distinct characteristic of high-performance firms as compared to others covered in his study. Among their findings, Gerwin & Barrowman (2002) included the need for more research to try to understand how to coordinate and integrate: 1) the hierarchy of teams in large-scale development projects – given that existing studies have focused on the lead team and have not discussed its relation to other teams, which necessarily exist in a large-scale project; 2) a company's portfolio of multiple partially concurrent projects; 3) the partners involved in the development effort.

Embraer found itself in a difficult competitive situation when it set out to develop the EMB 170. The competition was already ahead when the program was implemented, requiring completion in a record 38 months instead of 48. At the same time, Embraer could not afford to relinquish its goal of producing an aircraft with superior characteristics to those of the competition. Otherwise Bombardier would dominate the market, potentially forcing Embraer into insolvency. As if these problems were not enough, Embraer lacked sufficient capital to develop the product. Because of these competitive problems, it faced a dual challenge: firstly it had to raise the funds required for a large-scale project and secondly it needed to develop the product in a record time frame for the firm and for the industry as a whole.

The first problem was solved by resorting to a new method known as risk partnering. For the EMB 170/190 the firm decided to work with risk partners who would be responsible for 70% of the total investment, as well as the usual commercial and technological partners. Given the absolute need to produce an aircraft that customers would regard as superior, Embraer fostered greatly increased customer participation in product development. For the first time, all 17 partners in the development of the 170/190 family were involved from the very beginning and actively participated in the product definition stage.

After conception of the EMB 170 (Camargo Jr. et al, 2001), Embraer sought partners and customers to assure product feasibility. Starting with GE, which was to manufacture the engines, and Honeywell, which was to supply the avionics, Embraer brought together a constellation of 17 partners including Sundstrand, Parker, C&D Interiors, Gamesa, Liebherr, Hamilton, Mitsubishi, Latecoere, Kawasaki, and Akros. The first customer to buy the plane while it was still in the conceptual design stage was CrossAir, a Swiss airline.

Having chosen its partners and found the first customers, Embraer then embarked on joint definition of the EMB 170, as required for IPD. Inclusion of these partners and customers in the design process required a new organizational structure. The matrix structure created was devised to be capable of incorporating everything learned in the process of concurrent engineering, and to involve both partners and customers (regional airlines) to an extraordinary degree (Bidault, Despres & Butler, 1998).

Joint Definition is the design stage during which performance requirements are detailed for each of the aircraft's systems and for its functional and physical integration. It is a fluid stage which entails defining the requirements for each partner. After this stage it becomes increasingly more difficult to negotiate design changes. Requirements are defined for each system and subsystem and are converted into formal product development targets recorded in specification and interface documents.

Embraer adopted IPD to assure rapid and successful product development. The new approach was devised as a step beyond integrated multiproject management (Affonso & Campello, 1998) insofar as it built integrated treatment of all aspects into the product development program. The idea was to create integrated processes and broaden the scope of the work done by project teams to include more functions from the rest of the firm and outside it. According to this new approach, the stages of a program were formally defined as follows: preliminary studies, joint definition, detail design and certification, serialization, and phase-out.

Over the last decade, however, emphasis has been placed on the need for a multifunctional team to manage large-scale projects. A good example is the management of strategic truck programs by Daimler Chrysler Brazil. Instead of one lead team at the top of the program, why not have multifunctional teams at all the various levels of the organizational structure for the activities in question? In other words, why not have a multifunctional team responsible for each level and division of the structure?

3.2. *Structure of the EMB 170/190 program*

The EMB 170/190 program is managed by a cell (Program Management Nucleus) of 11 individuals that is led by the Program Director. For each aircraft (170, 175, and 195) there is a Chief Engineer. All chief engineers are members of the program management cell. In addition, representatives of Quality, Process Organization, and Planning & Control etc. are also members of the cell. In the future there will also be an individual manager for the EMB 190. This project had not started when the interviews were conducted for this paper. Above the management cell is the CEO of Embraer.

Each chief engineer heads a management cell for his aircraft, called the Technical Nucleus as shown in Figure 1. Various Product Development Managers (PDMs) are members of these cells. There is a PDM for Aeronautics, another for Structures, another for Electrical & Electronic Systems, and so on. Some of these product development managers are responsible for DBTs (Design Build Teams or aircraft parts), others for IPTs (Integrated Product Teams, or integrated systems such as aeronautics or structures). The job of DBTs is to make sure the aircraft can be assembled physically. Functional subsystems are located in all parts of the aircraft and someone must ensure that they all fit together and function as expected when assembled in the finished product. Each DBT is responsible for a physical piece of the aircraft (nose, tail etc). The IPTs manage the functional view. Their role is to guarantee effective design of subsystems for propulsion, electricity, hydraulics etc. Each PDM heads a cell for his own unit. For example, the Aeronautics PDM leads a cell made up of individuals linked to IPTs such as Aerodynamics, Flight Quality, Performance, etc.

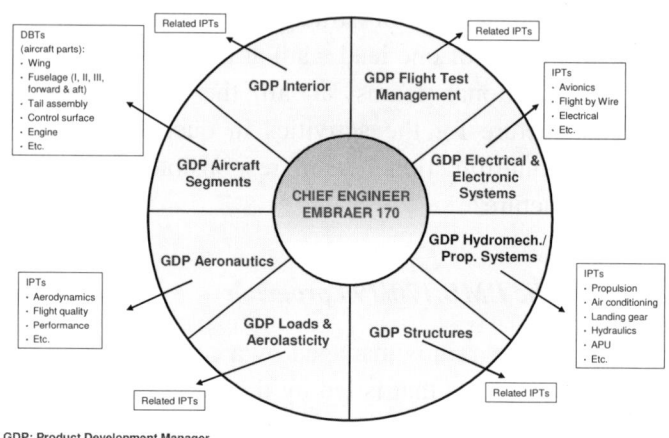

Figure 1. Technical nucleus.

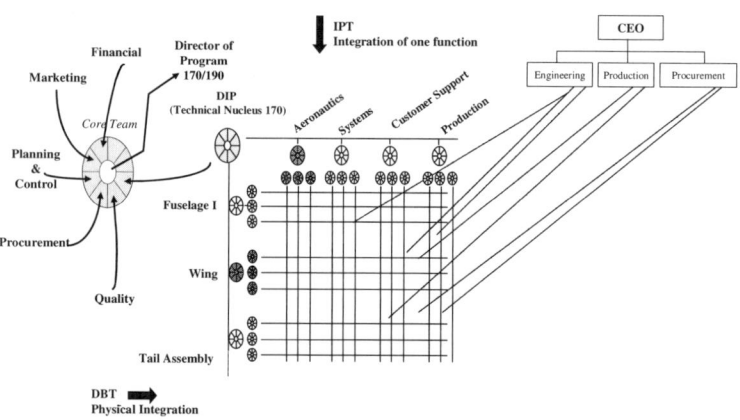

Figure 2. Organization DIP ERJ 170/190.

All these cells operate in a matrix structure, as shown in Figure 2. It is important to note that Figure 2 is a simplification of the actual structure because the 170 program involves 16 partners in several countries. For illustration only Embraer teams are shown in yellow and partner teams in blue. Thus the structure is comprised of cells in a matrix that extend beyond Embraer to interact with other structures in other firms located in different countries, with different languages and cultures. This structure has played a key role in tackling the challenges of the new program. The development team structures used for different aircraft parts must be similar to those of their counterparts in other levels of the product tree. This relationship is represented by the circles in Figure 2.

It is fundamental to grasp the idea that functional and physical integration does not take place only at the level of the program as a whole, i.e. at the root of the product tree. Each level – from the bottom up – in the tree must have the same internal structure for all teams. These structures share the same concerns with production, marketing, customer requirements, engineering, technical assistance, usability, maintenance, and physical integration of the aircraft. Every specialist reports to his functional manager as well as being committed to the program.

3.3. Matrix structure and management cells

Planning a new product is always a multidisciplinary process that involves functional areas such as engineering, R&D, marketing, human resources and production, amongst others. Matrix structures have been used for many years to integrate these areas effectively without needing to follow traditional hierarchical structures. They combine the advantages of project-based structures and functional structures, for example tighter budget control, more efficient use of human and material resources, better integration of product parts, and achievement of deadlines. However, the matrix alone is insufficient in the case of highly complex new products involving many technologies, a high degree of interdisciplinarity, very high levels of specification control, and a very large number of specialists. At points of intersection in the matrix decisions involve so many variables and are so complex that it is hard for any one person to take them in a timely manner and with the requisite quality.

The solution adopted by Embraer for combining a matrix and cell-based structure was an important step toward resolving the intersection decision problem. In particular, decisions are shared by a team comprised of the best suited people to deal with the problem. This group tracks the development of the new product at regular meetings. Chief responsibility remains with the cell coordinator, who can co-opt extra people if necessary to address specific issues. Before cells were introduced, it was up to each head to select appropriate specialists for every particular need. The current system enables a project memory to be constructed, reducing the likelihood of errors and increasing the group's learning capacity while shortening lead times and enhancing the quality of the end-product.

Each manager in the intermediate hierarchy belongs to a cell that is comprised of individuals who address the main intermediate performance targets, as well as the most relevant managerial and technical issues for product integration and performance. This provides a means to replace the impossible need for omniscient and ubiquitous supermen with the practicality of multifunctional teams of similar structure at each point in the hierarchy.

The nature of the concerns motivating the various hierarchical levels evidently varies from one level to another. At the top, business issues and customer satisfaction predominate. Technical and operational aspects of the project are paramount lower down in the product structure. Thus cell composition may vary according to the level and type of item involved. External and internal customers, as well as intermediaries who link the project to them, play a fundamental role in decision making on functional issues.

A critical point is the use of variable organizational forms depending on the specific needs of major project stages. In Embraer's case, the cell-matrix structure was of the greatest importance for the interface definition stage, during which 600 engineers interacted intensely. In the previous stage, the basic design was defined by some 20 people from various areas in the company, for which one multifunctional team was sufficient.

During the program effective use of a matrix structure combined with cells required an effort to train specialists in the necessary behavioral and project management techniques.

4. Conclusions

Joint definition of a complex product such as an aircraft requires a high degree of interaction among all the teams and firms involved. Prospective customers must also be included in the process, with all their requirements at varying levels of stringency. The specialties involved in the development of systems, equipment, and components are often similar. The organization that emerged at Embraer had to respond to these needs at all levels of the product tree. The result was a matrix structure combined with cells. The use of cells represents an important addition to the virtues of a matrix structure. A matrix promotes an adequate division of labor between routine and project requirements, while cells assure the integration of the sophisticated competencies necessary to achieve high performance in large-scale projects. At all levels in the product tree of a complex product, quality and time management require the participation of many individuals who have different specialties and belong to different organizations.

The following results were obtained: a) a reduction in total time taken to develop the 170 program; b) a reduction in the cost of reworking during the design stage, which is also expected to occur in the production stage; c) management concerns explicitly embodied in the cell structure for managing the division of labor. All of these results led to a more effective integration.

There are a number of key prerequisites for this form of organization to be successful. One key prerequisite is the need to prepare people for its use. They must be trained in leadership, teamwork, project management, and the usage of matrix and cell structures. Another important prerequisite concerns attitudes and behavior, in particular to improve team work and time spent on these topics. Complex structures lead to an increase in conflict and can therefore adversely affect performance. No amount of organizational charts or manuals can produce success if people do not have constructive and collaborative attitudes.

A great deal more research must be done to understand how complex projects can best be structured. The intention of this paper is to contribute to this effort through its presentation of a successful practical case that analyzed a highly specific organizational structure. Above all, the role of management cells in matrix structures with multiple axes is an important line of investigation. As noted by Gerwin & Barrowman (2002), this subject requires attention from researchers working in business administration.

References

Affonso, L. C. & Campello, A. C. (1998). Gestão Integrada de Multi Projetos em uma Indústria Aeronáutica. XX Simpósio de Gestão da Inovação Tecnológica. São Paulo, SP, Brazil. Nov. 17-20.

Bernasco, W., Weer-Nederhof, P.C., Tilema, H., & Boer, H., (1999). "Balanced matrix structure and new product development process at Texas Instruments materials and control division". *R&D Management*, Vol. 29, issue 2, pp. 121-131, April.

Bidault, F., Despres, C. & Butler, C. (1998). Leveraged Innovation: Unlocking the Innovation Potential of Strategic Supply. Macmillan Press, London, UK.

Camargo Jr., A. S. et al. (2001). "Desenvolvimento de produtos e processos: um estudo de caso do ERJ 170." Anais do III Congresso Brasileiro de Gestão de Desenvolvimento de Produtos. Florianópolis, SC.

Clark, K. B. & Wheelwright, S. C. (1993). Managing New Product and Process Development. Free Press. New York, USA.

Corey, Raymond E. & Star, Steven H. (1971). Organization strategy – a marketing approach. Graduate School of Business Administration, *Harvard University*, Boston, pp. 62-63.

Filkelstein, Clive (1991). "Visualizing How Corporate Change Affects Data Needs"; *Computerworld;* Vol.25, Issue: 50, December.

Frank, H., (1992). "Empowering employees through networking", *Networking Management,* Vol. 10, Issue 11, October.

Fremann, C. (1982). The Economics of Industrial Innovation. 2nd edition. Frances Pinter (Publishers). London, UK.

Gerwin, David et Barrowman, Nicholas J. (2002). An Evaluation of Research on Integrated Product Development. *Management Science,* Vol. 48, No. 7, pp. 938-953

Griffin, Abbie (1997). PDMA Research on New Product Development: Updating Trends and Benchmarking Best Practices. *Journal of Product Innovation Management;* 14; 429-458. Elsevier Science nc. New York, USA.

Larson, Erik W. & Gobeli, David H. (1987). "Matrix Management: contradictions and insights". *California Management Review,* Vol. 29, No. 4, Summer, pp. 126-138.

Shtub, A., Bard, J. F. & Globerson S. (1994). Project Management – engineering, technology and implementation. Prentice Hall, Englewood Cliffs, NJ, p. 217.

CHAPTER 22

A STUDY OF R&D OUTSOURCING IN THE MANUFACTURING INDUSTRY: A CASE STUDY OF PS2

Fujio Niwa* and Midori Kato**

*National Graduate Institute for Policy Studies, and National Institute of Science and Technology Policy, Tokyo, Japan; **Tokyo Keizai University, Tokyo, Japan

R&D outsourcing is becoming one of the most important strategies for manufacturing companies. We conducted a study of R&D outsourcing in the manufacturing industry to clarify and conceptualize the current circumstances. The research was conducted through a case study, namely, the PS2 of Sony, mainly for fact-finding purposes to clarify what we could learn from this cutting-edge case. Our findings can be summarized as follows:
(1) R&D outsourcing has been increasing quantitatively. Furthermore, the quality of R&D outsourcing will progress because it is based on building strategy. Timing is a condition for achieving the strategy.
(2) The organizational structure and management are conditions for achieving the strategy.
(3) R&D outsourcing is part of the strategy of acquiring technology.
(4) R&D strategy is locating above R&D outsourcing and business strategy is locating above R&D strategy respectively. R&D outsourcing is subject to such a strategic structure.

1. Introduction

1.1. Background to R&D outsourcing

The period around 2000 was a chaotic one for companies in Japan. During the structural recession, Japanese companies were unable to implement clear decision-making, as they strove to find their future trajectory. At the time, one challenge they faced was the conquest of Death Valley Japan, the construction of a scheme via which technical achievements would efficiently result in profitable new business. As background to this, we can cite not only external factors, such as the rapid globalization of the economy and scientific/technical activities and acceleration of business, but also decreasing competence, which is greatly dependent on the accumulation of closed and in-house technologies. The lack of corporate and business strategy, as well as the technical side, influenced this significantly.

On the other hand, some advanced companies anticipated the future, established a strategy, and engaged in definite actions. Such firms have positioned R&D outsourcing in their business and R&D strategy. R&D outsourcing is very strategic, requiring a target to be attained effectively by "selection and concentration." In addition, since the strategic options are very varied, new guidelines and a framework beyond those in the past is required, because the strategy itself will become more complicated.

R&D outsourcing differs essentially from other kinds of outsourcing. The reason is that R&D outsourcing means firms contracting externally for the creation of knowledge, and it is uncertain when they make a contract whether they will obtain the expected technology. It is not routine work and they do not repeat the same jobs. Moreover, R&D outsourcing generates many long-term relationships between firms and this changes as the phase moves from R&D to business. It is necessary to analyze R&D outsourcing by taking such points into consideration.

Although R&D outsourcing is becoming a powerful strategic method, it is a relatively new undertaking for Japanese manufacturers and we have few examples. Considering the 21st century as one of global competition and cooperation, R&D outsourcing is likely to become both important and necessary. With this in mind, we engaged in research on the R&D outsourcing of manufacturers through quantitative analysis (Niwa, 1999, 2001, Niwa & Seike, 1998) and case studies (Niwa & Kato, 2000). This paper conveys the results of our study.

1.2. Framework

Previous research on R&D outsourcing by company was very varied and had a long history. Here, we take the boundary, capability, learning/knowledge creation, and game theory to review the appropriate perspective on the emerging patterns of R&D outsourcing.

1.2.1. Boundary

The basis of R&D outsourcing is the selection and combination of the technologies that are made in house and those that are outsourced. Therefore, the boundary between the two has represented an important framework, such as transaction cost. The concept of transaction cost, as represented by Williamson (1985), covers a wide scope, though it is often associated with classical discussion.

Veugelers and Cassisman (1999) found that companies whose sources of innovation were internal information tended to link their internal and external knowledge, i.e., to use the means of "make and buy." Although discussion concerning the transaction cost generally requires an alternative selection rather than "make or buy." Admittedly, since such collaboration has become more diversified and complicated, a simple "make or buy" framework may be incapable of correctly analyzing their behavior.

Langlois and Robertson (1995) considered that the conventional theory of transaction cost had disregarded the aspect of strategy. Therefore, they integrated the capability approach and the transaction cost and arranged the influence exerted by the diffusion of knowledge on the boundary. They focused on boundary change as the transaction with the market or other firms, and advocated the concept of dynamic cost dealing.

1.2.2. Capability

R&D with outsourcing generally involves developing competitive technologies in house and outsourcing uncompetitive ones; hence firms should establish which of their technologies are competitive.

The core capabilities (Barton, 1992) and "Resource Based View, RBV" (Barney, 1991, Collis, 1998) make it possible to explain the typical behaviors of Japanese companies, which gain competitive advantage by building their own unique technology in the long term, based on the perspective that their aim is to gain resources that are hard to emulate.

Recently, dynamic capability has been discussed, from reflection to the static of RBV. In this discussion by Teece (1997) and others, there is no sustained competitive advantage, and the organizations should continue to establish new competitive advantages by rearranging the resources to hand. In particular, dependency on past knowledge decreases when subject to uncertain competition (Eisenhardt and Martin, 2000). This concept seems useful for considering R&D outsourcing because the external resources can render the rearrangement of resources faster and more drastic. The accumulated resources established via dynamic capability might also resemble "dynamic knowledge networking" (Rosenkopf, 2000). Whether or not inclusion of the external resources as an option facilitates bridging the structural hole (Burt, 1992) remains one of our crucial issues.

On the other hand, swift development is difficult without the effective use of past knowledge, because technology represents a very complex knowledge system. An ability to draw out the appropriate knowledge from the knowledge assets of the organization, as well as any dynamic aspect, remains necessary, as Barton (1992) states. (Does a firm practicing R&D outsourcing hesitate to rearrange its resources just because it is unfamiliar with the theory of dynamic capability?) Moreover, absorptive capability (Lane and Lubatkin, 1998) and complementary investment (Barton, 1992, Freeman, 1991, and Gambardella, 1991) of the external knowledge should still be necessary.

1.2.3. *Learning and knowledge creation*

In R&D outsourcing, aspects of knowledge creation and organizational learning are required, as well as alliance. Among them, we focus on the inter-organizational mutual learning (i.e., Lane and Lubatkin, 1998) and dynamism of knowledge creation. Ciborra and Andreu (2001) insisted that knowledge creation occurred during business practice as an aspect transcending the boundaries of firms and that these boundaries changed in line with the relations between organizations.

Mutual learning through products and patents, which is based on external monitoring, has been effective among Japanese companies. On the other hand, new types of mutual learning are premised on post-collaboration competition, based on the condition that each shares the technology or know-how. As a technology or product approaches completion, they share "Ba" (Nonaka and Konno, 1999) and exchange their knowledge directly with each other, especially in the forefront of R&D, where close interaction appears vital. However, they will return to indirect mutual learning through renewed observation of products and behavior after their collaboration. Combining hetero knowledge facilitates the generation of innovation, although unshared language, understanding, and meaning (Weick, 1984) at an early stage adversely affects the same (Kato, 1995).

1.2.4. *Game theory*

Emerging collaboration consists of a combination of cooperation and competition (Nalebuff and Brandenburger, 1996). Firms merge and cooperate in the meantime as if they were the same company, and return to becoming outsiders as competitors or complementers at a certain point. Therefore, the game theory approach, which deals with a long-term and interdependent relationship and focuses on how this relationship is changed, seems a useful framework.

However, a perspective based on a combination of multiple theories is more appropriate for the study of R&D outsourcing. Parkhe (1993) analyzed interfirm alliances with his framework, fusing game theory with transaction cost, and found that mutual trust built over time had a positive influence on the success or failure of strategic alliances. Larsson and Lars (1998) combined organizational learning and Das and Bing-Sheng (2000) engaged in RBV and game theory. We can find examples in which firms accord priority to timing over resources among recent cases. In other words, if firms do not have sufficient time to achieve their aims, they build close relationships with other companies, even if they have rich resources. In contrast, they are unable to get over the boundary of firms if they have time. Considering such cases requires both game theory and the boundary.

In terms of shared technology through collaboration, it is more difficult than before for each technology or product to have competence. Therefore, firms try to combine technology and products, namely, systematize them, in order to ensure their competitive advantage. Alternatively, in order to offer further differentiated goods to each customer, firms sell products packaged with added services. However, services are easier to imitate than products or technology, and meaning innovation is needed in the business structure. This is the reason why strategy becomes a more important determinant of R&D outsourcing than the number of resources. Business innovation is aligned with technical strategy inside firms and grasped as a business model from the outside. Therefore, it is also appropriate to discuss the business model in this study.

2. PlayStation 2 (PS2) of Sony Computer Entertainment

2.1. *Background of PS2 production*

We studied various cases of R&D outsourcing (Niwa and Kato, 2000). As a typical case we introduce the PlayStation 2 (PS2). This case involved a considerable amount of externally disseminated information, which was one of the reasons why we targeted it as a case study. All information introduced here was available in newspapers, magazines, lectures, and so on. We carried out several interview surveys concurrently, although the detailed information obtained from these is not introduced here. One of the purposes of the survey was to check the reliability of the available information. Subsequently, we selected the information. The information introduced here was gathered up to April 2000, except where otherwise indicated.

The PS2 is a games machine sold by Sony Computer Entertainment (SCE) and the successor to the PlayStation (PS). It was put on the market on March 4, 2000, and one million units were sold in the first two days. The main PS2 component is an MPU, called an emotion engine, and an image processing LSI called a graphic synthesizer. SCE developed the latter in-house, because it had sufficient technological ability. However, the former was developed jointly with Toshiba Corporation. This is an example of R&D outsourcing. (Incidentally, the R&D for the MPU of PS was also outsourced, and the partner was LSI Logic.)

Firstly, we discuss the initial pre PS2 launch situation. As mentioned above, the PS2 was the successor to the PS. The PS was launched on the market at a price of 398$ in 1994, and about 62 million units had been sold worldwide by April 2000. At that time, Nintendo and Sega dominated the game machine market. However, PS fought on bravely, in the U.S., rather than the domestic market, and achieved steady penetration, convincing SCE President Takeshi Takuraki of its potential future success. SCE developed a new business model for software sales to sell PS. It was a method whereby SCE bought all software developed for the PS by game software developers and bundled them with the unit. SCE grasped the game software distribution chain based on this business model and pioneered the packaged software market. Simultaneously, it could gain a high level of cash flow. The financial resources thereby earned were invested in the development of a next generation machine, PS2. As mentioned above, the MPU, the heart of the machine, was not developed in-house by SCE, but outsourced to LSI Logic. Outsourcing had already taken place in this field, and SCE considered the image-processing technology, rather than the MPU, to be its own core competence. The main PS2 components are as follows:

(1) MPU (microprocessor unit) "Emotion Engine"
(2) Image processing chip "Graphic Synthesizer"

2.2. Grand strategy of PS2

SCE confidently planned to sell the PS2 based on the public reaction to PS and a grand strategy:

(1) To dominate all rivals in the game machine market;
(2) To gain supremacy in the game software market;
(3) To establish and lead an "Integrated Soft Distribution Industry."

SCE had already achieved a significant position in the game machine market through the PS. However, in order to enhance this, it was necessary to take the new model to a higher market level. Its requirements included a high processing speed, excellent graphic functions, and many and varied games, the latter of which was deeply related to the second part of the grand strategy. To satisfy these requirements, the specification targets of the PS2 included a 128 bit type, a line width from 0.18 to 0.15μm, operation frequency of 300 MHz, and a processing speed of 6.2 gigaflops (floating-point operations per second). At the time, these represented considerable technological hurdles.

2.2.1. Why was the targeted high performance challenged?

First of all, the performance of the business game machine exceeded that of the home game machine because the area for parts equipped was not limited in the former case. PS2 leveled or reversed this relation and SCE opened up the PS2 hardware to business game machine companies in subsequent years to stimulate them to distribute PS2 games, and indirectly, the PS2 itself.

Secondly, the PS2 targeted a performance surpassing that of a supercomputer. The operation frequency of the high-speed MPU of the PS2 reached 294.912 MHz and 6.2 gigaflops, which were the usual capacity of the average supercomputer. For instance, the latest supercomputer released in 1998 by SGI, "Clay SV1" had a performance of 300 MHz and 1.2 gigaflops. It used "Pipeline (high-speed program execution technology)," which enabled a high level of floating-point operations per second, four times the operation clock frequency.

To achieve this function, SCE composed an MPU of 1.5 million transistors, connected between all circuits, via the wide width of the 128 bit data transfer. Furthermore, all the circuits, which had previously been scattered over two or more chips, were accumulated into a single chip.

Thirdly, the Emotion Engine that processed the images had five times or more the performance of the supercomputers of the day: 6.2 gigaflops. Its image-processing speed was several times faster than that of the

Pentium 3 released in March 2000 by Intel Corporation. It also exceeded that of the workstations used for CG (computer graphics) production.

2.2.2. Why was such high performance demanded of the game machine?

The reason concerned flexibility. Actually, the standard changed drastically because the technology used to deliver digital contents had been continuously evolving. It was hard for single purpose hardware to process the delivery and it became necessary to absorb technology evolution by software processing. Therefore, improved processing performance was needed.

The market accepted the image-processing technology very favorably. PS2's beautiful high definition images outshone those of existing game machines. The SCE slogan was to make it possible to enjoy theater movies at home, and it was successful.

One strategy adopted in game software was the pioneering of package software, as in the case of the PS. However, software development was overdue and it took considerable time to enrich the lineup of game software, because graphic software development required highly advanced technology. Only a small number of software makers were capable of such work, and the software inevitably become expensive. This was the reason why insufficient software was prepared immediately after the PS2 release.

The third component of the grand strategy, namely, the establishment and leadership of an "Integrated Soft Distribution Industry," was a future image drawn up by SCE and also shared by Sony Corporation. The nucleus was the idea that the PS2 was not only a game machine but also a home terminal, via which a large range of varied software could circulate via the Internet, in other words, making it a home set box. Since the PS2 was expected to be its first product, it was termed a wolf in sheep's clothing, or a Trojan horse (Microprocessor Report, 2000). The industry was capped by integrated software, because it intended to target various or all kinds of software. In concrete terms, this encompassed digitalized (amusement) software like games, movies, and music, as well as computer software. Such massive information could be delivered to homes via the Internet through high-speed cable television (CATV). In order to use such massive software, the DVD (digital versatile disc) was included. Furthermore, it was also assumed to have a massive external hard memory. Microsoft reacted immediately to this grand SCE strategy, announcing that it would develop and release a rival product in the same market, "Xbox," immediately after the announcement of PS2. This product naming correctly homed in on the characteristic of PS2, namely, its function as a domestic network terminal or home set box.

In concrete terms, SCE planed to gradually switch software, music, and movie sales from package sales, such as in the form of CDs (compact disc) at sales points, to net sales using the PS2 connective function. It formally declared the start of Internet delivery of game, movie, and music software in 2001 by making PS2 a receiving terminal. Because of the function of the home net connection equipment, it became a competitor to the personal computer. The leaders of Sony and SCE at the time were quoted as saying it was a "plot to enter the net delivery," "challenge Wintel (Microsoft and Intel)", and so on. The market had considerable expectations that that the PS2 could become the standard new amusement machine, combining music, movie, and games.

SCE advertised the fact that the PS2 could offer the following attractions to users:

(1) They could enjoy theater quality movies via the efficient MPU and graphic processing LSI;
(2) The DVD could reproduce massive software;
(3) The PS2 could receive various forms of software, such as videos, via high speed Internet through wideband cable television (CATV);
(4) The downloaded data could be stored in an external hard disk, connected by a USB terminal post.

The social infrastructure via which such a strategy was realized was being enhanced and the broadband infrastructure (Internet connection using the unused CATV bands) was rapidly developing. The Ministry of Posts and Telecommunications of the day planned to unfreeze this business to first-category communication corporations, such as Nippon Telegraph and Telephone (NTT), whereupon it was subsequently thought that about 10% of households would be able to use Broadband by around 2001.

2.3. R&D outsourcing by SCE

The essence of the grand strategy was timing. The possibility that SCE could develop a highly efficient MPU for PS2 was not zero, although it would take several years to complete. On the other hand, there was little time to begin the grand strategy, due to the strategies of rival companies, the rapid advancement of related technology, and also the rapid maintenance of the social infrastructure for the "Integrated Soft Distribution Industry." It was necessary to launch the grand strategy immediately under such circumstances, since SCE had no time to develop a highly-efficient MPU by itself. It is thought that SCE opted for R&D outsourcing rather than in-house development after comparing the company's future fruits obtained from the integrated soft distribution industry of the grand strategy with the huge and wide-ranging transaction costs involved. The company culture of Sony Corporation, to continuously pursue self-evolution, was also a major decisive factor in the decision to opt for R&D outsourcing. Excessive transaction costs (Williamson, 1985) would mean SCE giving up the grand strategy as well as the in-house development of the MPU. This was a comparison between the means of transaction costs and the target of the leader in the integrated soft distribution industry, and the time factor was assumed to limit the available room to maneuver.

SCE started to search for a semiconductor maker to serve as a partner in R&D outsourcing for the MPU. In the case of the PS, the partner was LSI Logic, and SCE sought the best partner worldwide based on this experience. Several potent semiconductor makers competed desperately to become the partner and at the time, the game machine market was growing enormously. Game machines evolved into palm size supercomputers as the market grew. The main battlefield of the MPU up to that point had been the personal computer and the market scale of game machines was too different from that of the MPU. However, when the PS was put on the market in 1994, 60,000,000 units were sold in four years. The game machine market had approached about one quarter of the personal computer market. The processor installed in game machines and STB was suppressed to a low unit price of $30 - 40 per unit, based on the assumption that it would be possible to ship more than one million units annually. The reason was that economies of scale were lost when the processor price exceeded one tenth of the product market prices ($300 - 400). However, if the market share could be secured, the software property stored, and the users captured, steady demand and a mass production position effect could be expected. This was the reason why the processor makers unhesitatingly ploughed their leading-edge technology into game machines and STB. The high potential for success of the PS2 was widely forecast and the R&D outsourcing of MPU was common in the game machine industry. The LSI made by NEC Corporation had been installed in the "Nintendo 64" of Nintendo Ltd. (released in June, 1996) and the LSI made by Hitachi Ltd. was installed in the "Dream Cast" of Sega Enterprises (November, 1998).

Toshiba and NEC Corporations battled to be the outsourcing partner for the MPU of PS2. Toshiba Corporation readied more than 100 scientists/engineers and won the battle. In order to implement enough human resources, Toshiba Corporation discontinued two of its own MPU development projects. Toshiba System LSI laboratory and Toshiba America Electronic Components Inc. (TAEC) took charge, and bases were established in Japan and the U.S. The engineer in charge of the basic design of the MPU was an American architect, who had been employed in the MPU development discontinued for this project.

2.4. *Toshiba Corporation as R&D outsourcing partner of SCE*

It was forecasted that Toshiba Corporation would obtain considerable profits by becoming the SCE R&D outsourcing partner and the company indeed strove very hard to do so, amid the ongoing semiconductor war. The major reason was to make profits in the semiconductor business, secondly to rapidly improve its semiconductor technology level, and thirdly to survive as a semiconductor maker in the fiercely competitive global market.

The technological targets requested by SCE were challenging. It was said that Takeshi Takuraki, SCE Vice-president of the day, insistently cited the necessity for a 300 MHz processor. The Toshiba leader could not understand why such a high-frequency processor was necessary for a game machine. However, in the end he, too, was persuaded of the absolute necessity for 300 MHz. The circuit line width of 0.18μ was also a high technological hurdle at that time.

On the other hand, SCE told its partner to treat costs with less regard. Although the chip size became about 20% larger than the original specifications, the Vice-President consented and Toshiba constructed an appropriate development system. First of all, the project team was established, based on the System LSI laboratory and following the SCE proposal, and started development in 1996 about three and half years before the release. This was a kind of "total war", in which more than 100 scientists and engineers participated, and development expenditure of $100 million was invested. Moreover, Toshiba set up an international collaborative development system spanning Japan and the U.S. System LSI Laboratory supplied 60 R&D personnel. In addition, several selected engineers from the MASIC (M = microprocessor) Division and the computer-related divisions of Toshiba joined the project from the processor users' standpoint.

Toshiba suffered from the existing workloads of talented personnel. To secure manpower, it had to discontinue two big joint projects: (1) MPU development for U.S. Silicon Graphics (SGI) Co. led by LSI Laboratory; and (2) a joint project with U.S. Chromatic in the media processor field.

The basic design of the processor, an art exceeding the available technology, was turned over to an American engineer. Very few Japanese could (and can) do it. Toshiba could not leave the basic design of the most important processor to Toshiba America Electronic Parts Corporation in San Jose. In its System IC Section there was one American architect (Dr. X) who had a particular talent. He had been employed from a venture company for 64-bit MPU development. The basic design of the processor was left to him, right from the conceptual ideas of the grand design to the circuit design. This was the most vital ingredient for the PS2's success.

Toshiba had a hard time with cultural friction between Japan and the U.S. Dr. X not only was American but also had an artist's temperament.

He requested a wide range of resources to accomplish the goal, to which the Japanese management could not react flexibly. He became frustrated, and the development schedule was long overdue. He demanded flexible management based on the American tradition. On the other hand, the Japanese established a relatively flexible management system compared with the norm in Japan. For instance, on-the-spot directors executed almost all their own decision-making and rarely consulted point by point with their headquarters in Japan. The confidence of the Japanese top managers in on-the-spot directors, and preparation by on-the-spot directors prepared for the worst, and so on were significant factors in the project. This is a very good example of Japanese corporate management being flexible, despite appearing to be very rigid. The perception of this preparation depended on the corporate culture and climate, meaning what was considered high flexibility from the Japanese perspective was very rigid from the American viewpoint. Dr. X left the office to begin consulting, only leaving word only that he wanted to resign when the project was almost completed.

Although Dr. X was a special case, Toshiba suffered from a larger communications gap with the American side. Toshiba initially proposed that the Japanese side share the development of manufacturing technology while the American side share the basic design. However, the difference between the U.S. contract culture and that of Japan exposed frequent credibility gaps during the project's execution. Consequently, the success factors were said to have been the clearly defined purposes and having the American side implement the basic design. Toshiba felt it gained significant advantages from the experience of managing the double-pole technological development involving Japan and the U.S., despite facing these cultural differences.

The time constraints of the development period (three and a half years) were very severe. Although many engineers were optimistic by nature, they often wanted to give up, due to the extraordinarily tough demands of SCE, such as the image processing performance, and because development experiences involving semiconductors were not readily available. Toshiba had no experience of large-scale and difficult development. Following the project, orders increased because its success proved it had the planning ability needed for the system LSI business.

The Oita Factory took charge of the mass production of the 128-bit CPU for the PS2. It designed the circuit from scratch and could study the cause of any failure instantly. If it had been manufacturing chips whose technology was introduced from the U.S. company, it would have been difficult to analyze failures because of the existence of black boxes in the design. Mass production in Oita Factory proceeded smoothly.

SCE gained hugely from the project. The first gain was the realization of the grand strategy and the second was the transfer of high technology from Toshiba. The technological specification level of the graphic synthesizer, namely, the image-processing LSI developed by SCE, had reached that of the MPU developed by Toshiba. The circuit line width was near 1.8µ, and the operation frequency was near 300 MHz.

3. Concluding Remarks

A quantitative analysis of Japanese manufacturers (Niwa, 1999) makes it clear that R&D outsourcing has increased and its range has diversified, which has led to a change in quality, the advancement of R&D, and competition among companies. Because increasing R&D outsourcing means the further disclosure of technology between parties, it requires new technical and business strategies. In particular, in co-development among competitors in the same industry, strategic collaborative relationships that focus on the difference between the synergy of business and technical strategy have become more widespread, since they will soon compete in the same market. Under such circumstances, the relationship among firms has changed.

The Sony case study shows that it prioritized the appropriate timing of entry into the market over their own development and intended to ensure the creation and domination of a new business. Therefore, the case of PS2 is a fine example of building a technology strategy based on one of business. The business strategy covered not only the products (game machines), but also the larger business of home networking. Toshiba, an R&D partner

Figure 1. Structure of strategy.

of Sony, also achieved the PS2 derivative technology. That is, Sony and Toshiba succeeded in building a win-win relationship.

We discuss such strategic policies of firms. In addition, we examine the fact that the relationship between firms differs significantly in the planning phase compared with the R&D phase.

3.1. *Summary of discussion*

(1) R&D outsourcing has been increasing quantitatively.
(2) The quality of R&D outsourcing will be developed because it is based on building strategy.
(3) Timing is a condition for achieving the strategy.
(4) Organizational structure and management are conditions for achieving the strategy.
(5) R&D outsourcing is a part of the strategy of acquiring technology. R&D strategy is located above R&D outsourcing.
(6) Business strategy is located above R&D strategy. R&D outsourcing subjects to such upper strategies. Figure 1 shows these relationships.

References

Barney, J. B. (1991). Firm Resources and Sustained Competitive Advantage, *J. Management*, 17, pp. 99-120.
Brandenburger, A. M. and Nalebuff, B. J. (1996). *Co-opetition*, Currency Doubleday, New York.
Burt, R. (1992). *Structural Holes*, Harvard Univ. Press, Cambridge.
Ciborra, C. U. and Andreu, R. (2001). Sharing knowledge across boundaries, *J. Information Technology*, 16, 2, pp. 73-81.
Collis, D. J. and Montgomery, C. A. (1998). *Corporate Strategy: A Resource-based Approach*, McGraw-Hill.
Das, T. K. and Bing-Sheng, T. A (2000). Resource-Based Theory of Strategic Alliances, *J. Management*, 26 Issue 1, pp. 31-62.
Eisenhardt, K. M. and Martin, J. A. (2000). Dynamic Capabilities: What are they? *Strategic Management J.*, 21, 10/11, pp. 1105-1121.
Freeman, C. (1991). Networks of Innovators: a synthesis of research issues, *Research Policy*, 20, pp. 499-514.
Gambardella, A. (1992). Competitive advantages from in-house scientific research: the US pharmaceutical industry in the 1980s, *Research Policy*, 21, pp. 391-407.

Kato, M. and Niwa, F. (2001). Strategy of R&D Outsourcing - Typology of R&D Using External Resources and Cost-Benefit Analysis – (in Japanese), *Proc. the 16th Annual Meeting of the Japan Society for Science Policy and Research Management*, pp. 93-96.

Lane, P. J. and Lubatkin, M. (1998). Relative absorptive capacity and interorganizational learning, *Strategic Management J.*, 19, 5, pp. 461-477.

Langlois, R. N. and Robertson, P. L. (1995). Firms, Markets and Economic Change : A Dynamic Theory of Business Institutions, Rout Ledge, London.

Larsson, R., Bengtsson, L., Henriksson, K., and Sparks, J. (1998). The Interorganizational Learning Dilemma: Collective Knowledge Development in Strategic Alliances, *Organization Science*, 9, 3, pp. 285-305.

Microprocessor Report (2000). *A Wolf in Sheep's Clothing*, 14, 10.

Nalebuff, B. J. and Brandenburger, A. M. (1996). Co-opetition, New York: Doubleday.

Niwa, F. (1999). An Inter-industrial Comparative Study of R&D Outsourcing, In L. Branscomb, et al. (eds.) *Industrializing Knowledge - University-Industry Linkage in Japan and the United States –*, the MIT Press, Boston, pp. 128-156.

Niwa, F. (2001). R&D Outsourcing of Companies - Its Present and Perspective –, *R&D Management*, January, pp. 6-10 (in Japanese).

Niwa, F. and Kato, M. (2000). R&D Outsourcing of Companies - Strategy Chang based on Outsourcing and Networking -, *Proc. 15th Annual Meeting of the Japan Society for Science Policy and Research Management*, pp. 209-212 (in Japanese).

Nonaka, I. and Konno, N. (1998). The Concept of 'Ba': Building a Foundation for Knowledge Creation, *California Management Review*, 40, 3, pp. 40-54.

Parkhe, A. (1993). Strategic alliance structuring: A game theoretic and transaction cost examination of interfirm, *Academy of Management J.* 36, 4, pp. 794-829.

Rohlfs, J. H. (2001). *Bandwagon Effects in High Technology Industries*, MIT Press, Cambridge.

Rosenkopf, L. (2000). Managing Dynamic Knowledge Networks, in G. Day and P. Schoemaker (eds.) *Wharton on Managing Emerging Technologies,* John Wiley and Sons, New York, pp. 337-357.

Shapiro, C. and Varian, H. (1998). *Information Rules: A Strategic Guide to the Network Economy*, Harvard Business School Press, Boston.

Teece, D. J., Pisano, G., and Shuen, A. (1997). Dynamic capabilities and strategic management, *Strategic Management J.*, 18, 7, pp. 509-533.

Veugelers, R. and Cassisman, B. (1999). Make and buy in innovation strategies: evidence from Belgian manufacturing firms, *Research Policy* 28, pp. 63-80.

Weick, K. E. (1984). The social psychology of organizing, 2nd ed. Reading, Addison-Wesley, Mass.

Williamson, O. (1985). *Market and Hierarchies*, Free Press, New York,

CHAPTER 23

A STRATEGIC SCIENCE AND TECHNOLOGY PLANNING AND DEVELOPMENT PROCESS MODEL

Grace Bochenek*, Carey Iler*, Bruce Brendle*,
Timothy Kotnour**, and James Ragusa***

*U.S. Army TARDEC, Warren MI, USA; **University of Central Florida, Orlando, USA; ***Independent Consultant, Titusville, USA

This paper contributes to the understanding of generalized strategic science and technology (S&T) planning and development through the application of a new process model to the research and technology initiatives of the U.S. Army Tank-Automotive Research, Development and Engineering Center (TARDEC). A reflective case study is used to document how the organization refocused and formalized its previous method of S&T planning and development to more relevantly and responsively support: (a) the present war on terrorism, (b) the need for near-term solutions to deployed military system operational capability gaps, and (c) maintenance of a future perspective and technology development competency. For model creation, a middle management steering group and action teams were formed (under change management sponsorship of a champion) to formulate and implement an improved process model considered essential to near- and longer-term organizational success and the ever-present goal of providing "Superior Technology for a Superior Army." A technology manager can use elements of this paper and its described approach and methodology, derived strategic S&T planning and development model, and identified implications (challenges, lessons learned, and success measures and evaluation criteria) to more effectively and efficiently review, assess, and revise as needed the S&T initiatives of other organizations.

1. Introduction

One of the many challenges of a science and technology (S&T) development organization is how best to focus and manage its mission, vision, goals, objectives, and customer/stakeholder needs within the constraints of budgets, human and physical resources, and schedule requirements. The need for continuous performance improvement is critical to technical organizations in an era of dynamic changes, economic constraints, and international competition. To these ends, organizations must focus on specific S&T portfolio planning and development to ensure that desired and timely results are achieved, and that customers' needs and requirements are satisfied within available resources. The primary question this paper addressed is: what are the elements of an overall philosophy and process to guide systematic S&T planning and development by organizations and technology managers?

This paper focuses on the entire strategic S&T management system and planning and development process from customer/stakeholder needs and requirements identification and funding allocations, to enterprise-wide results. Required for full S&T implementation is recursive S&T planning developed initially and revised as needed. Central to the process is the identification of all linkages, strategic functions, activities, roles, responsibilities, and deliverables. Essential for continuity are formalized S&T plans complete with roadmaps that visually portray why specific efforts are initiated, and what, when, where, how, and who will be responsible for customer/stakeholder needs and requirements satisfaction within identified funding/budget limits and developed schedules. This paper provides a specific application of a newly created process model to the research and technology development initiatives of a focus organization--the U.S. Army Tank-Automotive Research, Development and Engineering Center (TARDEC) of Warren, Michigan.

2. Research Approach and Methodology

To answer the primary questions of the paper, a review of relevant literature was initially undertaken. From these reviews, a variety of organizational and technology management challenges and required thrusts were identified.

2.1. Organizational challenges and thrusts

Technology organizations continue to face the challenges of ensuring that S&T efforts produce value to society, the economy, and their organizations (Geisler, 2001). To respond to these value challenges, organizations need to initiate two thrusts. The first is to ensure that research and development (R&D) activities are fully integrated and that full collaboration exists within the organization and with external stakeholders. Miller and Morris (1999) point out that a key element is the inclusion of a full range of stakeholders in the R&D process. These stakeholders include partners, customers, R&D, marketing, and production representatives. Their participation supports the development of a shared context (i.e., needs and values) for knowledge leading to the technology, the developed technology, and resultant products. For example, technology pull (from users) and push (from developers) satisfy both needs and values, and contribute to a shared context for all stakeholders. Chiesa (2001) further highlights the need for R&D activities to be fully integrated with competitors, suppliers, customers, and distributors.

The second thrust is to develop and execute an integrated management approach for multiple layers of strategies and best practices for R&D and portfolio management. The latter being the balance of projects and activities that best support the mission, vision, goals, and objectives of the organization and the needs of its stakeholders. Matheson and Matheson (1998) define the need and a series of best practices to connect a multitude of corporate, business, portfolio, and project strategies. The extent and scope of these best practices point to the need for a systematic approach to organizational technology management.

2.2. Technology manager challenges and thrusts

In the past, managers have used various organizational management tools to improve performance (Rigby, 2001). Today, S&T-focused organizations and technology managers are turning to an expanded and integrated set of initiatives such as strategic management, portfolio management, technology roadmapping, project management, and knowledge management to address the challenges they face. Technology managers are now finding that they must manage and function in an R&D environment pursuing two thrusts: (a) integrating core processes throughout the organization, and (b) implementing multiple strategy layers and best practices. These thrusts create challenges for technology managers that include: (a) strategic planning for technology products, (b) new product project selection, (c) organizational learning about technology, and (d) technology core competencies (Scott, 1998). These thrusts and challenges require the use of various methods and tools to develop and manage a project portfolio. Steps in the portfolio management process include: (a) identifying the R&D budget, (b) defining potential R&D projects, (c) evaluating projects, (d) selecting projects, (e) implementing projects, and (f) measuring and adjusting projects and the portfolio (Chiesa, 2001).

2.3. Case study method and focus

To better understand how organizations and technology managers can successfully implement the above core processes and manage challenges and thrusts, a reflective case study (Kotnour and Landaeta, 2004) focusing on a target S&T organization was initiated. The experience gained by the authors in the creation and implementation of an S&T planning and development process model offered a unique opportunity to align a technology management organization's challenge with a performance improvement development and implementation approach, and to share this experience with others. Findings and conclusions presented in this reflective case study are based on a one-year and continuing development and implementation effort by the authors and the focus organization.

3. Case Description

3.1. *Focus organization overview*

The U.S. Army Tank-Automotive Research, Development and Engineering Center (TARDEC) is the nation's laboratory for advanced military ground combat and support vehicle technologies. Its parent organization is the Army's Research, Development and Engineering Command (RDECOM). Because TARDEC is headquartered in Warren, MI (a part of metropolitan Detroit and the world's automotive capital), the organization is uniquely positioned to ensure that it remains committed to developing and delivering near- and longer- term advanced military technologies. The organization accomplishes its mission and vision through: (a) research, development, and engineering; and (b) leveraging and integrating advanced technology into ground systems and tactical (support) equipment throughout a system's life cycle. The organization is committed to increasing the Army's agility, versatility, responsiveness, deployability, lethality, sustainability, and survivability, through the development of advanced ground vehicle and support system technologies for a superior Army (*TARDEC Information Booklet,* 2004).

To sustain its mission, roles, and responsibilities, TARDEC recently developed and has begun implementation of an improved strategic S&T program management system and process model to ensure that the organization remains relevant and responsive to its customers--now and in the future. This system revision was initiated because of the organization's responsibility to continually improve its performance during the present war on terrorism, while concurrently supporting its ongoing mandate to provide technology development and supporting services for the Army in the long term.

3.2. *Interfaces and responsibilities*

As Figure 1 illustrates, TARDEC serves two masters with regard to providing S&T operational solutions and support services. They are: (a) RDECOM that provides first-level reviews, recommendations, and approval for TARDEC's S&T project initiatives such as Advanced

Technology Objectives (ATOs); and (b) its primary customers--Program Executive Offices (PEOs) and Project Managers (PMs) who have general program (e.g., ground combat systems) and specific project (e.g., armored Stryker Brigade vehicles) needs and requirements. This later group represents (from a TARDEC perspective) the ultimate customer--soldiers and other organizations that develop their doctrine and tactics, and provide training and logistics support. As such, TARDEC fits into and supports a Soldier and Ground Systems Life Cycle Enterprise--a life cycle system of Army commands, enterprises and alliances designed to function as a network of linked organizations that are integrated and function as an enterprise system of systems. While Figure 1 is a summary chart, numerous other Army and Department of Defense (DoD) elements (including support and development contractors) are involved in the complete S&T interface process model. While not complete, all essential elements are represented in Figure 1 for the purpose of this paper.

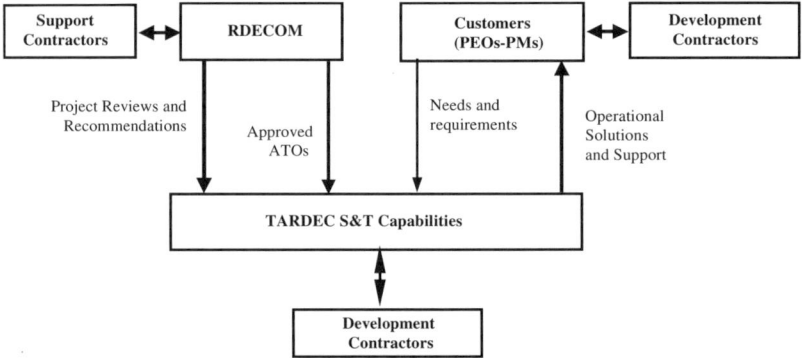

Figure 1. TARDEC S&T interface process model.

3.3. *The need for change*

In spite of established and understood S&T interface process model relationships (ref. Figure 1), and accepted TARDEC customer/stakeholder services, products, and deliverable responsibilities, several organizational problems and a need for change became evident. The first identified challenge was the need to improve the way the organization interfaced and collaborated with those external to the organization--namely its PM customers, stakeholders, the active military, and funding groups. Second, it was felt that these interface relationships should and could be improved by formalizing the organization's internal method and processes for S&T planning and development. The overall rationale for supporting these felt needs was to maintain and improve TARDEC's continued viability as a relevant, responsive, and ready organization through the effectively and efficiently management of its external and internal relationships and activities. The need for change was also made even more acute by the current war on terrorism. To be relevant and responsive to PMs who are responsible for current and operational military force combat and support systems, TARDEC needed to focus more on near-term S&T (less than 2 year) developments, balanced with their traditional longer-term activities. With improvement needs and drivers identified, it was decided that concerted action be initiated to address and resolve the identified problems and capitalize on this opportunity for organizational improvement.

4. Results Achieved

4.1. *The strategic S&T process model*

This paper focuses on the entire TARDEC strategic S&T program management model and its process and responsibility elements, which are illustrated in Figure 2. Indicated are the main functional activities, indicated by numbers to guide the reader through the various phases of the process. The basic idea is that initial S&T planning has taken place in an earlier period (usually annually with updates), and the process of S&T development follows. Requirements are identified as technology capability needs for solution, TARDEC system of systems engineering integration occurs, functional organizations provide S&T systems engineering, and technology scanning reviews occur throughout the process. Concurrent with these process steps and on a regular basis, S&T planning and replanning evolves and the process starts again.

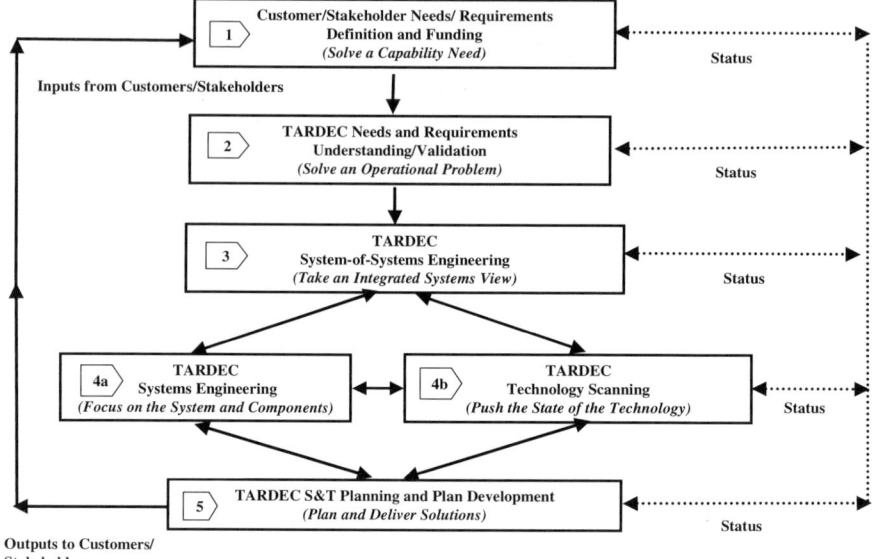

Figure 2. TARDEC strategic S&T program management model and process.

As Figure 2 indicates, the overall process follows the systems model that begins with inputs, in this case from customers and stakeholders, who seek solutions to capability and operational needs and requirements. Following the identification and input of needs, requirements, and funding implications, the process continues through various progressive steps. Needs and requirements are understood and validated, and in time solutions are identified and delivered, and planning and plans are finalized. A systems view is needed by specialists who take a system of systems engineering look at the entire system under consideration. In the context of TARDEC and Army environments, an entire system or platform would be an armored vehicle such as a tank, or a multi-purpose tactical truck used to support combat operations.

The systems perspective included in this paper and implemented by TARDEC is consistent with the development and implementation of DoD policy initiated in early 2004 with regard to systems engineering (Wiltsie, 2004). Systems engineering is now a guiding philosophy and methodology for TARDEC and others in the Army S&T community--by directive and necessity. Following system of systems engineering, organizational systems engineering and partners become the focal point of development activities in this process. Their responsibility, based on their specialized skills, is to "focus on systems" and eventually to perform various analyses at the systems level. In the case of TARDEC, and mentioned earlier, and critical to its mission are: mobility systems, survivability systems, intelligent systems, maneuver sustainment, and advanced software development. Continued in the process is the need for TARDEC to push the state-of-the-art through: (a) technology scanning (to understand what technologies are available), and (b) development of technologies. An important result of this awareness is an understanding of where technology gaps exist that prevents accomplishment of customer needs and requirements.

Finally, the process that began during an earlier period is again iterated. For the organization to deliver a solution, S&T planning and plan development must occur. S&T strategic planning, which is performed enterprise wide, results in an S&T plan and roadmaps that describe and visualize technology initiatives, responsibilities, and timelines. Throughout the process, status is provided to others including customers who have needs and requirements, and stakeholders who provide funding and program guidance.

4.2. *Phased process flow elements*

The following is a more detailed discussion of the major elements of the S&T strategic management model and process. Included are considerations for: the function, activities performed, the responsible party or parties, supporting elements, and resultant deliverables.

Customer/Stakeholder Needs/Requirements Definition and Funding. The first phase of the evolved process begins with various customers and stakeholders partners who are the focal point of and provide inputs to TARDEC's S&T strategic program management activities--their responsibilities being identifying operational needs, capability gaps and requirements, determining schedules and milestones, and establishing funding levels.

TARDEC Needs & Requirements Understanding. The second phase focuses on the actions initiated by TARDEC to understand customer needs and requirements. Also important is the identification of future requirements that may evolve from new technology developments. Required for the above activities are inputs and meetings with customers, stakeholders, funding agencies, and internal elements of TARDEC. Results of these activities are verification (where possible) and listings of identified capability gaps, and anticipated needs and requirements. Having a sanctioned, prioritized, and understood list of needs and requirements allows TARDEC to move forward to the next series of activities in this phase: developing schedules and beginning the process of identifying funding shortages. Other deliverables include a description of new capabilities, and an understanding and explanation of potential and future customer requirements that are based on recent developments. Important to this latter analysis is the concept of anticipated (technology-enabled) requirements, and the potential for possible new technology-enabled capabilities identified from recent internal and external technology breakthroughs.

System of Systems Engineering. The third phase of system activities is involves a system of systems engineering focus and perspective. Responsibilities of this team include: (a) the initial review of needs and requirements submissions and translations to technical specifications/metrics, (b) surveying and assessing candidate technologies, (c) developing and performing analyses of system (platform) alternatives, (d) developing computer-based models and performing simulations, (e) identifying roles and responsibilities for those within TARDEC and partners, (f) parsing actions to internal and external groups, (g) tracking activities and accomplishments, and (h) reporting status. Understandably, this activity is critical to the success of the S&T development and planning process. As should become evident, there are numerous activities performed during this phase that include the development of system (platform) concepts, and performing system analyses and trade studies.

Systems Engineering, Development, and Technology Scanning. The fourth phase of this system consists of two important S&T strategic functions: (a) TARDEC Systems Engineering and Development, and (b) Technology Scanning. Both functions are the primary responsibility of TARDEC's Associated Directors and their systems and components

A Strategic Science and Technology Planning and Development Process Model   351

technology development teams--namely Mobility, Survivability, Intelligent Systems, and Advanced Software Development. Specifics are included in the following figure and descriptions of activities and deliverables.

The first strategic function involves TARDEC Systems Engineering teams whose function is to develop systems and components that meet customer system requirements with regard to weight, volume, cost, and schedule parameters. A major initial activity performed by a variety of system organizational functional elements (e.g., survivability, mobility, and intelligent systems) is the definition of system and component: (a) performance requirements, and (b) concepts and alternatives. This involves the in-depth understanding of needs and requirements as defined initially by customers, and the limits of existing and known levels of technology.

Another important aspect of S&T systems engineering for new technology delivery is the "technology push" development of technologies and applications identified through the innovated efforts of organizational technologists. While these developments may not be needs or requirements driven, they can become significant contributions to customer organizations--near term or in the future. Lines of communications must exist to make these identified technologies or ideas known, with rewards and acknowledgement given to innovators.

The ultimate output of this phase of activities are the developments and transfer of technologies needed to satisfy operational gaps identified in the initial phase of the strategic S&T planning and development process--Customer/Stakeholder Needs/Requirements Definition and Funding. TARDEC does not have a production capability as mentioned earlier, but instead is a service organization.

The second strategic function involved in this phase of TARDEC's strategic S&T planning and development process involves technology scanning. This S&T scanning function, performed earlier by members of the system of systems team for full and integrated systems and platforms, is also performed by systems engineering team members oriented more to system and component-level technologies. This latter scanning activity requires a continual awareness and understanding of the state-of-the-technology relevant to ground vehicle combat and tactical support systems and components. An important aspect of scanning for existing and near-time technology developments is the identification of technology gaps that prevent or delay the development of solutions for operational gaps identified in earlier phase activities. The essence of this shared phase activity is to assist TARDEC technologists in pushing the state of the art through technology scanning (i.e. to understand what is "out there") combined with internal development, adaption, and adoption efforts.

S&T Planning and Plan Development. The <u>fifth and last phase</u> of the Strategic S&T Program Management System consists of a range of TARDEC S&T planning and plan development activities that span the life cycle of the process. Planning activities that occur during this phase include the definition of: (a) S&T maturity paths and decision points, (b) a transition plan from a present to future direction, (c) funding and partnering requirements, and (d) risk assessment. This planning is the responsibility of TARDEC's S&T Planning Team consisting of about twelve members with various functional experiences and skills that cover all functional TARDEC activities. Planning team members are responsible for facilitating, coordinating, and reporting status for all phases of the planning/replanning process, and the distribution of developed S&T plans inside and outside the organization.

Developed annual and periodic S&T plans serve as the basis for TARDEC Advanced Technology Objectives (ATOs), Advanced Technology Demonstrations (ATDs), other technology development activities, and the identification of unfunded capability and technology needs. In addition, resultant S&T plans and activities that go into its development serve as the basis for current and future budget submissions. This planning is an evolving and continuous process that results in the development of a family of roadmaps that visually portrays paths from identified customer needs and requirements to development and implement plans and schedules. Importantly, these roadmaps identify technology development issues of what, why, when (with milestones), how, and who (including partners) will participate in the process and their responsibilities.

5. Summary and Conclusion

This paper identified the ongoing need for an S&T development organization to focus and manage its mission, vision, goals, objectives, and customer/stakeholder needs within the constraints of human and physical resources, budgets, and schedules to produce value to society, the economy, and their own organizations. Also recognized was the need for continuous performance improvement that is critical to technical organizations in an era of dynamic changes, economic constraints, and international competition. A literature search identified the need for technology driven organizations to respond to value challenges by focusing on internal and external R&D collaboration and integration, and to develop a formalized approach to manage multiple layers of strategies and best practices.

Also identified was the need for S&T organizations and technology managers to improve performance by using an expanded and integrated set of initiatives such as strategic management, portfolio management, technology roadmapping, project management and engineering, and knowledge management to address the challenges they face. Technology managers must now manage and operate in an R&D environment pursuing two thrusts: (a) integrating core processes throughout the organization, and (b) implementing multiple strategy layers and best practices. A set of core processes important to achieve positive performance outcomes were identified that ranged from strategic management to learning/knowledge management. Identified best practices for technology strategy, portfolio management, and project strategy point to the need for a systematic approach to technology management.

While the focus of the research reported in this paper is on a single military S&T organization with a somewhat unique mission, numerous S&T organizations that also have the responsibility to develop and transfer technology to customers and stakeholders can benefit from the results of this paper. Other researchers and technology managers should be able to use elements of this paper and its described approach and methodology, derived strategic S&T planning and development model, and identified implications (challenges, lessons learned, and success measures and evaluation criteria) to more effectively and efficiently review, assess, and revise as needed the S&T initiatives of other organizations.

References

Bochenek, G., Kotnour T., and Ragusa J. (2005). Driving change from the middle in high-tech organizations: an approach and lessons learned from a military science and technology development organization. PICMET'05 Proceedings (to be published).

Chiesa, V. (2001). *R&D Strategy and Organisation.* Imperial College Press, London, England.

Geisler, E. (2001). *Creating Value with Science and Technology.* Quorum Books, Westport, Connecticut.

Kotnour, T. and R. Landaeta (2004). Writing reflective case studies for *the Engineering Management Journal (EMJ),* Proceedings for the American Society for Engineering Management 2004 Conference.

Kotter, J.P. (1996). *Leading Change.* Harvard Business School Press, Boston, MA.

Matheson, D. and J. Matheson (1998). *The Smart Organization.* Harvard Business School Press, Boston, Massachusetts.

Miller, W. L. and L. Morris (1999). *Fourth Generation R&D.* John Wiley & Sons, Inc. New York.

Rigby, D. (2001). Management tools and techniques: A survey. *California Management Review,* Berkeley, Winter.

Scott, G.M. (1998). The new age of new product development: Are we there yet?, *R & D Management*, 28:4.

TARDEC Information Booklet (2004). U.S. Army, Warren, MI.

Wiltsie, D.K. (2004). *Systems Engineering Policy in DoD*, Office of the Assistant Secretary of the Army Acquisition Logistics and Technology.

CHAPTER 24

INSTITUTIONAL MOT: CO-EVOLUTIONARY DYNAMISM OF INNOVATION AND INSTITUTION

Chihiro Watanabe

Department of Ind. Engineering & Management, Institute of Technology, Tokyo, Japan

Co-evolutionary dynamism between innovation and the institutional systems is decisive for an innovation driven economy. This economy may stagnate if institutional systems can not adapt to innovations and Japan's current economy is one example. Japan's system of MOT indigenously incorporates an explicit function which induces this co-evolutionary dynamism and enabled Japan achieving a conspicuous economic development.

1. Introduction — Institutional MOT

Innovative generation cycles that lead to emerging market innovation is highly dependent on institutional systems as illustrated in Figure 1.

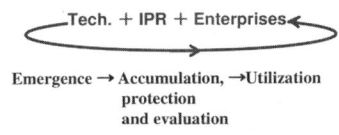

Figure 1. Innovative generation cycles.

Institutional systems are constituted by the following three dimensions (illustrated in Figure 2):

(i) National strategy and socio-economic systems,
(ii) Entrepreneurial organization and culture, and
(iii) Historical perspectives.

The prime role of MOT is management of cycles of technological innovation from the emergence to the final utilization. Activation of the innovation cycle depends largely on the co-evolution with the institutions. Change in innovation leads to change in institutions. The co-evolutionary dynamics between innovation and the institutional systems are decisive to an innovation driven economy. It may stagnate if institutional systems cannot adapt to evolving conditions as illustrated in Figure 3.

Japan's system of MOT indigenously incorporates this co-evolutionary dynamics and is respected as the most sophisticated system in the world. However, it changed to an opposite in the very last decade of the last century. This can be attributed to a conflict in the co-evolutionary dynamics due to the organizational inertia of the success story in the growth economy in an industrial society. On the one hand, a binding of the institutions (national strategy and socio-economic system, as well as entrepreneurial organization and culture) while on the other hand a historical perspective has shifted – from a mature economy into an information society.

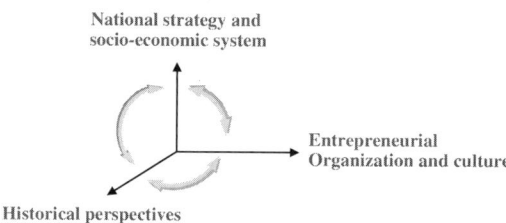

Figure 2. Three dimensions of institutions.

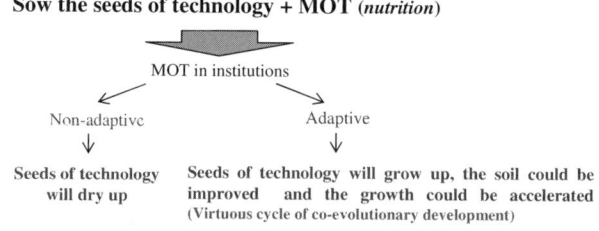

Figure 3. Science of institutional MOT — A conceptual illustration.

On the basis of this understanding and based on the three dimensional engineering approach (policy and strategy level, organizational level, and historical perspective) corresponding to the forgoing three dimensions of the institutions, an ambitious challenge was undertaken recently in Japan. It aimed at elucidating, conceptualizing and operationalizing the co-evolutionary dynamics between innovation cycle and institutions, leading to accruing global assets and thereby establishing new innovative science. This challenge, "Science of Institutional MOT" (SIMOT) aims at enabling any country with different institutions to effectively utilize its MOT.

This chapter demonstrates a concept of this challenge by providing an empirical analysis of Japan's co-evolutionary dynamics of innovation and institution, both success and failure. Section 2 outlines the significance of the co-evolutionary dynamics. Section 3 reviews Japan's co-evolutionary development cycle and leading-edge activities in reactivating such cycles. Section 4 analyzes reactivation efforts by exploring new frontiers of learning resources. Section 5 briefly summarizes the key implications and identifies the direction for further efforts.

2. Co-Evolutionary Dynamics between Innovation and Institutional Systems

2.1. *Co-evolutionary dynamics*

The innovation generation cycle is highly dependent on institutional systems. While institutional systems strongly shape emerging innovation, innovation may also change the underlying institutions leading to a self-propogating development trajectory as demonstrated in the upper side of Figure 4. Co-evolutionary dynamics may stagnate if institutional systems cannot adapt to evolving innovations as illustrated in the lower part of Figure 4. The system of MOT in most countries is suffering including Japan's system during its "lost decade" in the 1990s.

358 C. Watanabe

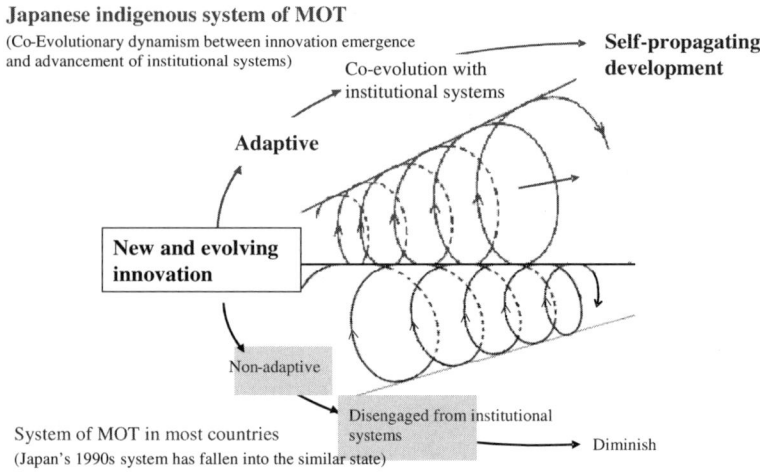

Figure 4. Co-evolutionary dynamics between innovation and institutional systems.

2.2. *Japan's indigenous system of MOT — X-Efficiency*

Japan's system of MOT indigenously incorporates an explicit function which induces this co-evolutionary dynamics and enabled Japan to achieve a conspicuous economic development.

Table 1. Trends in growth rate of GDP and TFP in Japan, the US and Germany (1960-2001) (% p.a.).

	1960 - 1973 GDP (TFP)	1975 - 1985 GDP (TFP)	1985 - 1990 GDP (TFP)	1990 - 1995 GDP (TFP)	1995 - 2001 GDP (TFP)	
USA	3.8 (1.5)	3.4 (1.0)	3.2 (0.9)	2.4 (0.9)	3.9 (1.5)	
Germany	4.6 (2.8)	3.8 (1.2)	5.2 (1.7)	1.5 (1.1)	1.1 (0.7)	
Japan	9.7 (6.2)	2.2 (1.4)	3.4 (2.8)	2.0 (-0.3)	1.8 (0.2)	
Direct effect of R&D investment	1.0	0.2	0.5	0.2	0.3	Techno-economic contribution
Indirect effect of R&D investment	2.2	0.4	1.0	0.4	0.5	Socio-economic contribution
Learning and spillover effects	3.0	0.8	1.3	-0.9	-0.6	

^a TFP and its components are estimated by the following equation:
$$\frac{\Delta TFP}{TFP} = \dot{TFP} = \underbrace{k^{-1}\eta \cdot \frac{\partial V}{\partial T} \cdot \frac{T}{V} \cdot \dot{T}}_{\text{Direct effect of R\&D investment}} + \underbrace{(1-k^{-1}\eta)\eta^2(\psi-1)k^{-1} \cdot \frac{\partial V}{\partial T} \cdot \frac{T}{V} \cdot \dot{T}}_{\text{Indirect effect}} + \underbrace{(1-k^{-1}\eta)\dot{F}_d - (1-k^{-1}\eta)\psi\eta \sum_i s_i \dot{p}_i}_{\text{Learning/spillover effects}}$$

where V: GDP; F_d: Final demand; T: technology stock; P: factor's price; s_i: $(P_iX_i)/(PV)$; X_i: factor i's quantity; η: production elasticity to cost; e: elasticity to production; $\psi = e/(1-e(1-\eta))$, k: profit ratio (=PV/C); and C: total cost.

Table 1 compares trends in growth rates of GDP and TFP (Total Factor of Productivity) in Japan, the US and Germany over the last 4 decades. It demonstrates that Japan's economic growth in an industrial society up until the end of the 1980s can be attributed to its conspicuous technological progress (TFP increase), which was enabled by a socio-cultural system (X-efficiency) rather than techno-economic contribution. Contribution of X-efficiency consisting of indirect effect of R&D investment as well as learning and spillover effects contributed more than 80% of Japan's TFP growth rate during the course of an industrial society.

3. Japan's Co-Evolutionary Development Cycle

3.1. *Japan's co-evolution and development cycle — Learning and assimilation*

Japan's conspicuous X-efficiency can largely be attributed to its intensive cumulative learning efforts with the following unique function:

(i) Motivated by xenophobia,
(ii) Cumulative learning stimulates assimilation of spillover knowledge,
(iii) Rich in curiosity, smart in assimilation, thorough in learning and absorption.

Based on this unique function, Japan's system of MOT successfully achieved a co-evolutionary development by learning and assimilating advanced innovation from the US and Europe, as well as advancement of its own institutional systems (as illustrated in Figure 5a). The US and Europe were inspired and improved their innovation and systems, from which efforts Japan learned and further developed its systems leading to a cycle as illustrated in Figure 5b.

Japan's cumulative learning minimized the impediments of the organizational inertia and it also accelerated assimilation of spillover technology.

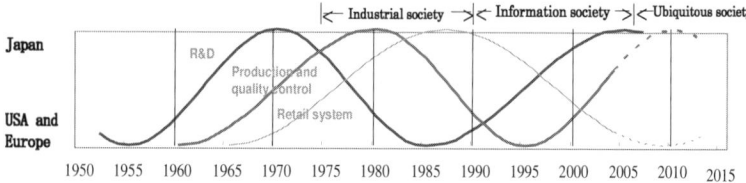

Figure 5a. Co-evolutionary development cycle of Japan's system of MOT.

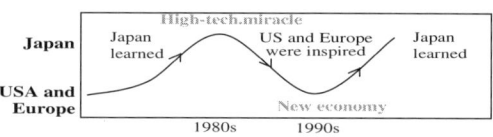

Figure 5b. Co-evolutionary development cycle by mutual inspire.

3.2. Contrast between co-evolution and disengagement

However, such learning efforts produced a negative outcome in the 1990s as demonstrated in Figure 6 due to:

- Xenophobia,
- Cumulative learning stimulates assimilation of spillover knowledge,
- Rich in curiosity, smart in assimilation, thorough in learning and absorption.

Facing such circumstances, the new innovation challenge in exploring new frontiers of learning resources has become indispensable (illustrated in Figure 7).

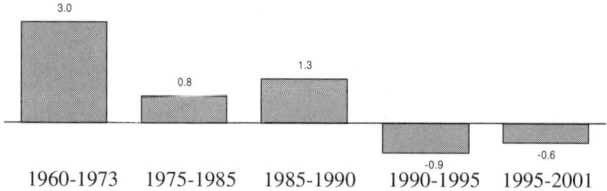

Figure 6. Trend in the contribution of learning to TFP and the consequent GDP growth rate in Japan (1960-2001) (% p.a.).

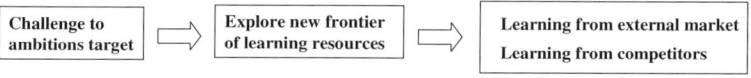

Figure 7. New innovative challenges for reactivation of potential learning ability.

Exploration of such new frontier can be expected by technological diversification effort, thereby gaining new experiences and assimilation of spillover technology. Figure 8 demonstrates correlation between technological diversification in 1995-1998 and operating income to sales (OIS) in 1999-2002 in Japan's leading electrical machinery firms. Figure 8 demonstrates that relatively smaller firms as Canon, Sony, Sharp and Sanyo endeavored in technological diversification efforts which resulted in increase their OIS.

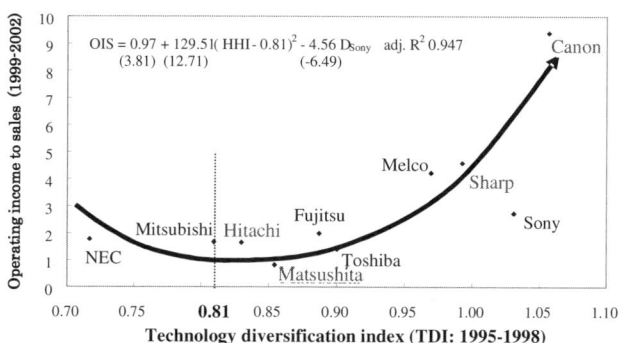

Figure 8. Correlation between technological diversification and OIS in Japan's top 10 electrical machinery firms (1999-2002).

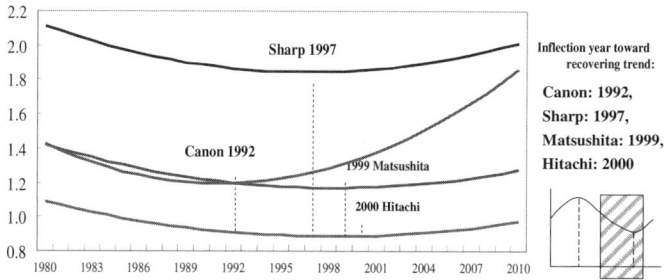

Figure 9. Trend in learning coefficients in Japan's four leading electrical machinery firms (1980-2003).

Table 2. Correlation between price of technology and governing factors of learning coefficients in Japan's leading machinery firms (1980-2003).

$$\ln P = \ln B - [(\alpha - \beta) + \beta \cdot b_1 \cdot t + \beta \cdot b_2 \cdot t^2 + \beta \cdot b_3 \cdot t^3] \ln T$$

	$\ln B$	$(\alpha - \beta)$	$\beta \cdot b_1$	$\beta \cdot b_2$	$\beta \cdot b_3$	adj. R^2	D W	1980-$b_1/{2 \cdot b_2}$
Matsushita	-9.94 (-5.61)	1.45 (4.92)	-0.030 (-5.01)	0.0008 (4.66)	-4.2*10⁻⁶ (-1.13)	0.994	2.81	1999
Hitachi	-7.47 (-3.43)	1.11 (2.99)	-0.024 (-2.73)	0.0006 (3.64)	7.9*10⁻⁶ (2.13)	0.997	2.34	2000
Canon	-7.95 (-3.00)	1.47 (2.31)	-0.045 (-2.05)	0.0019 (2.62)	-3.4*10⁻⁵ (-3.37)	0.989	1.98	1992
Sharp	-11.80 (-2.53)	2.15 (-2.07)	-0.034 (-11.42)	0.0010 (10.95)	-2.5*10⁻⁵ (-2.57)	0.989	2.01	1997

Learning Coefficient: $\lambda = -\frac{\partial \ln P}{\partial \ln T} = (\alpha - \beta) + \beta \cdot b_1 \cdot t + \beta \cdot b_2 \cdot t^2 + \beta \cdot b_3 \cdot t^3 \approx (\alpha - \beta) + \beta \cdot b_1 \cdot t + \beta \cdot b_2 \cdot t^2$ ⇒ Learning Coefficients of the leading firms follow concave trend with minimum level at time $t = -\frac{b_1}{2b_2}$

Such technological diversification efforts have born fruit recently in certain high-technology firms in improving their learning coefficients as demonstrated in Figure 9. Figure 9 demonstrates that while learning coefficients in Japan's leading electrical machinery firms continued to decrease in the 1980s and the early 1990s, Canon changed to an increasing trend from 1992, followed by Sharp in 1997. Contrary to these relatively smaller firms, while larger firms suffered longer years of decreasing learning coefficients, Matsushita changed to a slightly increasing trend from 1999 (similarly Hitachi from 2000). These demonstrate that cumulative learning efforts during the course of the lost decade in the 1990s have born fruit in recovering Japan's indigenously well functioning learning dynamics.

4. Reactivation of Learning by Exploring New Frontier

4.1. *Canon's technological diversification strategy*

Among leading high-technology firms that succeeded in revitalization of learning ability, Canon demonstrates a conspicuous achievement which can be attributed to the exploration of new frontiers in learning resources based on its unique technological diversification strategy. This strategy endeavors to develop new functionality by stimulating an inter-field technology spillover as demonstrated in Figure 10. This new functionality leverages co-evolution between indigenously developed or assimilated core technologies, and application of these technologies to new fields leads to a maximum return on the R&D investment.

Institutional MOT: Co-Evolutionary Dynamism of Innovation and Institution   363

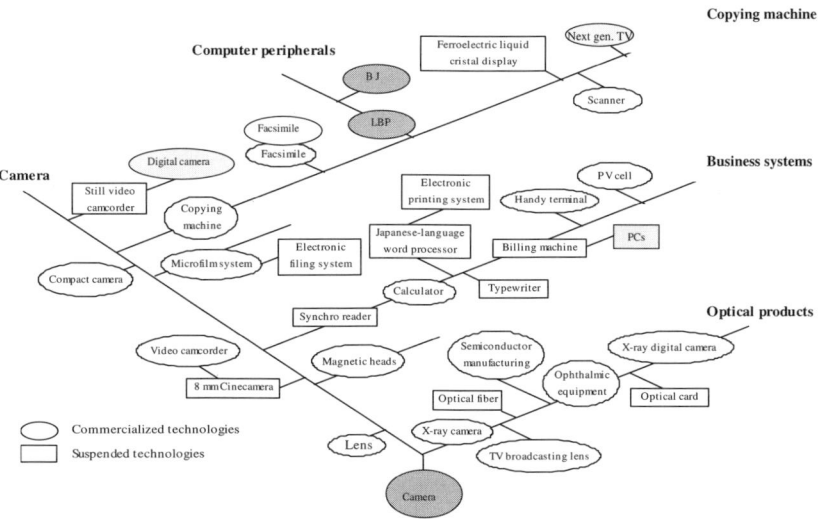

Figure 10. Canon's technological diversification paths demonstrated by its inter-technology web.

As a consequence of this technological diversification strategy, Canon constructed co-evolutionary trajectory between printers and PCs as demonstrated in Figure 11 leading to a virtuous cycle between them as demonstrated in Figure 12.

This virtuous cycle that is shared between price decrease, stock induction of technology, functional development, and price recovery by means of the addition of high-value innovative printers (depicted in Figure 13).

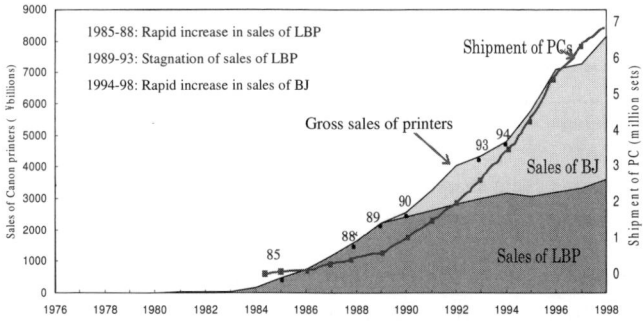

Figure 11. Co-evolutionary trajectory between Canon printers and personal computers (1976-1998).

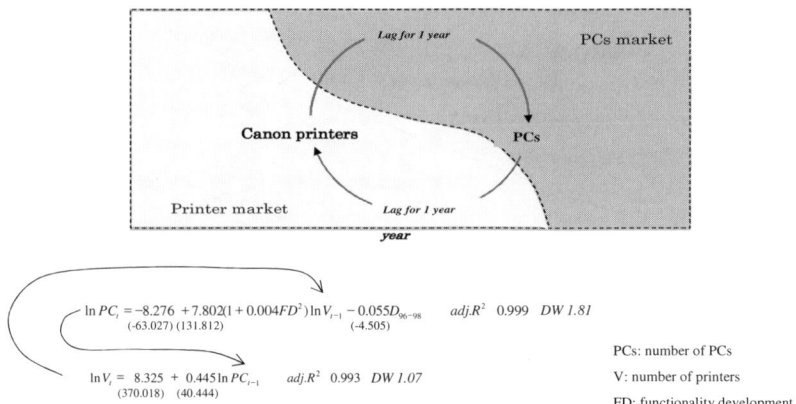

Figure 12. Virtuous cycle between Canon printers and PCs (1986-1998).

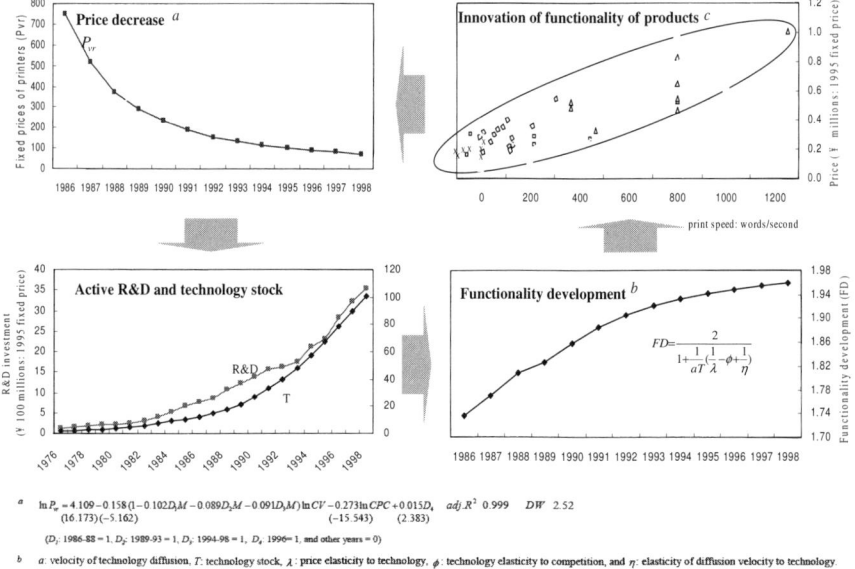

Figure 13. Virtuous cycle shared between price decrease, technological stock, functional development, and price recovery in Canon's printers.

In addition, Canon constructed a virtuous cycle between its printers and PCs producers such as NEC, Fujitsu, Sony, Toshiba as well as Dell and IBM, directly or through HP with OEM of LBP.

Canon's efforts in exploring new resources in learning by means of its technological diversification strategy suggest that one should reconstruct co-evolutionary dynamics between innovation and institutional systems in an information society:

(i) Challenge to ambitious target,
(ii) Learning from external market,
(iii) Learning from competitors, and
(iv) Harmony between cooperation and competition.

5. Conclusion: Accruing Japan's System of MOT to Global Assets

The co-evolutionary dynamics between the emergence of innovation and the advancement of institutional systems is decisive for an innovation-driven economy. A noteworthy surge in new innovation in leading edge activities in recent years by leading high-technology firms in Japan can be attributed to the resonance between indigenous strength in the Japanese firms incorporated during the course of an industrial society. The effects of cumulative learning actively absorbed their competitors in an information society and assimilated them in their business model. This suggests the significance of conceptualizing and operationalizing the co-evolutionary dynamics that accrue to global knowledge assets as illustrated in Figure 14.

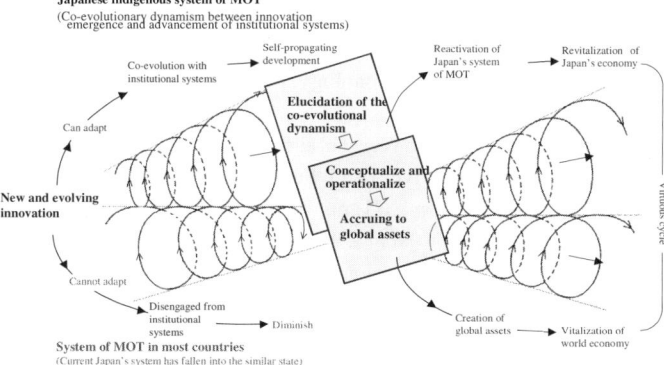

Figure 14. Co-evolutionary dynamics between the emergence of innovation and the advancement of institutional systems.

References

Binswanger, H. and Ruttan, V.W. (1978). *Induced Innovation. Technology, Institutions, and Development.* John Hopkins University Press, Baltimore.

Kondo, R. and Watanabe, C. (2003). The VIrtuous Cycle between Institutional Elasticity, IT Advancement and Sustainable Growih: Can Japan Survive in a Information Society? *Technology in Society*, 25, pp. 319-335.

Nelson, R.R. and Sampat, B.N. (2001). Making Sense of Institutions as a Factor Shaping Economic Performance. *Journal of Economic Behavior & Organization*, 44, pp. 31-54.

North, D.C. (1990). *Institutional Change and Economic Performance.* Cambridge University Press, Cambridge.

North, D.C. (1994). Economic Performance through Time. *The American Economic Review*, 84, pp. 359-368.

Ruttan, V.W. (2001). *Technology, Growth and Development: An Induced Innovation Perspective.* Oxford University Press, New York.

Watanabe, C., Santoso, I. and Widyanti, T. (1991), *The Inducing Power of Japanese Technological Innovation.* Pinter Publishers, London

Watanabe, C., and Clark, T. (1991). Inducing Technological Innovation in Japan. *Journal of Science & Industrial Research*, 50, pp. 771-785.

Watanabe, C. (1995). The Feedback Loop between Technology and Economic Development: An Examination of Japanese Industry. *Technological Forecasting and Social Change*, 49, pp. 127-145.

Watanabe, C., Kondo, R., Ouchi, N. and Wei, H. (2003). Formation of IT Features through Interaction with Institutional Systems: Evidence of Unique Epidemic Behavior. *Technovation*, 23, pp. 205-219.

Watanabe, C, and Ane, B.K. (2003). Co-Evolution of Manufacturing and Service Industry Functions. *Journal of Services Research*, 3, pp. 101-118.

Watanabe, C., Kishioka, M. and Nagamatsu, A. (2004). Resilience as a Source of Survival Strategy for High Technology Firms Experiencing Megacompetition. *Technovation*, 24, pp. 139-152.

Watanabe, C., Kondo, R. Ouchi, N. Wei, H. and Griffy-Brown, C. (2004). Institutional Elasticity as a Significant Driver of IT Functionality Development. *Technological Forecasting and Social Change*, 71, pp. 723-750.

Watanabe, C., Matsumoto, K. and Hur, J.Y. (2004). Technological Diversification and Assimilation of Spillover Technology: Canon's Scenario for Sustainable Growth. *Technological Forecasting and Social Change*, 71, pp. 941-959.

CHAPTER 25

FIRMS WITH ADAPTABILITY LEAD A WAY TO INNOVATIVE DEVELOPMENT

Akihisa Yamada and Chihiro Watanabe

Department of Ind. Engineering & Management, Institute of Technology, Tokyo, Japan

It goes without saying that the majority of the results of R&D have not necessarily commercialized. It's also pointed that R&D activities need to be diversified to produce high-added value products and services. However, firms face a dilemma about the simultaneous solution in efficiency and diversification of R&D because of the trade-off between them. In order to find the optimal solution to these questions, firms must choose their R&D subjects carefully. This will involve adoption of new and emerging R&D subjects and suspension of existing or irrelevant ones, derived from continually changing market needs and technology trends. This paper discusses the R&D activities of Japanese firms, based on the hypothesis that firms need to be adaptable when selecting R&D subjects to produce products and services effectively and continuously. R&D adaptability is defined as the capability to adapt businesses to the environment by responding to technological opportunities flexibly and by selecting R&D subjects correctly with self-assessment. Comparative empirical analyses are conducted focusing on the co-evolution between such adaptability and the institutions in selected countries. In addition, analyses are also conducted to identify the institutional factors governing the competitiveness of these countries.

1. Introduction

Many scholars and business people agree that although many firms invest huge funds in R&D, most of their investment doesn't lead to commercial success or contribute to commercialization based on the results of R&D (Andrew and Sirkin, 2003, Balachandra and Friar, 1997, Stevens and Burley, 1997). In other words, almost all R&D outcomes will die out at the Death Valley and the Darwinian sea. Therefore, it is a critical issue for firms over the world to improve R&D efficiency. Compared to other advanced countries, Japanese firms are particularly criticized that their R&D investments don't contribute to the growth rate of TFP sufficiently during the 1990s (OECD, 2001).

Nowadays, businesses have to satisfy customer's demands for not only low-price but also high added value products such as digital appliances and hybrid car. Those firms which can develop and integrate various technologies into a single product can accomplish a high revenue and growth rate (Gambardella and Torrisi, 1998, Gemba and Kodama, 2001, Granstrand *et al.*, 1997, Markides and Williamson, 1994, Watanabe *et al.*, 2003). Gambardella and Torrisi (1998) proposed that firms which focus on fewer businesses demonstrate better performance while firms which maintain greater technological diversification perform better. According to these arguments, the improvement of both R&D efficiency and technological diversification is required to accomplish a firm's growth and improvement of its revenue. However, managers will face a serious dilemma about the simultaneous achievement of efficiency and diversification of R&D because of the trade-off between them.

We hypothesize that R&D adaptability is required for firms to improve their R&D efficiency and technological diversification from the perspective of selecting R&D subjects. Empirical analyses are conducted to compare such adaptability and the external environments, especially the institutions of selected countries, which co-evolve with adaptability. In addition, analyses are also conducted to identify the institutional factors that affect the competitiveness of these countries.

2. Selection of R&D Subjects and R&D Adaptability

In terms of the operational level of R&D, the selection of R&D subjects is important for firms to improve the R&D efficiency and achieve technological diversification. The selection of R&D subjects includes adoption of new subjects and abatement of existing ones. This is based upon thorough consideration of the impact on the market, lead time required for commercialization, consistency with the corporate strategy, expectations of the market, and the effects on motivations and employment of researchers due to the change of R&D subject. From the viewpoint of efficiency and accomplishment, it is necessary to adopt R&D subjects that lead to advancement of commercialization and growth in profit as soon as possible. In addition, it is required to terminate current R&D subjects which obviously will never contribute to the commercialization or profit, and transfer the surplus gained by cutting down those subjects to others with higher potential. From the viewpoint of diversification, it is requisite to adopt more R&D subjects from technological fields while contending with the practice of business strategy and technological strategies.

Contrary to Japanese firms, US and European firms aggressively make use of M&A to select business fields. Moreover, they also facilitate actively the collaboration with public research institutions such as universities. The high mobility of human resources contributes to business environment where the adoption of new R&D subjects as well as the abatement of current ones can be carried out more easily. This implies the difference of government policies, corporate strategies and policies and cultures among all the countries have significant influence on the efficiency and diversification of R&D.

With the forgoing background, it is considered to be a requisite for firms to embody a certain capability to select appropriate R&D subjects in order to maintain the growth of profit and the progress of enterprise value. This capability implies the capability to adapt businesses to the environment by responding to technological opportunities flexibly and by selecting R&D subjects correctly with self-assessment. A business model for which such adaptation capability is embedded can be resilient against the exogenous environmental change.

3. R&D Adaptability and Institutions

It is generally realized that R&D adaptability should be considered from two viewpoints. One is firm's internal factors and the other is the firm's external environment. These two viewpoints don't exist independently but interact with each other. If the government offers preferable environment and institutional system, it is easier for a firm to adapt itself to external environment and as a result it can focus on its own projects/business effectively. On the other hand, a firm will have no choice other than making more effort to adapt to the external environment if the country is under unstable social condition or excess regulations. Firms can co-evolve with an external environment positively or negatively.

From the internal viewpoint of firms, subjects of analyses will be time series of actual R&D subject, reasons of selection, laboratories and researchers. From the external viewpoint of environment, each country's institutions, one of dominant factor affecting R&D adaptability, and the chronological trends will be subjects of analyses.

Many discussions have taken place and many definitions and opinions regarding institutions exist, but a clear answer has not yet been available (Williamson, 2000). These discussions concerning the definition and concept of institutions can chiefly be classified into three categories (Hall and Taylor, 1996, Kono, 2002). The first one is the economic view which considers institutions as constraints or rules and is helpful to explain the institutional changes. The second one is the sociological view which considers institutions as cultural and cognitive frame and is helpful to explain the continuousness of institutions. The last one is the historical view which treats institutions with "path dependence" as one of the key concepts and discusses how the previous historical processes affect on the creation, retention and declination of institutions. As these views are developed basically from each academic field, the overall picture of institutions has never been captured.

4. Methodology

Based on the foregoing discussion, first institutional factors which affect R&D adaptability are analyzed. In addition, the growth and maturity of IT and manufacturing technology that contributes to the economic growth and the components extracted as institutions are analyzed. Institutions are considered to be constituted by the following three dimensions: (i) national strategy and socio-economic system,

(ii) entrepreneur organization and culture, and (iii) historical perspectives. The three dimensions of institutions are postulated by the 21st Century COE (Center of Excellence) SIMOT (The Science of Institutional Management of Technology) program of Tokyo Tech (Tokyo Institute of Technology The 21st Century Program, 2004). The analysis is conducted with indicators of these three dimensions to extract the characteristics of each dimension by means of principal component analysis (PCA). The analysis encompasses 40 countries to represent different areas over the world and is composed of OECD, the original countries of ASEAN, NIEs[1], and BRICs as shown in Table 1.

Table 1. Forty countries and its regions in this analysis.

Europe		Asia	North America
1 Austria	13 Luxembourg	25 China	35 Canada
2 Belgium	14 Netherlands	26 India	36 Mexico
3 Czech	15 Norway	27 Indonesia	37 USA
4 Denmark	16 Poland	28 Japan	
5 Finland	17 Portugal	29 Korea	
6 France	18 Russia	30 Malaysia	
7 Germany	19 Slovak	31 Philippines	Others
8 Greece	20 Spain	32 Singapore	38 Australia
9 Hungary	21 Sweden	33 Taiwan	39 Brazil
10 Iceland	22 Switzerland	34 Thailand	40 New Zealand
11 Ireland	23 Turkey		
12 Italy	24 UK		

The first dimension of the institutions is national strategy and socio-economic system. In this dimension, 9 variables are selected in terms of national strategy, social system and economic system as tabulated in Table 2. National strategy represents the fundamental policy of government and nation like democracy, laws and regulations. Social system indicates the physical infrastructure, human resources and education. Economic system includes not only GDP, the most basic economic indicator, but also the level of activeness of trade and R&D.

---

[1] NIEs include Hong Kong and former Yugoslavia, but these countries are excluded in this analysis.

Table 2. Indices representing national strategy and socio-economic system (1st dimension).

1. National strategy and socio-economic system		
Detailed categories	Variables	Sources
1.1 National stategy		
1.1.1 Democracy	Need for economic and social reforms	IMD (2005)
1.1.2 Constitution, Law, Regulation, Standard and Manner	Efficiency of legal framework	WEF (2005a)
1.1.3 Separation of the three powers of administration, legislation and judicature	Risk of political instability	IMD (2005)
1.2 Social system		
1.2.1 Education system	Quality of the educational system	WEF (2005a)
1.2.2 Employment system	Skilled labor	IMD (2005)
1.2.3 Infrastructure investment	Overall infrastructure quality	WEF (2005a)
1.3 Economic system		
1.3.1 GDP and GDP per capita	GDP PPP per capita	IMD (2005)
1.3.2 Trade-based nation, Export and Import	Trade to GDP ratio	IMD (2005)
1.3.3 Tech-based nation, ICT and Government ICT	Total expenditure on R&D (%)	IMD (2005)

The second dimension is entrepreneur organization and culture. In this dimension, 12 variables are selected in terms of strategy and business model, employment, promotion and training, structure, doctrine, philosophy, and ethics as listed in Table 3. Strategy and business model imply the business strategy and capabilities used to implement its strategy into an operational level. Employment, promotion and training indicate firms' human resources management, and structure indicates organizational structure. Doctrine, philosophy and ethics reflect firms' governing values and ethics.

The last dimension is historical perspectives. 12 variables are chosen for this dimension to indicate geographical structure, culture and tradition, state of development and paradigm and phase of industrial society as tabulated in Table 4. Geographical structure shows the geographical structure including population and the variety of races. Culture and tradition indicate each country's customs, religion and culture. States of development and paradigm and phase of industrial society represent the development and maturity of countries, respectively.

Table 3. Indices representing entrepreneurial organization and culture (2nd dimension).

2. Entrepreneurial organization and culture		
Detailed categories	Variables	Sources
2.1 Strategy and business model		
2.1.1 Vision and Business strategy	Nature of competitive advantage	WEF (2005a)
2.1.2 Business model and Market policy	Extent of marketing	WEF (2005a)
2.1.3 R&D and ICT	Capacity for innovation	WEF (2005a)
2.2 Employment, promotion and training		
2.2.1 Appointment	Hiring and firing practices	WEF (2005a)
2.2.2 Promotion	Reliance on professional management	WEF (2005a)
2.2.3 Traininig	Extent of staff training	WEF (2005a)
2.3 Structure		
2.3.1 Entrepreneurial organization	Large corporations	IMD (2005)
2.3.2 Affiliated firms	Small and medium-size enterprises	IMD (2005)
2.3.3 Foreign capital	Ease of hiring foreign labor	WEF (2005a)
2.4 Doctrine, philosophy and ethics		
2.4.1 Business doctrine	Degree of costomer orientation	WEF (2005a)
2.4.2 Philosophy and Ethics	Ethical behavior of firms	WEF (2005a)
2.4.3 Corporate governance	Efficasy of corporate boards	WEF (2005a)

Table 4. Indices representing historical perspectives (3rd dimension).

3. Historical perspectives		
Detailed categories	Variables	Sources
3.1 Geographical structure		
3.1.1 Geopolitical environment	Urban population	IMD (2005)
3.1.2 Population	Total population	WEF (2005a)
3.1.3 Homogeneity and Heterogeneity	Gini index	World Bank (2005)
3.2 Culture and tradition		
3.2.1 Culture, Custom and Common idea	National culture	IMD (2005)
3.2.2 National spirit, Moral, Ethic, Manners and Customs	Protectionism	IMD (2005)
3.2.3 Religion	Justice	IMD (2005)
3.3 State of development		
3.3.1 Rapid economic growth	Real GDP growth per capita	WEF (2005a)
3.3.2 Mature economy	Life expectancy at birth	WEF (2005a)
3.3.3 Diminish population and Aging	Population over 65 years	WEF (2005a)
3.4 Paradigm and phase of industrial society		
3.4.1 Industrial society, Information society and Post-information society	Market environment ICT	WEF (2005b)
3.4.2 Heavy and chemical industrial structure	Productivity in industries PPP	IMD (2005)
3.4.3 Knowledge-intensified industrial structure	Productivity in services PPP	IMD (2005)

5. Results

5.1. *Results of PCA*

Table 5 demonstrates the results of PCA by utilizing 9 variables of the 1st dimension in 40 countries. Cumulative variance proportion is 78.8% by taking three principal components (PC) with Eigen values over 1.0. By evaluating the high level of loading of variables in PCs, 3 PCs: PC_{11}, PC_{12} and PC_{13} can be interpreted as "quality of traditional development base," "manufacturing oriented socio-economic system" and "commodity trade dependency," respectively.

Similar to the 1st dimension, the result of PCA conducted by utilizing 12 variables of the 2nd dimension is summarized in Table 6. Cumulative variance proportion is 74.4% by taking two PCs with Eigen values over 1.0. By evaluating the high level of loading of variables in PCs: PC_{21} and PC_{22} can be interpreted as "high qualified managerial system" and "liquidity of work force," respectively.

Similarly, PCA is conducted by utilizing 12 variables of the 3rd dimension and its result is summarized in Table 7. Cumulative variance proportion is 74.7% by taking three PCs with Eigen values over 1.0. By evaluating the high level of loading of variables in PCs, PC_{31}, PC_{32} and PC_{33} can be interpreted as "productivity seeking nationality," "elasticity of heterogeneous nations" and "brain resources supply potential."

Table 5. Statistical results of selected principal components (1st dimension).

Variables	Loading of variables		
	PC_{11}	PC_{12}	PC_{13}
Need for economic and social reforms	0.27	0.83	0.02
Efficiency of legal framework	0.92	-0.15	0.04
Risk of political instability	0.82	-0.18	0.09
Quality of the educational system	0.85	0.23	0.05
Skilled labor	0.54	0.54	-0.32
Overall infrastructure quality	0.91	-0.21	-0.01
GDP PPP per capita	0.78	-0.33	0.11
Trade to GDP ratio	0.32	0.22	0.86
Total expenditure on R&D (%)	0.78	-0.05	-0.43
Eigen value	4.75	1.29	1.05
% of common variance explained	52.83	14.35	11.63

Table 6. Statistical results of selected principal components (2nd dimension).

Variables	Loading of variables	
	PC_{21}	PC_{22}
Nature of competitive advantage	0.86	-0.15
Extent of marketing	0.87	-0.16
Capacity for innovation	0.87	-0.17
Hiring and firing practices	0.18	0.73
Reliance on professional management	0.93	0.03
Extent of staff training	0.96	-0.08
Large corporations	0.65	0.01
Small and medium-size enterprises	0.83	0.05
Ease of hiring foreign labor	0.09	0.74
Degree of customer orientation	0.92	-0.06
Ethical behavior of firms	0.95	0.13
Efficacy of corporate boards	0.87	0.19
Eigen value	7.70	1.23
% of common variance explained	64.14	10.24

Table 7. Statistical results of selected principal components (3rd dimension).

Variables	Loading of variables		
	PC_{31}	PC_{32}	PC_{33}
Urban population	0.65	-0.14	-0.44
Total population	-0.53	0.21	0.67
Gini index	-0.69	0.14	-0.38
National culture	0.09	0.80	-0.38
Protectionism	0.78	0.43	-0.10
Justice	0.82	0.42	0.20
Real GDP growth per capita	-0.62	0.31	0.08
Life expectancy at birth	0.89	-0.18	-0.05
Population over 65 years	0.72	-0.48	0.15
Market environment ICT	0.68	0.50	0.29
Productivity in industries PPP	0.87	0.04	0.11
Productivity in services PPP	0.88	-0.14	0.03
Eigen value	6.14	1.71	1.11
% of common variance explained	51.13	14.29	7.40

5.2. *Multiple regression analysis*

Based on the forgoing PCA, a cross-country multi-regression analysis on 40 countries is conducted between the 8 PCs of institutional factors. The factors are selected by the forgoing PCA, the development of MT (Manufacturing technology) and IT (Information technology) - which are represented by Production Process Sophistication (IMD, 2005) - and NRI (Network Readiness Index, WEF, 2005b), respectively. By means of Backward Eliminating Method (BEM) with a 5% significant level criteria, the PCs which have significant influence on both MT and IT are identified. Table 8 demonstrates its statistical results.

Looking at Table 8, it's noted that the structure of institutional systems that influences the development of MT and IT is represented by 6 PCs out of all 8 PCs as follows.

Table 8. Multiple regression analysis between PCs of institutional factors and leading technologies.

	Explained variable	Constant	PC_{11}	PC_{12}	PC_{13}	PC_{21}	PC_{22}	PC_{31}	PC_{32}	PC_{33}	adj. R^2
MT	Production process sophistication	4.952	0.293	0.108		0.719	0.075		-0.239		0.938
		(130.73)	(2.49)	(2.16)		(6.27)	(2.20)		(-4.83)		
IT	NRI	0.705	0.446	-0.092	-0.113		0.075		0.102		0.935
		(26.64)	(4.93)	(-2.55)	(-3.35)		(2.20)		(2.65)		

Note: Positive (+) and negative (-) indicate PCs which provide significant impacts on respective technologies. While + indicates that high-score countries provide positive impact in inducing technology, - indicates that low-score countries provide positive impact in inducing technology.

5.2.1. Manufacturing technology (MT)

Four PCs as "quality of traditional development base" (PC_{11}), "manufacturing oriented socio-economic system" (PC_{12}), "high qualified managerial system" (PC_{21}), and "liquidity of work force" (PC_{22}) contribute to the advancement of MT leading to increase the global competitive advantage of manufacturing industry. However, "elasticity of heterogeneous nations" (PC_{32}) impedes this development. Institutional factors inducing Japan's high level MT can be attributed to the well-balanced level of PC_{11} (ranked as 15 out of 40 countries), PC_{12} (similarly ranked 14), as well as PC_{21} (ranked as 11), and its non-heterogeneous nations (ranked as 37 in PC_{32}).

5.2.2. Information technology (IT)

Similar to MT, "quality of traditional development base" (PC_{11}) and "liquidity of work force" (PC_{22}) induce IT development. Contrary to MT, "elasticity of heterogeneous nations" (PC_{32}) also contributes to this development in the advantageous countries, while "manufacturing oriented socio-economic system" (PC_{12}) impedes this development in these countries. Furthermore, "commodity trade dependency" (PC_{13}) also impedes this development in the advantageous countries.

6. Conclusion and Future Works

The analyses extracted and analyzed the institutional systems comprehensively by adding factors such as historical perspectives and cultures that have been relatively abstract from the perspective of R&D adaptability (i.e. the capability to deal with the trade-off between R&D efficiency and diversification of technology). Based on these extracted factors, institutional factors that influence MT and IT that played a leading role in an industrial society and in an information society, respectively are identified.

The factors that affect MT and IT most positively are "high qualified managerial system" (PC_{21}) "quality of traditional development base" (PC_{11}), respectively. However, both "quality of traditional development base" (PC_{11}) and "liquidity of work force" (PC_{22}) provide a positive impact on MT and IT. These indicators, from the perspectives of economics and management theory, demonstrate the level of performance of MT and IT based on the regulations of the government as well as firms' strategies and operations.

The noticeable point of the analyses is the impact of "elasticity of heterogeneous nations" (PC_{32}) on MT and IT. While "elasticity of heterogeneous nations" (PC_{32}) influences MT negatively, it affects IT positively. In other words, MT depends on the concentration and integration based on the homogenous institutions while IT emphasizes more on relatively heterogeneous and diversified ideas. MT can work well in the case of integrated architecture product or service, while IT works better on module architecture. Moreover, this correlation affects the competitiveness of each country (Fujimoto and Oshika, 2006).

This chapter conducted analyses to elucidate the relationships between external environment (including national-level institutions) and

R&D adaptability from a macro perspective. These analyses and findings urge a further analysis on internal R&D adaptability.

References

Andrew, J.P. and Sirkin, H.L. (2003). Innovating for cash, *Harvard Business Review,* 81, 9, pp. 76-83.

Balachandra, R. and Friar, J.H. (1997). Factors for success in R&D projects and new product innovation: a contextual framework, *IEEE Transactions on Engineering Management,* 44, 3, pp. 276-287.

Fujimoto, T. and Oshika, T. (2006). Empirical analysis of the hypothesis of architecture-based competitive advantage and international trade theory, MMRC Discussion paper, University of Tokyo. Retrieved July, 24, 2006, from http://www.ut-mmrc.jp/dp/index.html.

Gambardella, A. and Torrisi, S. (1998). Does technological convergence imply convergence in markets? Evidence from the electronics industry, *Research Policy,* 27, 5, pp. 445-463.

Gemba, K.and Kodama, F. (2001). Diversification dynamics of the Japanese industry, *Research Policy,* 30, 8, pp. 1165-1184.

Granstrand, O., Patel, P. and Pavitt, K. (1997). Multi-technology corporations: Why they have "distributed" rather than "distinctive core" competencies, *California Management Review,* 39, 4, pp. 8-25.

Hall, P.A. and Taylor, R.C. (1996). Political science and the three new institutionalisms, *Political Studies,* 44, 5, pp. 936-957.

IMD. (2005). The World Competitiveness Yearbook 2005, Lausanne, Switzerland.

Kono, M. (2002). Institutions: Theory and model for social science, University of Tokyo Press, Tokyo, Japan (in Japanese).

Markides, C. and Williamson, P.J. (1994). Related diversification, core competences and corporate performance. *Strategic Management Journal,* 15, pp. 149-165.

OECD. (2001). Science, technology and industry outlook, OECD Publications, Paris.

Stevens, G. and Burley, J. (1997). 3,000 raw ideas equals one commercial success. *Research Technology Management,* 40, pp. 16-27.

Tokyo Institute of Technology the 21st Century Program SIMOT (Science of Institutional Management of Technology). (2004). Science of Institutional MOT - Elucidation of Japan's Co-evolutionary Dynamism Accruing to Global Assets, Retrieved November, 13, 2005, from http://www.me.titech.ac.jp/coe/detail_e.pdf.

Watanabe, C., Hur, J.Y. and Matsumoto, K. (2003). Technological Diversification and Firm's Techno-economic Structure: An Assessment of Cannon's Sustainable Growth Trajectory, *Technological Forecasting and Social Change,* 72, 1, pp. 11-27.

Williamson, O.E. (2000). The new institutional economics: taking stock, looking ahead, Journal of Economic Literature, 38, 3, pp. 595-613.
World Bank. (2005). World Development Indicators 2005, World Bank, Washington DC.
World Economic Forum (WEF). (2005a). The global competitiveness report 2005-06, Palgrave Macmillan, New York, NY.
World Economic Forum (WEF). (2005b). The global information technology report 2004-05, Palgrave Macmillan, New York, NY.

CHAPTER 26

TOWARD PROJECT STRATEGY TYPOLOGIES: CASES IN PHARMACEUTICAL INDUSTRY

Peerasit Patanakul*, Aaron J. Shenhar*,
Dragan Z. Milosevic**, and William Guth***

*Stevens Institute of Technology, New Jersey, USA; **Portland State University, Oregon, USA; ***New York University, New York, USA

As part of our continuous studies on the strategic approach to project management, this study focuses on combining the two worlds— project and business— by deriving specific and appropriate project strategies in line with business strategies or objectives. In this preliminary study, we centered our attention on the strategies of projects in pharmaceutical industry. The results of this research illustrated that project strategies existed in an implicit way and were used in managing projects for better business results. Two types of project strategies were found. One is the "Product advantage/Schedule driven strategy," which leads a project team to focus on the product quality/competitive advantage and project schedule. The other type is the "Efficiency driven strategy," focusing on the project schedule to increase the efficiency and productivity of the company. These research results are very encouraging since they are a basis for our future study on project strategy typologies. In addition they can be used in building the foundation for adding project strategic planning to traditional project planning practice, which are still focused on operations and mostly on "getting the job done."

1. Introduction

Although project management has become a central activity in many organizations, it is rarely approached as a critical business function. Project success is often perceived as meeting time and budget goals, and project planning is focused to a large extent on building scope, budgets, and schedules. However, to be effective, projects must be in line with the overall business strategy and projects should be perceived as engines driving business results (Cooper *et al.* 1998; Pennypacker and Dye 2002). With this perception, project management should shift its focus from the traditional operational view of "getting the job done" to "a strategic approach to project management." To strategically lead projects, project managers should develop a strategic mindset. This means that the project managers should understand the strategic directions of their organizations so that they can lead projects accordingly. They should also develop an explicit and formal project strategy that they can use to guide their project management efforts. This project strategy should be aligned with the organization's business strategy.

While a number of authors have written in supporting of this normative theory, it has not yet been subject to systematic empirical examination, development or testing. This study is a part of our longer-term empirical study dealing with the strategic approach to project management. Its objective is to empirically document the existence of project strategy (either explicitly or implicitly) as an element of project management, and to verify that there is variation in project strategies that may support and enhance different types of business strategy. In sum, the objective of this research was to explore the existence of project strategies in order to develop project strategy typologies. In this preliminary research, we studied projects in pharmaceutical industries and we were able to identify the initial typologies of project strategies. Our preliminary results were presented here, yet further research should be conducted.

2. Literature Review

In the general management literature, several scholars extensively conducted research on organizational strategy, resulting in numerous definitions and frameworks. For example, Wright *et al.* (1992) defined organizational strategy as "the top management's plans to attend outcomes consistent with the organization's missions and goals."

Mintzberg (1994) offered five different definitions for strategy, calling them "the five "P" framework," which includes plan, pattern, position, perspective, and ploy. Porter (1980, 1985, 1996) describes strategy as the creation of a unique and valuable position, involving a different set of activities, and categorized generic strategies into cost leadership, differentiation, and focus. Miles & Snow's strategies typology includes reactors, defenders, analyzers, and prospectors (Miles and Snow 1978). Moore (1995) recognized three types of strategies— product leadership, customer intimacy, and operational excellence. Besides the definitions and frameworks of organizational strategy, many scholars have studied the alignment and performance relationships across organizational hierarchy: corporate, business, and function (Youndt *et al.* 1996; Papke-Shields and Malhotra 2001). They argued that a good fit between business strategy and functional strategies can improve the organizational performance (Chan and Huff 1993; Luftman *et al.* 1993). The functional strategies include, e.g., manufacturing, information technology, and R&D strategies.

In project management, studies with regards to strategy are rather limited. Several works related to strategy were in the context of project selection. Even though the studies suggested that projects should be selected (as parts of portfolio) to support business strategy, how to strategically manage these projects after selecting them was rarely mentioned. It was not until recently that researchers centered their works on strategic issues in project management. This includes the development of a theoretical framework for managing projects in a strategic way (Shenhar, 2004), a framework for aligning project management with business strategy (Milosevic and Srivannaboon 2004), and a strategic framework for project manager appointment (Milosevic and Patanakul 2004). In particular, Shenhar (2004) studied over 120 projects in various industries and concluded that a more strategic approach was needed to project management. Project managers should be perceived as leaders, who must manage their projects for better business success, and for winning in the market place. A similar conclusion was reached by Morris (2004). In their studies, Shenhar (2004) and Shenhar *et al.* (2005) also argue that to be able to lead a project for better business success, a project manager needs a project strategy. They describe a project strategy as "the project perspective, position, and the guidelines on what to do and how to do it, to achieve the highest competitive advantage and the best value from the project outcome." This means that project managers, including project team members should have a strategic mindset. They should understand the business objective of an organization, the definition of project products and their competitive advantage/value to the organization. Further, the project description, success and failure criteria, and the guidelines or procedures for project management (Shenhar *et al.* 2005), see also Table 1. Even though several researchers conduct their studies on strategic issues in project management, the research in this area is still in its early state. More research needs to be done.

Table 1. The elements of project strategy.

Project Strategy Components	Description
Business Perspective	The business motivation for implementing this project or producing project products
Objective	The ultimate business goals of the project
Product Definition	The description of project products including their specifications
Competitive Advantage/Value	The reasons why the project products are better than other products and the value the products create
Success and Failure Criteria	The perspective and expectations that the organization has for the product/project
Project Definition	The project boundaries, scope of work, project deliverables, and project type
Strategic Focus	The mindset and guidelines for behavior to achieve the product's competitive advantage and value

Source: Shenhar *et al.* (2005).

3. Research Methodology

The objective of this study was to explore project strategies in order to develop project strategy typologies. In this study, we found it appropriate to employ a case study research methodology since we found it appropriate in such a study, where the area of focus is not well explored and current perspectives seem to be inadequate (Eisenhardt 1989). To guide our research activities, the following underlying hypotheses were developed.

Underlying Hypothesis 1: *Project strategy is implemented so that project is led for better business success.*
Underlying Hypothesis 2: *Different projects are managed with different project strategies.*

In this preliminary research, we started with the study of four projects from four different companies in pharmaceutical industry. Even though we planned to study projects in various industries, we began with cases in pharmaceutical industry because of the availability of project information. The descriptions of each project are in Table 2.

Table 2. Project description.

Case name	QR	RBA	PX	EDT
Project definition	Develop non-prescription medication for better and safer treatment	Develop premium product offering, which goes beyond basic generic products	Manufacture special skin care products in high volume for traditional market	Develop/Implement new process that will shorten product development time
Project duration	10 months	18 months	6 years with series of products	Ongoing since 2000
Project budget	Confidential	$270,000 (w/o clinical study)	Confidential	Confidential
Project product	Seasonal medication	Non-seasonal medication	Non-seasonal medication	New development process

By following the case study research methodology, a content analysis including a within-case analysis was performed in order to gain an understanding of each case. Then, we conducted a cross-case analysis to identify the similarities/dissimilarities among these cases on the issue of our interest (Yin 1984; Eisenhardt 1989). In particular, we used the definition and framework of project strategy (Shenhar, *et al.,* 2005) as shown in Table 1 in our case analysis.

4. Research Results

We found from our preliminary research that in most cases, projects were managed by using implicitly project strategies. However, these project strategies were not part of the formal project plan and documentation. Again we defined a project strategy as "the project perspective, position, and the guidelines on what to do and how to do it, to achieve the highest competitive advantage and the best value from the project outcome." In this study, the project teams used project strategies to lead their project management toward a focus on product quality and product development time. The following are detailed discussions starting with a brief summary of each case followed by the results from the cross-case analysis.

5. Brief Case Summary

Case QR: The objective of the QR project was to develop a non-prescription medication for a consumer market. This project was implemented as an effort to gain the company's market share and revenue in this product category. In doing so, the team had to develop a high quality cream for better and safer treatment of cold sore. In addition, since the project produced a seasonal product, the project team had to complete the project and launch the product before the winter arrived. If the project was delayed, the company would lose its shelf space, which would lead to a loss in revenue. With a proactive project management approach, including a focus on product quality and project schedule, the team was able to launch the product ahead of schedule, resulting in a significant gain in market share and revenue. The product became the No. 1 pharmacist recommended cold sore treatment in the United States.

Case RBA: RBA was a development project of a non-seasonal medication. Its focus was on developing superior heartburn-relief tablets that go beyond the consumers' basic needs. By offering new product value and thanks to the product differentiation, the company intended to gain more profitability due to premium pricing. This was, in fact, an effort to regain the company's product leadership that suffered from a

20% loss in market share in the past 4 years. With the product superiority in mind, the team attempted to launch the new product to the market as early as possible. In fact, the project was completed 6 months ahead of schedule and the product has gained a high consumer acceptance.

Case PX: PX was the development project of non-seasonal, external used products. Its objectives were 1) to develop a series of specialty lotions, launched in a six-year period, and 2) to use its development process as a framework for future product developments. These objectives were aligned with the company's strategies. For the PX project, the team used a traditional project management process. Since the series of products were scheduled to be launched every other year, the project milestones were strictly defined. These milestones were highly visible and tied to specific product launched schedules. In addition, with an emphasis on product quality, specific quality targets were set and monitored formally.

Case EDT: EDT was a project created to implement a new operating approach that would enhance drug development processes— with the new approach, the pre-clinical R&D time would be decreased by 50%. This would lead to potential gains in benefits and profitability, the cost containment of the company, and a 30-50% increase in the productivity of pre-clinical development. In general, the project was implemented with a strict timeline. The pace of the project was driven with a business aspect— the amount of dollars that could be saved and earned by the company each year. The project milestones were set in consistence with the number of development projects that were successfully implemented by the new operating approach. So far, several projects have been successfully implemented with this new approach.

6. The Existence of Project Strategies

The findings of the cross-case analysis indicate the existence of project strategies in our four studied projects (see Table 3). However, these strategies do not exist explicitly. In all four cases, project strategies were not parts of the formal project plan and documentation.

As parts of project strategies, the product and project definitions were well documented and understood by the project teams in our study. These definitions helped the project team understand the big picture of their projects including their products and deliverables. For example, in the QR case, the project was defined as the producing and launching of

Table 3. Short summary of project strategy.

Case name	QR	RBA	PX	EDT
Business perspective	A market opportunity exists in this product category, leading to a potential gain in sale revenue	A potential to regain market leadership and revenue by introducing premium products	A possibility to gain revenue by introducing special products to global traditional market places in high volume	A potential gain in revenue and cost reduction by shortening product development time
Objective	To gain market share in this product category and to gain sale revenue	To regain market leadership from a 20% loss in the last 4 years and to increase revenue	To increase revenue by focusing R&D and marketing on products of the leading brands	To increase revenue by launching new products to market 2-year faster
Product definition	Non-prescription cream for cold sore treatment	Heartburn-relieved tablets	Skin care lotions for traditional market place	New development process
Competitive advantage/ value	Effective cream for better and safer treatment, launched before winter arrives	Premium tablets with effective ingredients that go beyond basic Generic products	Special skin care products, responding to the customers' needs and being a platform for future products	New development process, focusing on increasing process efficiency/speed

Table 3. (Continued)

Case name	QR	RBA	PX	EDT
Success/ failure criteria	*Project:* - Product quality - Project schedule tied with window of opportunity *Business:* - Revenue gained	*Project:* - Product quality - Project schedule, time-to-market *Business:* - Revenue gained - Market leadership	*Project:* - Product quality and future product offerings - Project schedule, time-to-market *Business:* - Revenue gained	*Project:* - Project schedule - Have freedom on budget overrun *Business:* - Revenue gained
Project definition	Manufacture and launch non-prescription medication	Develop, manufacture, and launch premium product offerings	Manufacture and lunch products, globally, in high volume	Implement new process
Strategic focus	Proactive project management approach to launch a quality product within the window of opportunity	Product superiority and business mindset- the faster the product launched, the more revenue gained	Traditional project management process with a product quality focus and clear milestones	Clear implementation timeframe with a predefined expected outcomes (No. of projects using new process)

non-prescription medication of cold sore treatment with FDA approval. The product was the FDA-approved cream containing active ingredients for better and safer treatment.

Less documented, although the project teams were aware of it, were the competitive advantage and success/failure criteria. In all four cases, the teams recognized what was made their product better than other products and what the expectation of management on their project. In the RBA case, the team understood that they had to develop, produce, and launch premium heartburn-relieved tablets. To be able to compete in the market, the team realized that they had to use effective ingredients to make their tablets better than the basic generic products. In terms of success/failure criteria, the project team knew that the product quality and project schedule were the main criteria that management expected.

Even though, the business perspective and objective of the project were not obviously documented in the formal project-level documents, they were articulated by the project managers. These business perspectives and objectives helped the project team understand the business aspect of their project including its product. We found in the RBA case, that the team realized the potential to regain market leadership by introducing premium heartburn-relief tablets to the market. In fact, the understanding of the business perspective and objective helped the project managers in our study choose an appropriate approach to manage their project.

Strategic focus is the least documented component of project strategy. Ironically, the project managers in our study practice it appropriately. These project managers had the right mindset in managing their project. They used appropriate guidelines that led the project team to develop a competitive product. In the PX case, a traditional project management process with clear milestones was used in accordance to the focus on product quality. On the other hand, a proactive project management approach was used in the QR case.

Based on the existence of all elements of project strategies, we conclude:

Proposition 1: *Project strategies implicitly exist when managing projects in the pharmaceutical industry.*

In all four cases, the implicit existence of project strategies set a tone for project management. The understanding of the business perspective and objective helped define the products possessing competitive advantages. This led to the definition of the project scope, including the success/failure criteria. Then, the project manager set the strategic focus that guided the team to produce competitive products or products with a high value. This approach, in turn, helped the organization achieve its business success. Thus, we propose:

Proposition 2: *Project strategy is needed to guide project management towards better business success.*

6.1. Various types of project strategies

When analyzing the similarity and dissimilarity among project strategies, we noticed in three out of four studied projects (QR, RBA, and PX) that the common project strategy used was based primarily on the product competitive advantages and project schedule. In this study, we refer to it as the "Product advantage/Schedule driven strategy." This strategy drives project management toward a focus on developing high quality products with compressed development time. In the EDT project, the strategy used was the "Efficiency driven strategy," which drove the project team toward schedule focus. We noticed from all four cases that their project strategy was aligned with the business objectives of their company. In particular, these companies' objectives were e.g. to gain profitability and revenue, to gain market share, to respond to the customers' needs, and to increase productivity. In each company, its project strategy enhanced these objectives. In sum, we suggest:

Proposition 3: *To be consistent with business strategies, various types of project strategies are used when managing projects in the pharmaceutical industry.*

Product advantage/Schedule-driven strategy: With the "Product advantage/Schedule-driven strategy," product quality has a high priority in the project. In the QR, RBA, and PX projects, the teams had to develop high quality medications to differentiate their products from those of their competitors. In fact, in the pharmaceutical industry, a project strategy emphasizing on product quality seems to be appropriate. Some of the reasons are:

- A medication that provides safer and better treatment often receives high acceptance from the consumers and brings financial benefits to the company. Time is therefore spent on searching for the most active ingredients appropriate for the products. This includes non-prescription medications that have to compete in commodity markets.
- Some medications are required to have an approval from the Food and Drug Administration (FDA) for safe and effective treatments.

With the nature of the industry and the business objectives/strategies that the companies were pursuing, we were not surprised to find that a project strategy centered on product quality was common to the three studied cases (QR, RBA, and PX). However, focusing the project strategy only on product quality may not be sufficient in all competitive environments, especially in those of non-prescription medications (QR and RBA cases). To truly gain the benefits from the products, the companies therefore concentrated also on shortening the development time.

In the QR, RBA, and PX cases, the project schedules were strictly attached to the launch schedule of the products. The project milestones were clearly defined. The project teams understood that the project schedules were compressed. In all three cases, management made the teams realize that "time was money"— the faster the product was launched, the more revenue and profit the company would gain. From our case study, we noticed that there were two types of product launches: "window of opportunity" and "as early as possible."

- The "window of opportunity" type is applicable to the products whose launch dates are attached to a specific time frame (e.g., seasonal medication: the QR case). If the project team cannot launch the product to accommodate this time frame, the company may lose the opportunity of revenue generation from the product.
- The "as early as possible" type is used when the team has to get the project done quickly for the product to gain an early market profit. This type of launch schedule is rather common to product development in various industries.

The evidence from the case showed that for the product with the window of opportunity launch dates, the project team had a high concentration on project schedules and milestones since the deadline was visible to the team. Further study should be conducted to investigate the difference in project management between these two types of launch schedules. In sum, we propose:

Proposition 4: *A project strategy with a focus on product quality and project schedule is rather common for projects in the pharmaceutical industry.*

Efficiency-driven strategy: We found evidence of the "Efficiency-driven strategy" when we studied the EDT project. In particular, this strategy drives the team toward the efficiency and productivity of the company by focusing on project schedule. One notation that is worth mentioning is the difference between the EDT project and the QR, RBA, and PX projects. While the other three projects developed products for external customers, EDT was an internal project. As in the EDT case, the team realized that a successful implementation of the new process would lead to potential gains in benefits and profitability, a cost containment of the company, and an increase in productivity of pre-clinical development. They therefore attempted to implement this process as early as possible. In addition, the project outcomes, which were measured by the number of projects using the new process per implementation timeframe, were predefined. Thus, we propose:

Proposition 5: *An efficiency-driven strategy with an emphasis on project schedule is used for managing internal projects in the pharmaceutical industry.*

Does project cost matter? We also noticed from our study that even though project cost was an important issue, it was a secondary focus in most of the cases. A possible reason was that the company's management in the QR, RBA, and PX projects made it understood to the project teams that an early launch of high quality products would bring high financial benefits to the company and would in turn cover extra project cost. As in the EDT project, the faster the team could implement a new effect process, the sooner the company could reduce the development cost. In sum, we state:

Proposition 6: *Project cost is a secondary focus when managing projects in the pharmaceutical industry.*

Since we studied only four cases, at this stage, it may be too soon to generalize our results. However, this preliminary study raised several questions that need further analysis. Some of the questions are: 1) Are the Product advantage/Schedule and Efficiency driven strategies unique to the pharmaceutical industry? 2) Are there any other project strategies that the teams pursue? The further investigation of these questions will lead us to the next stage of our research — the identification of project strategy typologies.

7. Implications

The results of this preliminary study bear several significant implications. It provides six research propositions that can be explored further in future empirical research. Even though the propositions were developed based on projects in the pharmaceutical industry, they could also serve as guidelines for the future development of propositions and hypotheses to study project strategies in different industries. In addition to research implications, some managerial implications can also be drawn from the results of this study.

Firstly, to successfully manage projects, project managers and team members should have a strategic mindset. Secondly, a project should be managed with a formal project strategy, which is documented and is a part of formal project plans and reviews. Having a project strategy will help a project team act appropriately according to different project situations, make the right trade-off decisions, develop a common project spirit and culture, etc. The documentation will help in terms of communication; facilitate some discussion and clarification, including the improvement of project strategy if needed. Thirdly, the project strategy should be in line with the business strategy/objectives of the organization. Lastly, with the nature of the industries and to satisfy customers, the Product advantage/Schedule-driven strategy is perhaps an appropriate project strategy for managing projects in the pharmaceutical industry. In addition, the Efficiency-driven strategy may be appropriate for an internal project, implemented to increase the company's efficiency and productivity. However, in managing projects, project teams should develop appropriate project strategies contingently according to the specific situations.

8. Conclusion

From this preliminary study, we found the project strategies of pharmaceutical projects. Even though, these strategies were not explicitly defined, the project teams understood them and managed the projects with their strategic mindset. We refer to these strategies as the "Product advantage/Schedule-driven and Efficiency- driven strategies." The former strategy leads a project team to focus primarily on the product quality/ competitive advantage and project schedule. The later strategy focuses on project schedule to increase the efficiency and productivity of the company. At this stage of our research, it is too early for us to draw any conclusion and determine whether these strategies are unique to the pharmaceutical industry. Further research needs to be done. However, this study empirically validated the concept of project strategy and variation in project strategies of pharmaceutical projects. With the next steps of our research, we believe that, perhaps, we can identify additional project strategies, which will in turn help us develop future propositions regarding project strategy typologies. First, we will focus our study on the pharmaceutical industry, and then we will expand it to various industries. In addition, for our longer-term research effort, we will examine the relationships between project strategy, business strategy, and project performance.

References

Chan, Y. E. and S. L. Huff (1993). "Strategic information systems alignment." *Business Quarterly,* 58(1): 51.
Cooper, R. G., S. J. Edgett, et al. (1998). Portfolio Management for New Products. Reading, M.A., Perseus Books.
Eisenhardt, K. M. (1989). "Building theories from case study research." *Academy of Management Review*, 14: 532-550.
Luftman, J. N., P. R. Lewis, et al. (1993). "Transforming the enterprise: The alignment of business and information technology strategies." *IBM Systems Journal,* 32(1): 198.
Miles, R. E. and C. C. Snow (1978). Organizational Strategy, Structure and Process. New York, McGraw-Hill.
Milosevic, D. and P. Patanakul (2004). A model for assigning projects to project managers in multiple project management environments. Innovations-Project

management research 2004". D. P. Slevin, J. K. Pinto and D. I. Cleland. Newtown Square, PA, Project Management Institute.

Milosevic, D. and S. Srivannaboon (2004). The process of translating business strategy into project management actions. Innovations-Project management research 2004". D. P. Slevin, J. K. Pinto and D. I. Cleland. Newtown Square, PA, Project Management Institute.

Mintzberg, H. (1994). The Rise and Fall of Strategic Planning: Reconceiving Roles for Planning, Plans, Planners. New York, Free Press.

Moore, G. A. (1995). Inside The Tornado. New York, Harper Business.

Morris, P. W. G. (2004). Moving from corporate strategy to project strategy: Leadership in project management. Project Management Institute Research Conference 2004, London, UK., Project Management Institute.

Papke-Shields, K. E. and M. K. Malhotra (2001). "Assessing the impact of the manufacturing executive's role on business performance through strategic alignment." *Journal of Operations Management*, 19(1): 5.

Pennypacker, J. S. and L. D. Dye (2002). Project portfolio management and managing multiple projects: Two sides of the same coin? Managing Multiple Projects. J. S. Pennypacker and L. D. Dye. New York, Marcel Dekker Inc.: 1-10.

Porter, M. E. (1980). Competitive Advantage: Creating and Sustaining Superior Performance. New York, Free Press.

Porter, M. E. (1985). Competitive Strategy: Techniques for Analyzing Industries and Competitors. New York., Free Press.

Porter, M. E. (1996). "What is Strategy?" *Harvard Business Review* (November-December): 68.

Shenhar, A. J. (2004). "Strategic Project Leadership®: Toward a strategic approach to project management." *R&D Management*, 34(5): 569-578.

Shenhar, A. J., D. Dvir, et al. (2005). "Project strategy: The missing link." 2005 Annual meeting of the Academy of Management: A new vision of management in the 21st century, Honolulu, Hawaii.

Wright, P., C. Pringle, et al. (1992). Strategic Management Text and Cases. Needham Heights, MA, Allyn and Bacon.

Yin, R. (1984). Case study research: Design and methods. CA, Sage Publications.

Youndt, M. A., S. A. Snell, et al. (1996). "Human resource management, manufacturing strategy, and firm performance." *Academy of Management Journal*, 39(4): 836.

CHAPTER 27

TECHNOLOGY KNOW-HOW PROTECTION: PROMOTE INNOVATORS, DISCOURAGE IMITATORS

Christoph Neemann* and Günther Schuh**
*Dep. Technology Management, Fraunhofer IPT, Aachen, Germany;
**Director Fraunhofer IPT, Aachen, Germany

As the economy in many emerging countries is on the rise, innovative enterprises are confronted with technology know-how theft and product imitations from those countries. Especially SMEs are facing problems in enforcing their legal and also legitimate claims on their intellectual property and know-how. In addition, technology firms are under pressure due to fierce competition with competitors in technological eye height. As intellectual property rights seem to become a blunt sword, a bigger picture on know-how-based competitive advantages in technology-driven firms is needed. To answer this challenge, a holistic approach to technology know-how protection is recommended.

1. Introduction

The competitiveness of European and other industrial nation's enterprises is based on innovation (Orgalime, 2001). Hence, technology-driven enterprises need to be able to protect their innovations from being counterfeited by competitors to assure a fair return on their research & development spendings. Otherwise – in the macroeconomic view – the incentives for R&D collapse, and – in the microeconomic view – the most innovative firms vanish from the markets or stop their activities in R&D (Hussinger, 2004).

Intellectual property rights create incentives for companies to engage in R&D, but they are not applicable to all kind of innovations. Moreover, the enforcement of legal rights, especially abroad, is difficult to obtain for small and medium-sized enterprises (Kingston, 2001).

This results in alarming effects on the world economy: Investigations on product piracy and counterfeiting revealed that up to 10% of the world trade volume account for illegally imitations and counterfeit products – equaling a sum of 500 billion Euros (ACG, 2004). As the prices for imitated products are usually lower than the original product prices, the actual loss comes at a much higher figure. Still, many imitations are never discovered as they are only offered on local markets (Orgalime, 2001). Estimates are that all these effects cause a job loss of 200,000 in the European Union (EU Taxation and Customs Union, 2004). This represents a tremendous and even rising danger for the well-being of the innovative companies and economies.

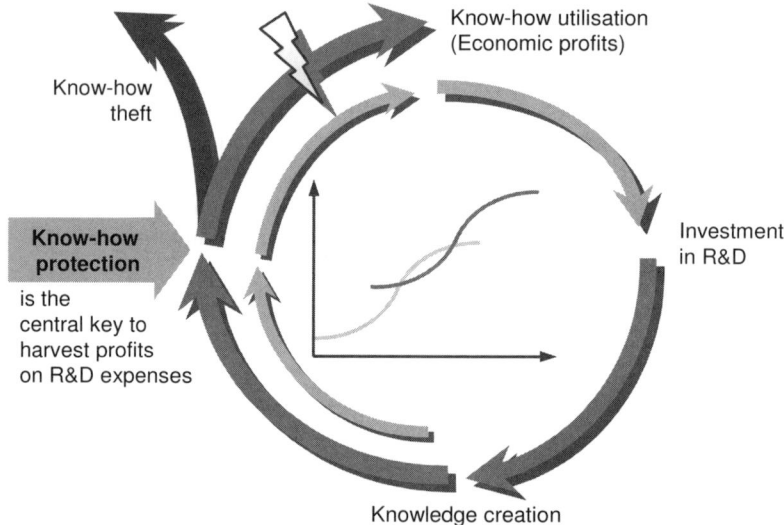

Figure 1. Product counterfeits break the healthy cycle of know-how creation, economic profits and investment into research & development (R&D).

Additionally to the loss of business volume that lacks to contribute to the companies coverage of R&D expenses, the companies are confronted with product liability claims for products they are not responsible for. This causes unnecessary efforts to prove the non-responsibility for these claims, while the damage on reputation and brand value is still not accounted for. But not only low quality imitations are giving innovators a headache - as the state of technology rises in the emerging countries, they are able to produce imitations at equal quality, hence building up a real threat on eye height to the innovator's product while still saving the cost for the development and marketing of the products.

As products have a high share of embedded technology know-how that is to grow evermore in future, this threat is crucial for innovative companies' competitiveness as product counterfeits break the healthy cycle of know-how creation, utilisation in terms of economic profits and their re-investment into new R&D projects (Idris, 2004).

2. Decision Model for Technology Know-How Protection

The decision model presented in this paper aims to solve this problem with a holistic approach to know-how protection. The decision support takes the type and characteristics of know-how, possible threats of know-how spill-overs and their mechanisms into account. It aims to deliver possible solutions for know-how protection on basis of this information, resulting in recommendations of know-how protection mechanisms.

Following the systems engineering approach (Haberfellner, 1999), the decision model is divided into six sub-models, each comprising a certain aspect of technology know-how protection (See Figure 2).

The *Company model* represents all company information that is relevant for know-how protection. It is introduced to identify this information in a manner that is familiar to the usual perception of a company, e.g. thinking in current and future products, strategies, implemented and future production processes etc.

The *Context model* structures all information that needs to be considered in regarding the competitive, but also legislative surrounding of the company. This includes the assessment of current and potential competitors and imitators, markets and applicable legislations.

The *Technology know-how model* abstracts the relevant know-how aspects from the Company model, regarding the properties that are relevant for its protection. It takes the inside view of the company and does an assessment against the know-how level of competing imitators.

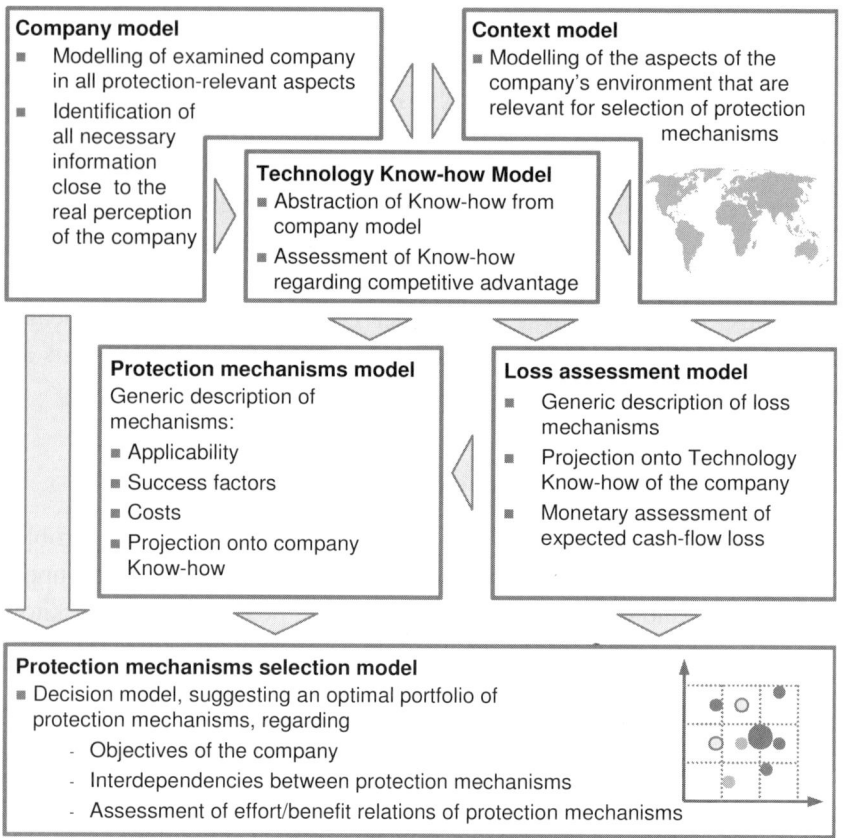

Figure 2. Concept of a decision model for technology know-how protection.

The *Loss assessment model* describes what kind of damage might occur, once the know-how has found its way to potential competitors and how the know-how is transferred to the imitator.

The *Protection mechanism model* describes possible generic actions for know-how protection which are matched to the use case during application of the model.

All this information is input for the *Protection mechanisms selection model* which eventually assesses the applicability and effectivity of the protection mechanisms, resulting in a recommendation of a portfolio of know-how protection mechanisms for the company under analysis. Additionally to the mentioned constraints, also strategy implications of the company are considered for the identification of the optimal protection portfolio.

Selected elements of the model are explained in detail in the following sections.

3. Analysis and Assessment of Technology Know-How

At first, it is necessary to identify the company's technology know-how in a structured way that makes know-how accessible for assessment and evaluation. This is realised in the so called Technology know-how model.

The technology know-how model derives information from the Company model and the Context model by assessing the value the products and the processes create for the company. In case of the products, this is done by assessing their contribution to turnover and profits. Those contributions are then split to the product functions and the product technologies realising those functions.

For the process technologies, the current and future process chains within the company are assessed regarding their value-add and their contribution to differentiation.

By matching process and product technologies to know-how components, the connection between technologies and the know-how becomes visible. An assignment of percentage weights that illustrate to what extent a technology contributes to the know-how value allows to calculate a monetary value for the company know-how elements. This proceeding is illustrated in Figure 3.

4. Identification and Assessment of Loss Threats

After analysing the valuable know-how of the company, it is necessary to identify the effects that occur if parts of this technology know-how find its way to potential competitors.

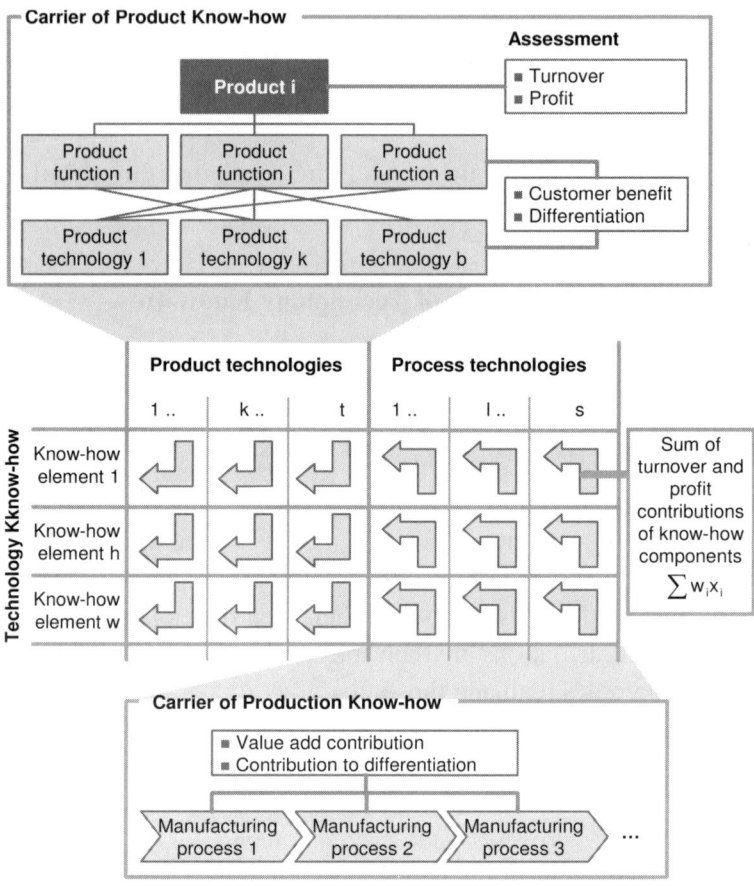

Figure 3. Assessment of technology know-how elements.

First, the company has to identify the potential ways through which know-how might leave the company. Based on the company's know-how assessment, the mechanisms that might occur are selected from a generic classification of spill-over effects (Figure 4). The proposed classification takes into account the know-how carrier, the willingness to transfer the know-how and the awareness of know-how transfer. These three dimensions have an general impact on the importance of the spill-over mechanism, and they also provide a framework to classify the effects like the examples given in Figure 4.

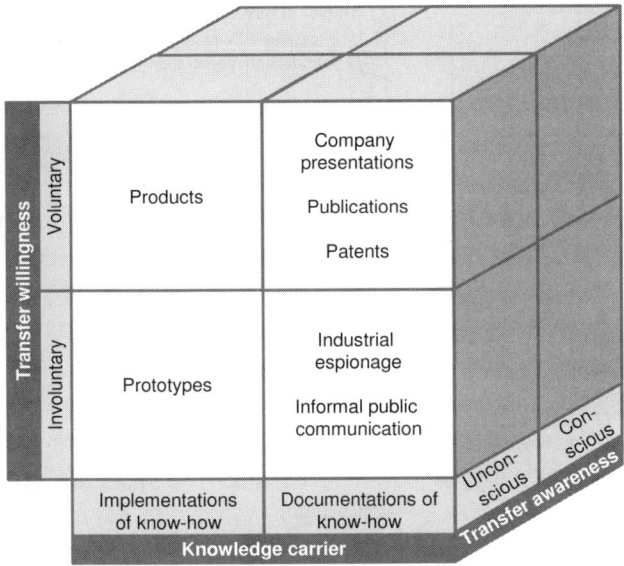

Figure 4. Classification of know-how spill-over mechanisms.

The determinants of such losses are various, and they are covered in an assessment that estimates the loss in cash-flow for the company. The loss in cash-flow is determined by the probability of loss and the height of loss in case of appearance. For the assessment of the turnover at risk, the product life cycle model is used to identify the parts whose margin contribution might be threatened from the point when the first imitations appear. The rate of turnover loss is mostly determined by the rate of product users that will switch to the imitation product, and by the price decline caused by the appearance of imitations (Figure 5).

5. Know-How Protection Mechanisms

The know-how protection mechanisms model describes different protection mechanisms, their success factors and their limits in detail. A categorisation of protection mechanisms allows to see in what phases they might be applicable. This categorisation is derived from a game theory model that describes the process how an innovator and an imitator act and react.

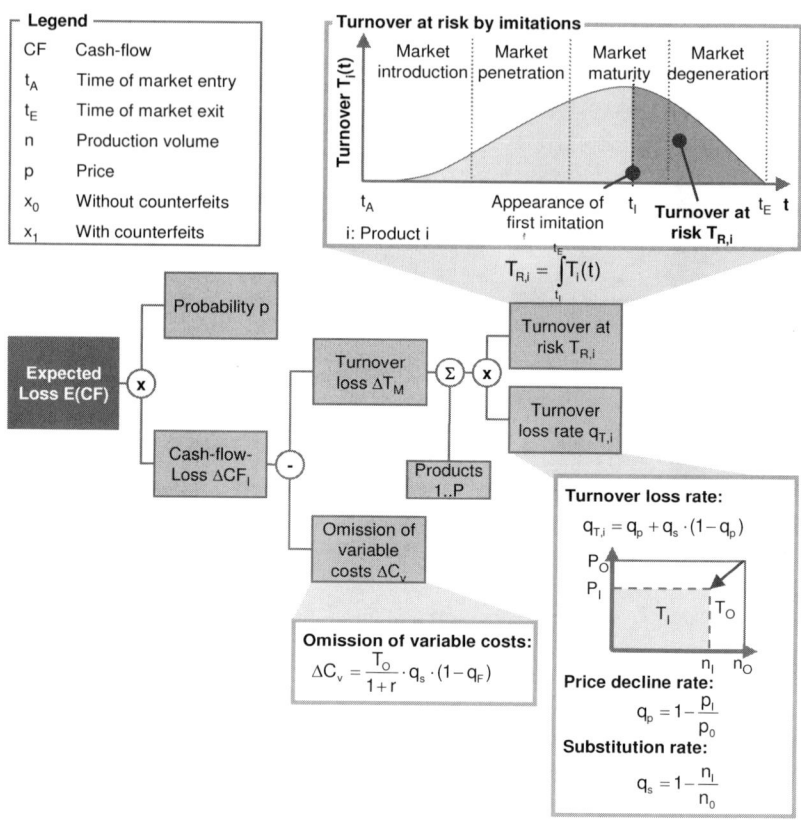

Figure 5. Assessment of potential loss caused by imitations.

First, the imitator decides whether to imitate or not. This decision can be influenced by protection mechanisms that lower the imitation attractiveness for the imitator. Alternatively, threatening with persecution, for example in case of violation of intellectual property rights, can move an imitator to withdraw from his plans. Second, the imitator tries to acquire the necessary know-how for realising the products which is another approach for protection mechanisms. Examples are actions that actively protect information or alternatively make Reverse Engineering of the product difficult for the imitator. If he still succeeds, actions that hinder the reproduction of the necessary components come into play. This could be the use of non-standard parts that are not widely available on global supply markets, or the use of processes with high fix costs which imitators usually avoid in the fear of being convicted to discontinue production.

Hindering the sale of imitations is another possibility for protection that can be realised by offering additional services the imitator can not deliver, e.g. lifetime service, direct distribution to end customers etc.

Enforcing persecution, usually with legal means, is another option as soon as the imitations reach the market. After successfully fighting the imitator, communication of this consequent action can help to deter other potential imitators. Those last two activities show effect only in the next cycle of the 'imitation game' that is shown in Figure 6.

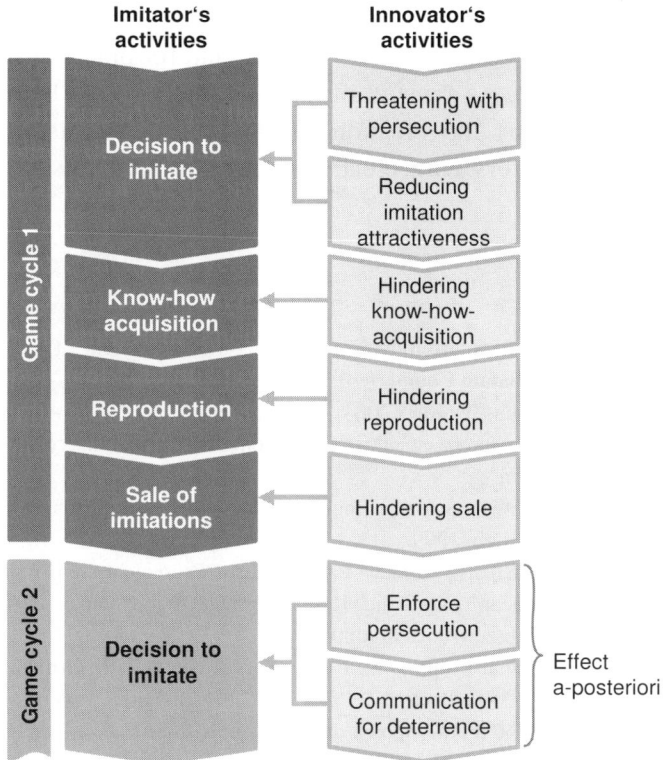

Figure 6. The 'imitation game': Action and reaction between innovator and imitator.

6. Summary and Outlook

The presented framework should allow an decision support for enterprises who are facing copies of their technological products and processes, leaving them unable to earn a fair return on their own R&D expenses.

However, many of the challenges are still to be found in details of the process and especially in the implementation of such counter-measures against know-how theft. However, while the implementation of single actions is still an effortful task, the application of such a framework is necessary to have a common and powerful approach as most actions deliver their full effect in combination with others – and on the other hand might even conflict with each other if not carefully selected and customised.

As a conclusion, we still see a high demand for practical solutions in this field of research. As a matter of fact, the protection of know-how has to be as creative as the creation of the know-how itself. Only by surprising the imitators with creative protection solutions, we can stay ahead of them.

References

ACG (2004). The Anti-Counterfeiting Group: Outcomes Statement of the First Global Congress on Combating Counterfeiting, 2004.

EU Taxation and Customs Union (2004). Counterfeiting & Piracy – The Economic Consequences. http://europa.eu.int/comm/taxation_customs/customs/counterfeit_piracy/counterfeit1_en.htm

Haberfellner, R. et al. (1999). Systems Engineering. Published by Daenzer, W.F/ Huber, F., 10. edition, Zuerich, 1999.

Hussinger, K. (2004). Is Silence Golden? Patents versus Secrecy at the firm level. ZEW – Centre for European Economic Research, Discussion Paper 04-78. Internet: ftp://ftp.zew.de/pub/zew-docs/dp/dp0478.pdf

Idris, K. (2004). Intellectual Property: A Power Tool for Economic Growth. Publication of the World Intellectual Property Organization (WIPO). Internet: http://www.wipo.int/about-wipo/en/dgo/wipo_pub_888/index_wipo_pub_888.html

Kingston, W. (2001). Innovation needs patents reform. In: *Research Policy,* Vol. 30, pp. 403-423.

Orgalime (2001). Combating Counterfeiting. A practical guide for European engineering companies. October 2001. Internet: http://www.orgalime.org/pdf/counterfeiting_guide_en.pdf.

CHAPTER 28

COMPARISON OF TECHNOLOGY FORECASTING METHODS FOR MULTI-NATIONAL ENTERPRISES: THE CASE FOR A DECISION-FOCUSED SCENARIO APPROACH

Oliver Yu

San Jose State University, San Jose, California, USA

This paper reviews and compares major technology forecasting methods for multi-national enterprises (MNEs) and then presents a case for using a decision-focused scenario approach to manage the inherently irreducible uncertainties for forecasting the complex technology innovation and adoption process in a volatile MNE business environment.

1. Introduction

In the past two centuries, technology innovations, such as production mechanization, scientific management, advances in transportation and communication, and developments of weapons of mass destruction, have been dominant driving forces in economic development and societal change, including the *rise of multi-national enterprises* (MNE). As a result, accurate technology forecasting, including *the timing of development, the rate of adoption, and the impact of applications*, is vitally important not only in helping business organizations formulate responsive strategies and gain competitive advantages, but also in assisting governments and the public at large understand the long-term effects of technology on the economy, society, and environment. This paper provides a critical review of the major forecasting methods and presents a case for a scenario-based approach.

2. Comparison of Major Technology Forecasting Methods

There are many ways to classify technology forecasting methods. One way is to focus on the specific *underlying assumptions or hypotheses* about how the technology innovation process works or can be predicted. By this classification, we can concisely summarize and compare major technology forecasting methods into the following categories:

2.1. *Power of collective wisdom*

Underlying assumption: There is *power* in the collective wisdom of experts or senior executives or the professional community in forecasting the technology development, adoption, and application impact process.

Typical examples: Delphi, expert opinions, executive judgments, consensus.

Major advantages: General credibility; low cost.

Serious pitfalls: Inherent bias; blind leading the blind.

Basic applicability: Far-out technologies with little knowledge and experience.

2.2. *Potential leading indicators*

Underlying assumption: There are potential *early warning signs or leading indicators* about the technology development, adoption, and application impact process.

Typical examples: Environmental scanning and monitoring, patent analysis and citation and innovation search, survey of technology early adopters.

Major advantages: Plausible evidence; relatively low cost.

Serious pitfalls: Signs or indicators may be misleading; may miss isolated developments.

Basic applicability: Relatively slow-paced and gradual technology developments.

2.3. Continuation of historical patterns

Underlying assumption: Historical patterns or trends will *continue due to inherent nature or momentum* of the process.

Typical examples: Trend extrapolation, such as Moore's law (Moore, 1965); Substitution models, such as Fisher-Pry model (Fisher and Pry, 1971).

Major advantages: Support of empirical data and validity of short-term momentum; statistical analysis can be used to reduce random errors.

Serious pitfalls: Patterns or trends may have significantly changed and not continue as assumed.

Basic applicability: Short-term forecasting with ample historical data to ensure validity of the process momentum.

2.4. Analogies to well-known phenomena

Underlying assumption: The technology development, adoption, and application impact process is *similar to* some well-known physical or biological phenomena.

Typical examples: Technology life cycles, in which the technology development is analogous the life of an organism going through the stages of birth, growth, maturity, and death; Diffusion models, in which the market penetration of a technology is analogous to the physical diffusion process.

Major advantages: General plausibility and credibility.

Serious pitfalls: The underlying assumption for the analogy to a well-known phenomenon may be invalid.

Basic applicability: General technology development and adoption.

2.5. Structural relations

Underlying assumption: The technology development, adoption, and application process follows *plausible structural relations of various influencing factors*.

Typical examples: Relevance tree, cross-impact matrix (to identify the correlations), analytic hierarchy process.

Major advantages: Systematic; logical; descriptive.

Serious pitfalls: Difficult to include feedback loops and identify effects of lower order factors.

Basic applicability: Longer term technology forecasting based on a system perspective and framework.

2.6. *Causal models*

Underlying assumption: Causal models can be developed for the technology development, adoption, and application impact process.

Typical examples: Various techno-economic models, such as the classic economic supply and demand model (Samuelson and Nordhaus, 2004); System dynamics models (Forrester, 1973).

Major advantages: Sophisticated and impressive as the models provide mathematical models of the technology development, adoption, and impact process that give the appearance of deep insight and accuracy.

Serious pitfalls: Convoluted theories, expensive research requirements, and often incomprehensible to laymen and even incorrect because of the great inherent difficulties in attaining true understanding of the extremely complex socio-economic, political, and business aspects of technology development, adoption, and application impact process.

Basic applicability: As a continuing attempt to the ideal goal for technology forecasting.

3. Difficulties in Developing Accurate Technology Forecasts

Unfortunately, the record of technology forecasting has often been *poor* in comparison to predictions of physical phenomena or engineering endeavors. History is replete with faulty analysis and incorrect predictions (Schnaars, 1989). In examining the technology development, adoption, and application impact process, one may identify a number of major inherent sources of difficulties in accurate forecasting as follows:

- The technology development, adoption, and application impact process is generally highly *complex*, involving interactions among many related technology developments and myriads of social, political, and cultural factors; chaos and contingency theories even suggest that the process may be intrinsically intractable.
- There is a general lacking of *controlled experimentation to test and validate* the various assumptions and hypotheses underlying the traditional technology forecasting methods as there hav been in forecasting for physical sciences and engineering undertakings.
- Given all the complexity and lack of understanding, the popular interest for forecasters to focus on a single realization of the technology development, adoption, and application impact process would be *practically impossible to be accurate or even correct*.

Facing these difficulties, a case can be made for using a *decision-focused scenario approach* as an effective method for technology forecasting, *especially for the vastly uncertain business environment of the MNEs.*

4. Overview of a Decision-Focused Scenario Approach

The following description is based on an approach developed by SRI International, which has been widely applied in technology forecasting around the world since the 1980s (Wilson and Ralston, 2006).

Scenario approach is *conceptually different* from traditional forecast or sensitivity analysis methods. Strictly speaking, it does *not* develop a single forecast but a set of *structurally different but plausible alternative* scenarios that provides an *envelope to uncertainty* in the future business environment affecting the development and adoption of technologies of interest. Specifically, in contrast to the general characteristics of many traditional technology forecasting approaches, decision-focused scenarios are:

- not predictions, but rather are descriptions of alternative plausible futures
- not variations around a mid-point or base-case forecast, but rather are significantly, often structurally different views of the future

- not generalized views of desired or feared futures, but rather are specifically decision-focused visions of the future
- not products of outside futurists, but rather are the results of management and senior staff's insights and perceptions about the future

Furthermore, the special characteristics of this approach are:

- It *emulates* effective human thinking and decision-making process.
- It is *total system-oriented, context-based, inclusive* of all other methods and *integrative* of all available knowledge.
- It uses a *hierarchical, logical* approach to identify *key factors* in the business environment affecting decisions.
- It uses *collective diverse human judgment* (structured brainstorming and integration of experts from diverse background) supplemented by *detailed studies of important topics with uncertainty* (focus papers).
- It uses *systematic clustering and condensing of factors at each stage* to aid in human comprehension.
- It uses *repetition and redundancy* to ensure comprehensiveness; throughout the scenario development process, there is *always room* to recapture a missing factor or modify existing factors.

In a complex and dynamic business environment like that of the MNEs, the decision-focused scenario approach can be an effective technology forecasting method because it offers the following advantages:

- *Focus on decision objectives*
- *A total system view*
- *Rich context on plausible alternative futures*
- *A basis for managing uncertainty in the business environment*

On the other hand, local system-oriented single realization point forecast, even with sensitivity analysis, is almost always not only wrong but also often misleading.

5. The Decision-Focused Scenario Development Process

The decision-focused scenario development process consists of the following six iterative steps:

1. Identify the **Decision Focus**, which pinpoints the technology choices that need to be made.
2. Identify **Key Decision Factors**, which are the *key issues in the external environment* that directly affect the decision to be made and need to be forecast—for the MNE environment, they often include: technology development, market demand growth, industry structure, government regulations, resource requirements, international relations.
3. Identify **Micro and Macro Forces**, which are *major drivers of changes* in the external environment and the major causes of future uncertainty. In particular, those forces with high importance and low uncertainty will be the *dominant forces* present in all future scenarios.
4. Consolidate the *critical forces,* i.e., those forces with high importance and medium to high uncertainty and those forces with high uncertainty and medium to high importance to form **Axes of Uncertainty**.
5. Use the extremes of these axes to develop *plausible, structurally different, and internally consistent* **Scenarios of the Future**.
6. Assess **Scenario Implications**, which are *preliminary assessment of the general impacts* of the scenarios on Key Decision Factors and eventually the Decision Focus.

These Scenarios of the external environment may be refined by modifying the key Decision Factors and additional iterations of the scenario development process.

6. Major Resource Requirements for Decision-Focused Scenario Forecasting

Decision-focused scenario forecasting is highly *resource intensive*. Formal scenario development for a large business organization or government agency generally requires:

- *Commitment by top decision maker* for a development process involving 3 or 4 facilitated meetings of major stakeholders in technology development and adoption for a period of 3 to 6 months.
- *Continuous participation* of 8 to 12 *major stakeholders with diverse background*, as any interruption will seriously damage consensus building that is essential for scenario development.
- An effective *facilitator* to stimulate discussions and provide structure for the scenario development process, and a competent *recorder* to faithfully capture the contents and insights of the scenarios.
- A professional *research staff* to provide additional research needed to reduce the uncertainty as much as possible for the major driving forces.

It is important to note that although a full scenario development process requires significant resources, the general principles of the process can apply to all forecasting endeavors, regardless of the amount of available resources.

7. Description of an Actual Application

An application of the decision-focused scenario forecasting process by the author was to the decisions for the development of virtual reality (VR) technology for an Asian country X, presented here as an illustrative example.

7.1. *Decision focus*

This scenario forecasting effort is designed to assist Country X's information technology (IT) industries in making technology investment decisions for the period of 1996-2003. The focus of the scenario technology forecasts is:

> "To enter the international VR related products market and develop add-on value in hardware products for Country X's IT industry by the year 2000, and realize the return on investments by the year 2003."

Scenario scope: Development and adoption of VR hardware and software that provides intuitive and natural interactions with computer generated data.

Major elements of the decision focus:

- What are the opportunities in terms of application and product focuses?
- How should these opportunities be accessed, for example, through market strategy, technology strategy, key alliances, and what actions and timing needed to realize the desired returns by 2003?
- What will be the core technologies?
- What should be the government and private industry budgets?
- What should be the return on investment?

7.2. Key decision factors

The most important external factors affecting the decision were:

- How will the VR technology evolve, in terms of time frame, industry standards, and performance requirements?
- What will be the biggest hardware and software market demand?
- What will be the industry structure for virtual reality, in terms of competitors, leading vendors, and profit locations?
- What will be the growth dynamics of Country X's IT industry?
- What will be computing power on the PC platform?

7.3. Examples of dominant forces

High importance and low uncertainty forces:

- Computer power on PC (Moore's Law will continue)
- Internet development (will continue to expand)
- Bill of industry upgrades (will happen)
- Computer graphics (will evolve)
- Market information availability (will be available)

- Need for diversions & leisure activities (will grow)
- 3D graphics investment by game machine and content providers (will occur)
- Desire for fantasy with real feel and touch (will be important)
- Entertainment applications (will be the main applications)
- Patent trend/barriers (protectionism will increase)
- Country X government sponsored VR related projects (will be made)
- Long term R&D commitment (will be there)
- Country X peripheral manufacturing (will occur)
- Need for stress relief by sensual stimulation (will be important)
- International technology sources (no accessible to Country X)
- Hazardous situation training applications (will occur)
- U.S. agency (will take strong health and safety measures on product safety)

7.4. Examples of critical forces

High importance/medium uncertainty forces: Key components development, killer applications, market size of hardware and software, market size of entertainment applications, niche products, profitability of players.

High importance/high uncertainty forces: Interest of venture capitalists, rival countries' technology investment policies, internet development, value of VR building blocks, worldwide technology diffusion, interrelationships among companies, cost/performance, health and safety factors.

Medium importance/high uncertainty forces: Commercial time frame, VR research in US/Europe/Japan, country X's internal political situation, Japan, Singapore and other Asian country policies and actions, VR standards, interest of content providers, country X's system integration capabilities.

Note that the above lists of Critical Forces were obtained after *in-depth research* had been conducted to provide additional understanding and *reduce* as much *uncertainty* as possible for the following important topics:

- Key Components of Virtual Reality
- Needs and Market Potential for Virtual Reality
- Key Player Activities and Status
- Virtual Reality Technology Roadmap and Application Potentials
- Killer Application Characteristics
- Market Value of Virtual Reality Business
- Virtual Reality Standards

7.5. *Axes of uncertainty*

For this application, the Critical Forces were organized into four major *Axes of Uncertainty* for the Virtual Reality business:

1. *Industry structure*: Investment and source, openness of structure, market barriers, opportunity access, success potential.
2. *Asia-Pacific econo-politics*: Political conflicts, economic competition.
3. *Technology evolution*: Technology standards, technology integration, technology access, manufacturer dominance.
4. *Market demand*: Cost/performance, leverage, applications, enabling device, safety issues.

For each axes of uncertainty, two plausible extremes were developed from specific rationale. The following is an example:

Plausible extremes of VR industry structure axis
Extreme 1: Combative and fragmented, due to
- Low investment mainly from corporations
- Protective structure with high market barriers
- Limited access to opportunities

Extreme 2: Cooperative and integrated, due to
- Heavy investments and many from venture capitalists
- Open structure with international cooperation
- Full access to opportunities and many small companies have major successes

The two extremes for each of the other three axes are:

- *Asia-Pacific econo-politics*: Open vs. Closed
- *Technology evolution*: Stuck and/or Disjointed vs. Breakthroughs
- *Market demand*: Expensive Specialized vs. Cheap Mass Market

These extremes result in 16 possible combinations of candidate scenarios, which needed to be reduced for effective management by human mind.

7.6. Final scenario selection

Scenarios were selected based on the guidelines that each scenario should be:

- *"structurally" different*
- *internally consistent* to naturally fit into a "story line"
- *plausible*
- with *decision making utility* as a "test bed" for assessing alternative future actions

and together, the selected scenarios should span the realm of plausible alternative futures to provide an *"envelope of uncertainty."*

Through extensive discussions among the participants, three final scenarios were selected together with their respective short title as shown in Table 1.

Incidentally, the titles of the selected scenarios are important as they can serve as powerful memory triggers for participants to easily recall the details of the scenarios.

7.7. Example of the detailed narrative: The Life in Hell scenario

Uncertainty axes: Combative/fragmented industry structure, closed Asia, disjointed technology development, expensive specialized applications.

Table 1. Final scenarios.

Final Scenario	Axis of Uncertainty			
	Industry Structure	Asia-Pacific Econo-Politics	Technology Evolution	Market Demand
Life in Hell	Combative & Fragmented	Closed*	Stuck and/or Disjointed	Expensive & Specialized
Left Behind	Cooperative & Integrated	Closed*	Breakthrough	Cheap Mass Market
Waiting for Technology Spring	Cooperative & Integrated	Open**	Breakthrough	Cheap Mass Market

* Closed econo-politics makes it difficult for Country X to access Asia-Pacific technology developments and markets.
** Open econo-politics allows Country X to fully access Asia-Pacific technology developments and markets.

Summary: VR is perceived as an emerging technology with important applications. Major countries such as UK, US, Japan invest vast resources in VR related research in 90's. Although many breakthroughs, the cost performance of VR is still far from the reach of the mass market. Systems can only be afforded in special applications such as flight simulation and industry design. Country X's IT industry becomes interested in VR technology because it provides integration of many technologies and offers good profit potential for a well-structured PC industry. However, Country X has troubles obtaining key technologies from other countries or developing its own due to patent barriers and the competitive situation. Besides, relations with its political rival country get worse after 1996. Foreign capitalists are reluctant to invest in Country X owing to unstable political future. IT industry suffers because of the lack of investments and struggles to survive.

Detailed description: A detailed novelette-like description of the scenario was then developed to reinforce the plausibility and realism of the scenario among the participants as well as future reviewers.

8. Conclusions

Country X used the application results to develop a *robust* VR technology investment strategy that could effectively respond to all three final scenarios. In addition, the process was so well-received that a training program has been established to promote the process for other technology investment decisions.

The experience gained from this application also provides the following insights:

- Because they are derived from *unresolvable* uncertainties, the final scenarios should all be about *equally probable*.
- Powerful *facilitation skills* are necessary for effective implementation of the approach.
- The approach is basically a *democratic process to integrate diverse views*, which is key to a holistic understanding of the future; thus any dominance of views by rank or personality of the process participants must be avoided.
- The scenario development should integrate as *much information* and be as *rigorous* as practical.
- To be effective, the scenarios must be eventually *linked to decisions and strategies that provide robust responses to the final scenarios.*

Finally, from this application and many others, it can be confidently concluded that the decision-focused scenario approach is an effective method for technology forecasting in a highly complex and uncertain business environment like that of a MNE.

References

Fisher, J. and Pry, R., (1971). "A Simple Substitution Model of Technological Change," Technological Forecasting and Social Change, Vol. 3, pp.75-88.
Forrester, J., (1973). *World Dynamics*, 2nd Edition, Wright-Allen Press.
Moore, G., (1965). "Cramming More Components Onto Integrated Circuits," *Electronics*, April 19, (1965).
Samuelson, P. and W., (2004). Nordhaus, Economics, 18th Edition, McGraw-Hill/Irwin.

Schnaars, S., *Megamistakes: Forecasting and the Myth of Rapid Technological Change*, Free Press, (1989).

Wilson, I. and Ralston, B., (2006). *Scenario Planning Handbook*, South-Western Education PublicationI. Wilson and B. Ralston, *Scenario Planning Handbook*, South-Western Education Publication, (2006).

SECTION IV

STANDARDS AND EVALUATIONAL METHODS

CHAPTER 29

STANDARDS OF QUALITY AND QUALITY OF STANDARDS FOR TELECOMMUNICATIONS AND INFORMATION TECHNOLOGIES

Mostafa Hashem Sherif*, Kai Jakobs**, and Tineke M. Egyedi***

*AT&T, USA; ** Aachen University, Germany;
*** Delft University of Technology, The Netherlands

In this paper, we adopt the project management methodology to provide a checklist for managing the risks that could affect the standard development process as well as the quality of the final standard. We show how the adoption of these management techniques can help improve the quality of standards.

1. Introduction

The definition of quality of standards is not a straightforward task because standardization involves many stakeholders, each with one or more objectives. As a result, the production of a standard, while grounded in technical facts, is influenced by many other factors: business considerations, social interactions, ideological principles, etc. Yet in the end, successful standards have to respond adequately and timely to societal needs in terms of user requirements, regulatory constraints and unstated assumptions, whether collective or individual. Despite these difficulties, there is a need to define a way to evaluate and measure the quality of standards in the telecommunications and information technologies (ICTs).

A standard that is too late for market acceptance is a considerable waste of effort, while the over-specification of an emerging product or service can stifle innovation. Defective standards have significant costs in terms of incompatibilities, reworks or vulnerabilities; just consider the damage that denial of service attacks have produced. Could one identify potential risks while the standard is being processed? Can the quality of a standard be assessed while it is still under development so that appropriate corrective actions can be taken in time to fix the standard? In other words, is there a way to detect and correct faults in standards before they end up in products? In the absence of an objective methodology, statements on the quality of a given standard or the superiority of one standardisation process over another can only be subjective. However, the experience gained from the Total Quality Management (TQM) movement as well as from the project management literature can provide a useful starting point to derive more objective criteria.

The structure of the paper is as follows. In section 2, we define the problem and clarify the terminology used. In section 3, we review the basic elements of TQM and show that they are not readily applicable to standardisation. From section 4 onwards, we explain how project management methodology can provide elements to think about standards management. In particular, we address the following elements: scope, time, quality, cost, documentation, and resources and suggest telltale signs of trouble. In the concluding section, we identify areas that need further research.

2. Problem Definition

The list of stakeholders in standards development is quite large. It includes members of the technical committees that develop the standards, the technical organisations that carry the expenses of developing technical contributions and/or sending their representatives to these standards bodies, the standard development organisations (SDOs) that host the meeting space and approve the specifications, the product or service organisations that will use the standard, the end-users of the products incorporating the standards, the public and political institutions concerned with the outcome, etc. One way to organise these various elements is to use typical marketing terminology. Here, the "suppliers" are the participants that bring in technical contributions, thereby shaping the direction and content of the standards. The "producer" is the technical committee that merges various contributions into a coherent text to be approved by the attendees. The "owner" of the standard, at least from the copyright perspective, is the standards body. Members of the committee are expected to the interests of the "sponsors", the various companies or official entities that finance their participation. Once the standard document is published, the developers "consume" it to design products or services that will embody the specifications and put them in the hands of "end-users:" these are the consumers that use the product or service but not the standard per se. Figure 1 summarizes the main ideas in this paragraph.

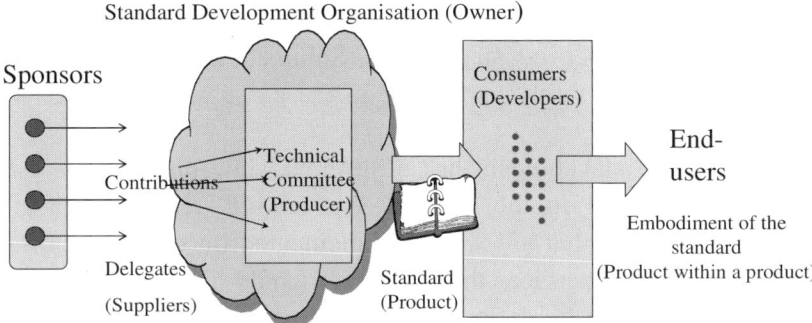

Figure 1. Value chain in standard development.

With this description in mind, it may be easier to see that none of the stakeholders have direct control over the process or the outcome. The sponsors only have loose control over their representatives and what they agree to. Not all members in the technical deliberations are actively involved in the decision-making; many are just interested observers. The suppliers and the consumers may come from different divisions of the same company, which means that their agendas may not be totally congruent. Furthermore, not all attendees have explicit mandates form their "sponsors" in which case their views many not necessarily represent those of their employers (Jakobs *et al.*, 2001). This creates a problem for the committee responsible for producing the standard in that they will not readily know whether the positions advocated are those of the sponsoring institution or the pet projects of some participants.

The standard owner, i.e., the SDO, has no direct control over the product in terms of timing, quality, etc. It can only monitor the standards process to verify that the agreed rules are adhered to. Another problem is that the owner of the standard is distinct from the patent holders, which could lead to contradictory strategies between the goals of standardization across organizational boundaries and the desire to exploit intellectual property rights. Specifically, the patent strategies and the technology-licensing practices of some participants can lead to the so-called "submarine patent" problem, whereby the patent holder can take advantage the continuous application mechanism of the U.S. patent system to submit additions to the original patent application and keep the invention secret until the technology in question matures. Once the technology is stable, the patent holder tries to enforce its claimed property rights on other industry players (Soininen, 2005; 2007). Finally, the "end-users" only rarely participate in the process, either directly or indirectly.

Each of these stakeholders has a different view on what standards quality entails. Even in the case of "successful" telecommunication standards, i.e., those that are widely implemented by vendors and used by operators to offer services that end-users would like to have, it is not clear how quality can be defined and, therefore, who is to safeguard it. For example, operators tend to select standards that are implemented by more than one vendor, are as transparent to the end-user as possible, flexible in their evolution, simple to maintain and support and are able to interwork with other operators' networks and end-user equipment and devices. They attempt to differentiate their service offers through pricing, coverage, quality or range of service options. On the other hand, equipment vendors would like to dominate markets with unique products (Sherif and Sparrell, 1992; Baskin et al., 1998) and would like to market their products as quickly as possible.

Deficient standards can cause significant damage to operators. For example, link state protocols have been defined without a robust mechanism to recover from a widespread loss of topology database information or overload conditions. This has lead to well-publicized service outages (Ash and Choudhury, 2004). Also, it is well known that

Multi-Protocol Label Switching (MPLS) and the associate Label Distribution Protocol (LDP) have been developed without taking into consideration the requirements for operations, administration and maintenance (OAM) in terms of service reliability including intrinsic means for detecting and locating failures (Cavendish *et al.*, 2004; Fang *et al.*, 2004). This means that network elements and services developed according to the standards have to be retrofitted to accommodate future mechanisms for failure detection.

Attempts at improving the governance of standards have thus far focused on reducing cost, shortening time intervals and enhancing collaboration among various standard organizations. For example, automation and electronic document submission have been adopted to enhance collaboration and to reduce costs. Processes were changed to respond to market conditions faster and to shorten the "time to market". To this end, most standards bodies have streamlined their processes and introduced new "lightweight" processes leading to new forms of deliverables. Yet, we need to go one step further and address the quality of the content of the final standard. In the next section, we try to see if TQM can provide some guidance on how to approach this task.

3. Standards and Total Quality Management (TQM)

According to Phillip Crosby (1979), quality is measured by conformance to requirements in terms of reliability, effectiveness, durability, maintainability, etc. This measure is not very useful in the case of standards. First, it is the standards development organisation that writes its own terms of reference based on participants' contributions. Second, the commercial and strategic goals of the participants are never explicitly stated, so the technical requirements may be silent on other constraints such as the time needed for standard development, performance improvement, or cost reduction.

Deming's approach (1986) emphasises the uniformity and the predictability of the production process to reduce cost and increase reliability. "Uniformity" and "predictability" would then mean that the rules are constant across all standards, that all standards pass through the same approval process and that the layout and organization of the final documents are the same. It is not clear, however, what advantages these characteristics that apply to mass production would give to the one-of-a-kind operation that characterizes standards production.

Juran's (1992) measure of quality stresses fitness for use. This measure was generalised by the second edition of the ISO 9000 standards series to the satisfaction of the needs and expectations of stakeholders. Unfortunately, not all stakeholders are equally involved in the standardization process. Even when "user requirements" are available, they are usually high-level expectations that need to be translated into requirements for sub-systems or components.

To summarise, the traditional criteria for TQM are oriented towards internal processes within a firm to ensure mass delivery of products or services. They do not address collaborative efforts performed outside the firm. In contrast, project management literature defines a project as a unique effort to produce a new product or service that meets certain requirements (Project Management Institute, 2000). In the next section, we argue that the production of a standard can be treated as a project and that the methodologies used for project management could guide us in addressing the quality of standards.

4. Standards as Projects

It is well established that there are many categories of standards (De Vries, 1999; 2006). For example, standards can be internal to an organization or external, i.e., originating from a standards development body. The purpose of internal standards is to streamline the operations to increase operational efficiency and improve the response to emergencies. External standards in ICT relate to several levels. At the level of the infrastructure, they define the shape and form of transmission of signals from one end user to another other and the way the various media can interconnect. Network providers depend on standards to reduce uncertainties concerning equipment interoperability as well as to manage these elements with specialized information systems for configuration, maintenance, account management, security, etc. Standards help them avoid the monopoly of a single supplier or the dependency on a rare technical expertise. Likewise, service providers — such as wide area operators or Internet service providers — benefit from standards in reducing the complexities of day-to-day operations to focus on their service delivery process. They also depend on these standards in their negotiations with other carriers or virtual operators. It should be noted, however, that the more the technology is standardized, the more important is the service delivery process in the competition among service providers. Content providers and content managers depend on standards to evaluate the performance and quality of the various service providers on one hand and to present their products to businesses and consumers on the other. Finally, users need standards to minimize the risk of a technology lock-in and to reduce the dependence on individual vendors.

In the case of external standard development, the standard body appoints a rapporteur (or convener) who effectively acts as a project manager. The rapporteur's authority is of an informal nature. It is based on the respect of other participants gained through experience and knowledge, and also through fairness in conducting the proceedings. The rapporteur organises the meeting agendas at his or her discretion but has very little formal authority over the contributions. Almost the same may be said about the sponsors. Even though they typically approve the technical contributions and select their delegates, they have limited influence over decisions taken in the corridors, etc. Finally, the main control that a standard organisation has is to revoke the mandate of the rapporteur. To sum up, in standards development, the overall chain of authority is very weak.

Fortunately, project management methodologies allow for this type of weak control (Project Management Institute, 2000). We consider each work item on the standards body agenda as a project. This project may involve the development of a new or the revision of an existing standard. Let us now use the project management methodology to see how the following items are managed in a standardization project: the scope, the time in terms of the speed of standardisation, the quality of the project conduct, the cost, the resources and the project documentation.

5. Scope Management

Good scope management is essential for the success of a project. The main issue is to prevent scope creep, i.e., to avoid being side tracked by new issues and to remain focused on what has been agreed to. If changes need to be made, e.g., to reflect changes in the external environment, changes in the scope of the standard must follow a rigorous process to evaluate various courses of action and their impact on the overall project. Many standards organization recognize the need to change the original scope but do not specify a formal procedure for change control.

As an example of scope creep, consider IP technology. Originally, it was used to carry best effort traffic through a network of pipes with all the intelligence concentrated in the end terminals. To carry voice, and to support mobile and wireless networks, basic architectural modifications were needed that include quality of service guarantees. These modifications introduce intelligence into the network, which goes against the Internet's basic architectural assumptions. Another impressive example would be ITU-T's X.500 Directory Service. Here, the number of pages of the specification – which may be used as a proxy measure for scope creep – increased from 70 to over 180 within a time period of around three years (and five different draft versions). And that was before the standard was even published (Jakobs, 1989).

The scope and the technical requirements should take into account the nature of the items to be standardized and the maturity of the technology involved, so that the details of the standard are commensurate with intended uses. For example, a specific category of standards, typically those that deal with algorithmic techniques such voice coding or encryption, might require more detail than a higher-level category of standards. Also, at the early stage of a technology, unnecessary details may be burdensome for future innovations.

In general, ICT standardization reduces the variety of interfaces, thus easing the burden on equipment vendors, operators and users. However, the degree of standardization sought varies with the stakeholder intent.

Signs that could indicate that the scope is not well managed are:

- There are no – or very superficial – terms of reference (ToR), which may indicate a lack of consensus among the stakeholders.
- The standard's scope is not consistent with the intended use of the standard and the phase in the life cycle of the technology. If the scope includes too many details at an early stage of the lifecycle, the wrong aspects may be included. If standardization takes place late in the life cycle but does not respond to specific end-user demands, it may be irrelevant.
- When the decision making process requires consensus, compromises can result in which opposing or partially conflicting requirements co-exist. This may lead to an overabundance of options. In the case of emerging technologies, for example, the inclusion of many options could mean that the standard's scope is too broad, or that market conditions are not well understood.
- Changes to the terms of reference are carried out to reflect changes in stakeholders' needs or composition without considering the impact on the program schedule.
- There is no formal process to re-evaluate the need for the standard or to change the scope in response to changes in the environment (e.g., new or competing technologies).
- Existence of other standard groups with similar or overlapping activities.
- Frustrated stakeholders leave because they feel that what their needs will not be properly reflected in the standard. The original scope might have been (too) broad to the point of including incompatible sets of requirements. Another possibility is that the process excludes groups that do not share the dominant market considerations. In either case, it is important to understand what is going on and how to react. In case legitimate stakeholders are being excluded, they may start a competing standards setting activity. In such a case, the duplication of effort will lead to fragmented markets.

Consider the case of potential standard wars with China, which is caught in the so-called "technology trap." Lacking advanced technology of their own, its manufacturers have to license foreign technology at high cost. To avoid paying royalty fees, China has developed its own DVD standard, wireless communications standards and so on. (De Jonquieres, 2006).

6. Time Management

ICT standards attempt to meet a multiplicity of needs that have different time horizons (time to market, time to scale, time to profitability). For example, there is a difference in the time windows of interest of equipment vendors and service providers that use the same technology as shown in Figure 2 (Sherif, 2003b).

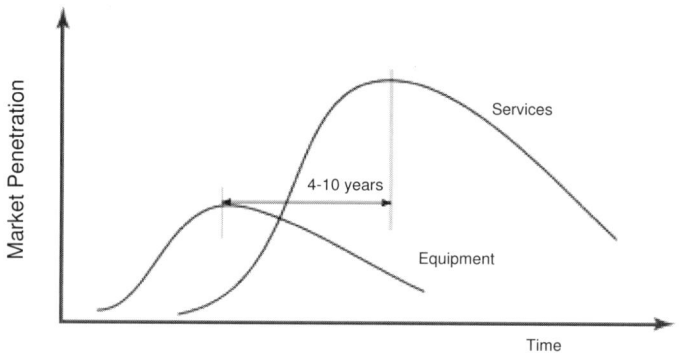

Figure 2. Timing windows for equipment and services based on a given technology.

In recent years, it has often been asserted that the "new breed" of standard organisations and consortia are capable of turning around specifications at a faster rate. There has been a strong emphasis on delivering standards as quickly as possible even though technologies with long lead times for development and deployment, such as for infrastructure projects, have a different time horizon than desktop projects. For these infrastructure projects, shaky technical agreements for the sake of quick consensus may backfire in the long run. From an end-user's perspective, it is likely that the quality of the final product or service counts more than the time the standards process took. From project management literature it is known that the triple constraints of cost, quality and time form a triad: improvement of one element comes at the expense of at least one other element in this triad. There is a danger that an overemphasis on standardisation speed may lead to quality reduction, increased cost or both (Jakobs and Blind, 2006; Sherif, 2003b). In the 1990s, the race to increase the rate of standard production encouraged many companies to establish or join consortia of likeminded parties interested in promoting a given solution. Although dropping out of the traditional standard bodies was effective in some situations, the indiscriminate application of this principle resulted in the exclusion of many interested parties. The excluded parties, in turn, banded together to form their own consortia. As a consequence, there was an avalanche of consortia that companies needed to monitor. This increased the cost of their standard participation.

In other words, because standards need to be available at the right time for the right products, the speed of development will vary according to the life cycle of the technology. One way of doing so is to match speed to technology maturity (i.e. technology S-curve) as well as market needs (Sherif, 2003a). The urgency of a standard varies on whether the innovation is incremental, architectural, radical or platform innovation. In general, service providers rely on incremental innovations to increase revenues while the profits of manufacturers are mostly associated with platform innovations.

Figure 3 illustrates how the type of standards needed at each phase of the technology life cycle depends on the technology maturity. Anticipatory standards are needed to specify the production system of the new technology provided they allow for errors. The associated specifications cannot be as robust as when the technology is mature and the market for the service is well defined. Enabling standards relate to refinements in the production system as well as improved embodiment of the technology. They are usually generated while the performance of the innovation is improving exponentially and initial products are commercialized. Standardization in this phase provides an avenue to the dissemination of knowledge about the technology. Formal standards organization or implementation consortia can equally develop enabling standards: formal specifications on the one hand and on the other, development tools, conformance verification processes and/or the promotion of the technology. Responsive standards relate to the manifestation of the technology in a service system. To define a responsive standard, the initial innovators may have to release technical information earlier than anticipated or shift product differentiation to areas such as quality, customer support or maintenance services.

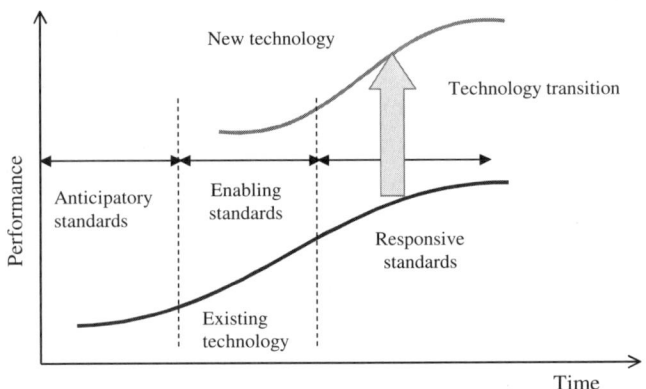

Figure 3. Timing of standards in relation with the technology S-curve.

7. Quality Management

From a process point of view, management of the quality of the output relates to its internal logical consistency and its conformity to the scope and the technical requirements.

Complex specifications are written over a period of time. Inevitably, ambiguities creep in because of the way technical experts use or misuse human languages. Internal inconsistencies could eventually lead to erroneous and/or incompatible implementations, which is problematic, particularly in communication applications. Some of the tools of structured development of software have been extended to standardization. For example, The European Telecommunications Standards Institute (ETSI) recommends that reviews (or "walk-throughs") be conducted to examine a specification and identify inaccuracies and inconsistencies, and that formal languages be used for protocol descriptions. Another option is to use technical editors (other than those who developed the standards). Lastly, there is considerable evidence that the use of bilingualism early in the drafting process can be of great help in the preparation of clear and unambiguous text, not only in English, but in the other languages as well. At a minimum, translating international standards into different languages before final adoption has a positive side-effect because translation involves new parties that go through a thorough review of the full text, so that they can uncover ambiguities and inconsistencies (Teichmann *et al.*, 2006). In addition to the associate costs, the linguistic issue is unfortunately trapped in non-productive debates on hegemony and cultural imperialism.

Other implementation-related problems that may later hinder interoperability are missing details — due to inattention to details, to excessive reliance on tacit knowledge or to keep competitors at bay—, lack of clarity about how options should be treated during implementation, and what consequences partial implementation of the specifications have (Egyedi, 2006). Specifications with pseudo-code or formal languages are clear and unequivocal. There are automatic tools to check the consistency of formal descriptions and detect potential problems, such as deadlocks in communication protocols. Once the content has been validated, formal specifications can be the starting point for automating the generation of the software or firmware, thereby reducing the chance of error. The protocol performance can be also evaluated. Another advantage is that the generation of test sequences for conformance testing can be automated to cover certain usages within predefined constraints (Linn and Uyar, 1994; Sherif and Krishnakumar, 1990).

Conformance testing procedures help determine the extent to which an implementation conforms to the technical requirements by interoperation with a reference implementation built according to the specification. In the past, the ITU supplied test procedures for its voice coders (G.726, G.729, etc.). ISO and the ATM Forum developed conformance procedures for their specifications. ETSI has made a Protocol and Testing Competence Centre (PTCC) available to its technical bodies to verify their deliverables. However, to reduce expenditures and shorten the development time, many protocols are finalized before these quality steps are undertaken.

A complementary approach to manage specification quality is to test the interoperability of different implementations to determine their end-to-end functionalities before final release. Ideally, a new specification should interoperate with the legacy ones, particularly those that have wide usage. Also, coordination with other standard bodies can ensure that emerging standards do not end up incompatible. This is not as difficult as it sounds, because the sponsors usually participate in several standards organizations. So, a minimum of coordination can already be achieved if each company harmonizes its contributions to the various standard bodies.

Accordingly, some of the signs of impending trouble are:

- No formal description
- Many options
- Missing information
- Inconsistency among different sections of the same document
- No tests for conformance or interoperability
- No pre-implementations

8. Cost Management

The principle that the costs of developing a standard should not exceed the expected benefits may sound reasonable, but it is very difficult to implement. There are direct costs in developing standards such as the salaries of the people involved and the cost of participation (subscription fees, travels, etc.). There are also other indirect costs, such as the cost of revealing proprietary information or the royalties paid for Intellectual Property Rights (IPRs), as well as the opportunity cost of delaying investments until the standard materializes. There are also the costs of too many specifications, of incompatibilities, etc.

The difficulty in cost estimating arises from two main problems. First, some costs are difficult to estimate and historical data is difficult to come by. Second, standard development is a shared endeavour and financial record keeping for the expenditure is not centralized.

In general, what is needed is to establish a measure to assess the costs and benefits of the following:

- Backward compatibility
- Additional options
- Development of a reference implementation, conformance testing suites and end-to-end service testing before the standard is published.

What is needed for project planning and decision does not have to maintain the rigor of accounting standards. Potential measures could be the development effort in terms of man-hours.

A more challenging area is to develop the cost of incompatibilities or defective standards. How to do so is not obvious at all. One of the difficulties is that vendors and service providers are usually very reluctant to admit any problem with their specific implementations.

9. Resource Management

Standard setting requires the availability of many resources: logistics, coordination, etc. However, the most important resource by far is the availability of experts, i.e., knowledgeable, motivated, and committed individuals who come up with the specifications. Moreover, these individuals need to be willing to co-operate with the project manager within the scope of the project.

What makes the standards setting different from many other projects is that participants are usually detached by various project sponsors, even though these sponsors are often competitors. As a consequence, not all the necessary expertise may be publicly available. Furthermore, for many of these participants, standards setting is a side activity that takes them away from their functional responsibilities. In practice, except perhaps for cases of statutory standardization, the process for technical standardization is more oligarchic than democratic.

Signs of trouble include, but are not limited to:

- One or more strong participants take over the agenda and drive it beyond the scope of the project.
- Competing interests of the sponsors lead to a deadlock in the deliberations.

- Individual rivalries among participants prevent progress: this happens, particularly, if the sponsors do not give clear mandates to their representatives or employees.
- "Group think" and the rejection of new ideas is a problem common to technical committees.

Being dependent on voluntary contributions makes resource management a difficult task. Few painless "corrective" measures are available. One such a measure is to define the decision criteria in the scope including an escalation procedure.

A more significant measure would be to include in the standardization budget of the standard organization an item to allow the hiring of the necessary technical expertise, on a project basis, to act as facilitators and/or independent mediators answerable only to the standardization organization.

10. Documentation Management

Readability of the document is an important quality of the standard in terms of its fitness for use. International standards production uses natural languages (mostly English) and other non-verbal tools such as equations, pseudo-code, drawings, etc., to describe the technical content. Yet, technical/scientific experts are not necessarily good technical writers, especially if English is not their native language. The deficiency in writing skills may end up in creating a gap between the intention of the standards producers and the interpretation of the standard consumers, as they develop products or services (Teichmann *et al.*, 2006).

Although the final draft of the specifications is published for comments, the various contributions that justify the choices made during the standards process as well as any supporting data are generally not available. Some technical editors solve this problem by including explanatory material; web-centered document management may offer another route by connecting the text with tutorials and other ancillary documents (Purper, 1999; Spring *et al.* 1999).

A typical sign of trouble in document management is that the standard specification is not well organised or indexed, making it difficult to retrieve information. Even though most standardization bodies have a basic form and style for their published documents, very few have editors to check whether a reader which did not participate in the deliberations could get the necessarily information easily. This is particularly problematic in international bodies because the working language (English) is often the second or third language of the technical experts.

Usually, some of these problems are found ex-post, when the missing or "hidden" information lead independent teams to develop incompatible implementations.

11. Stakeholders in Standards Quality

Table 1 depicts how various stakeholders in the standardization approach the issue of standards quality. The table shows that end-users are mostly interested in the outcome and not the technical process. As a consequence, they are not represented at the initial phases of the project when the scope and the requirements of the future standard are being discussed. Another reason for the absence of end-users is that some standards bodies (e.g., ISO) consider concrete implementations to be commercial realizations that lie outside the scope of ICT standard projects. This, in turn, poses structural problems for attempts to understand legitimate end-user concerns and include them from the start in the standards process.

Some non-governmental organizations may attempt to fill that gap and try to represent the "consumers" at large. For example, the European Commission and the *European Free Trade Association* (EFTA) fund the *European Association for the Co-ordination of Consumer Representation in Standardisation* (ANEC) to contribute to the standardization process. However, the cooptation of an unelected group does not insure adequate representation of the wide range of end-user interests across the Union. In other words, the absence of the citizenry and the lack of a democratic mandate to the Association and its mission puts the whole approach to end-user representation very much in doubt.

It is quite clear from this study that, at any instant of time, none of the stakeholders has complete information or direct control over the outcome of the standardization process. More explicitly, the assumption of transparency on which market mechanisms are built is not satisfied. The implication is that market forces alone cannot lead to improved quality in standards development.

These two aspects, the lack of democratic representation for end users' interests and the inadequacy of market mechanisms in ensuring standards quality, call for a re-evaluation of the approach taken in the last two decades which has de-emphasized the role of governments and of non-commercial institutions on the quality of standards.

Table 1. Stakeholders' interest in standards quality.

Stakeholder	Angle of Interest	Quality Emphasis	Relevant Project Management Aspect
Owner (standards body)	Legitimacy Due process	P	Resource
Producer (technical committee)	Technical Due process	O, P	Quality, resource, time
Supplier (committee participant and standard developer)	Technical Due process	O	Resource, quality, documentation
Sponsor (Companies financing participants)	Marketing Financial (possibly technical)	O	Time, cost, resource
Consumers (implementers of standard)	Technical Ease of implementation	O	Quality, documentation
End-users (users of standard-compliant product)	Usability (interoperability and functionality) of standard-compliant product or service	O	Quality
Regulators	Legitimacy Due process	O, P	Quality, documentation

O = Outcome, P = Process.

12. Conclusion

To the best of our knowledge, we have been the first to attempt to define what quality of standards means taking into account the multidimensional nature of standardization. In this paper, we adapted the principles of project management to the standards setting process in the ICT area to provide a way of managing risks.

The diverse interests that affect standardization, the distributed nature of its management process and the time lag between a standard and its implementation in products and services mean that there is no clear accountability in terms of profit and loss responsibilities caused by deficiencies in an ICT standard. In some cases, those who pay the cost of the lack of quality are not those who made the decisions. Thus, market mechanisms will rarely provide the driving incentive to carry out the intensive planning and coordination across organizational boundaries that is needed to produce a quality standard.

While process quality does not always lead to product quality, the checklist of potential troublesome signs can be used to manage the risks associated with the adoption of a given standard, i.e., identification of risk factors, the likelihoods of their occurrence and the evaluation of various contingencies. Recognizing these telltale signs at an early stage of the process could greatly improve the quality of the standardization process. Risk identification and management can be done individually or collectively, by the rapporteurs, the standard bodies or the participant companies.

The project management methodology adopted here was accepted after a long gestation period that allowed for data collection to validate its prescriptions. Likewise, we need to explore the correlation between the standards setting process as such and the quality of its output. Such a study requires long term tracking of several standards and their implementations.

References

Ash, J. and G. Choudhury. (2004). "PNNI routing congestion control." *IEEE Communications Magazine,* 42 (11), 154–160.

Baskin, E., K. Krechmer, and M. H. Sherif. (1998). "The six dimensions of standards: Contribution towards a theory of standardization." In: Lefebvre, L. A., R. M. Mason, and T. Khalil (Eds.), Management of Technology, Sustainable Development and Eco-Efficiency, Elsevier, Oxford, U.K., pp. 53–62.

Cavendish, D., H. Ohta, and H. Rakotoranto. (2004). "Operation, Administration, and Maintenance in MPLS networks." *IEEE Communications Magazine*, 42 (10), 91–99.

Crosby, P. B. (1979). Quality is free. The art of making quality certain, New York, McGraw Hill.

De Jonquieres, G. (2006). "To innovate, China needs more than standards." Financial Times, July 13, p. 11.

De Vries, H. (1999). Standards for the Nation. Analysis of national standardization organizations, Kluwer Academic Publishers, Boston, USA.

De Vries, H. (2006). "IT standards typology." In: Jakobs, K. (Ed.), Advanced Topics in Information Technology Standards and Standardization Research, Vol. 1, Idea Group, Hershey, PA, USA, pp. 1–26.

Deming, W. E. (1986). Out of the crisis, Massachusetts Institute of Technology, Center for Advanced Engineering Study, Cambridge, USA.

Egyedi, T.M. (2006). "Experts on causes of incompatibility between standard-compliant products." In: Doumeingts, G., J. P. Mueller, G. Morel and B. Vallespir (Eds.), Enterprise Interoperability. Springer Verlag, Berlin, Heidelberg, New York.

European Telecommunications Standards Institute (ETSI). "Making better standards: Practical ways to success," http://portal.etsi.org/mbs.

Fang, L., A. Atlas, F. Chiussi, K. Kompella, and G. Swallow. (2004). "LDP failure detection and recovery." *IEEE Comm. Mag.*, 42 (10), 117–123.

Jakobs, K. (1989). "The Directory—Evolution of a standard." In: Bojanov, K.; Angelinov, R. (Eds.), Proc. IFIP TC6/TC8 Open Symposium on Network Information Processing Systems, North Holland, Amsterdam, pp. 281–289.

Jakobs, K., R. Procter, and R. Williams. (2001): "The Making of standards." *IEEE Communications Magazine*, 39 (4), 102–108.

Jakobs, K. and K. Blind. (2006). "Co-ordination in ICT Standards Setting." In: Bolin, S. (Ed), The Standards Edge: Unifer or Divider? The Bolin Group, Menlo Park, CA.

Juran, J. M. (1992). Juran on quality by design. The new steps for planning quality into goods and services. New York, Free Press.

Knutson, J. and I. Bitz. (1991). Project management. How to plan and manage successful projects, American Management Association, New York.

Linn, R. J. and M. Ü. Uyar. (1994). Conformance testing methodologies and architectures of OSI protocols, *IEEE Computer Society Press*, Los Alamitos, CA.

Moseley, S., S. Randall, and A. Wiles. (2003). "Experience within ETSI of the combined roles of conformance testing and interoperability testing." In: Egyedi, T. M., K. Krechmer, and K. Jakobs (Eds.), Proceedings of the 3rd IEEE Conference on

Standardization and Innovation in IT, SIIT 2003, IEEE Press, Piscataway, NJ, pp. 177–190.

Project Management Institute. (2000). Guide to the Project Management Book of Knowledge (PMBOK® Guide), Newtown Square, PA.

Purper, C. B. (1999). "An environment to support flexibility in process standards." In: Jakobs, K. and R. Williams (Eds.), Proceedings of the 1st IEEE Conference on Standardization and Innovation in IT, SIIT'99, IEEE Press, Piscattaway, NJ, pp. 189–198.

Sherif, M. H. (2003a). When is standardization slow? *J. of IT Standards & Standardization Research,* 1 (1), 19–32.

Sherif, M. H. (2003b). "Technology substitution and standardization in telecommunication services." In: Egyedi, T. M., K. Krechmer and K. Jakobs (Eds.), Proceedings of the 3rd IEEE Conference on Standardization and Innovation in Information Technology, SIIT 2003, IEEE Press, Piscataway, NJ, pp. 241–252.

Sherif, M. H. and A. S. Krishnakumar. (1990). "Evaluation of protocols from formal specifications: A case study with LAPD," Proceedings of GLOBECOM'90, IEEE Press, Piscattaway, NJ, Vol. 2: 879–886.

Sherif, M. H. and D. K. Sparrell. (1992). "Standards and innovation in telecommunications." *IEEE Comm. Mag.,* 30 (7), 22–29.

Soininen, A. H. (2005). "Open standards and the problem of submarine patents." In: Egyedi, T. M. and M. H. Sherif (Eds.), Proceedings of the 4th IEEE Conference on Standardization and Innovation in IT, SIIT'2005, pp. 231–244.

Soininen, A. (2007). "Patents and standards in the ICT sector: Are submarine patents a substantive problem or a red herring?" To appear in: *J. of IT Standards & Standardization Research,* 5 (1).

Spring, M., R. Andriati, and V. Vathanophas. (1999). "Usability of a collaborative authoring system for standards development: Preferences, problems, and prognosis." In: Jakobs, K. and R. Williams (Eds.), Proceedings of the 1st IEEE Conference on Standardization and Innovation in IT, SIIT'99, IEEE Press, Piscattaway, NJ, pp. 211–226.

Teichmann, H., H. J. de Vries, and A. J. Feilzer. (2006). "Linguistic qualities of international standards," *J. of IT Standards & Standardization Research,* 4 (2), 70–88.

CHAPTER 30

A METHODOLOGY TO MEASURE THE INNOVATION PROCESS CAPACITY IN ENTERPRISES

José Ramón Corona-Armentats, Laure Morel Guimaraes, and Vincent Boly

I.N.P.L.- E.R.P.I., Nancy, France

Innovation represents a major strategic element, critical to the development of the company. Management has to routinely evaluate company's capability with respect to innovation. Even if innovation systems adopted by enterprises are specific, a methodology for measuring the innovative process is warranted. In this paper we propose an evaluation methodology regardless of the industry specifics. It is based on an innovative pilot model, which contains 13 practices, developed in our laboratory. In this paper, a comparison is presented between an in-house methodology called IPI (Innovation Potentiality Index) and two multiple criteria methods (Electre and AHP) to measure the innovation process capacity of companies. IPI defined their strong and weak points regarding to several indicators. The result of our research leads to an evaluation tool allowing both a local and global comparison between companies issued from the same industrial field or not.

1. Introduction

Innovation represents a major strategic element of development policies at macroeconomic and microeconomic level of an organization. It relates to the new techniques, products, processes or services (Afuah, 1999). It results in the improvement or the technological change in companies (Escorsa *et al.*, 2001). At the beginning, innovation is, at microeconomic level, a search to adapt a new activity, and then the company will have to ensure its stability and expansion on the market (Dert, 1997; Afuah,

1999; Turriago, 2002). At the macroeconomic level, countries implemented actions in order to activate, to encourage, to develop innovation, and to make the necessary adjustments needed to increase assistances and stimulation for it's begin (Amable *et al.*, 1997). Consequently, the development of the "capacity of innovation" of the companies is a major stake (European Commission, 1995). As a consequence, in the one hand, executives and leaders develop adapted management actions, and, on the other hand, the actors of the economic development engage stimulating operations. Thus, it becomes essential to evaluate the actions carried out to increase the capacity of innovation in terms of efficiency and effectiveness. Managers need to assess if the resources brought for the deployment of the innovation in a company prove to be justified. Our research has been thus related to the measurement of the capacity to innovate of the industrial systems.

2. Research Problems

Our attention is directed toward technological innovation within manufacturing companies. The aim is a better understanding of the management practices, the methodologies and the organizational schemes developed at one given moment in a company. Finally the objective is to determine if this potential is favorable for innovation. We focus on the activities supporting the process of innovation and not on the results of this process. Based on an innovative pilot model, which contains thirteen practices, developed in our laboratory (Boly, 2004), and taking in account that innovation processes adopted by enterprises are specific, we propose a methodology for measuring the innovative process capacity of a given company. Both a local analysis and global comparison between companies issued from the same industrial field are possible thanks to the methodology. In this paper, this methodology called IPI (Innovation Potentially Index) is described and later a comparison between IPI and two multiple criteria methods (Electre and AHP) is developed. Experimental results allow the definition of the assets and limitation of this approach.

3. Measurement Systems and Measuring Innovation

Through time, human beings sought to understand the environment around them, and to manage it with methods of comparison and control. In other words, it was often considered that measuring can permit controlling, that controlling can permit managing, and finally managing can permit improving the system studied. However to compare, measurements require a common denominator for all sizes. Metrology is the science of measurement that uses a universal language whose definition, evaluation, resulting expression of measurements, units, and uncertainties are its starting point. Measuring is concerned with the comparison of results expressed through a numerical value and obtained with a recognized method (whose characteristics of the reference frame are clearly established) by several partners (Himbert, 1998). Consequently, measurement uses specific tools (Morel *et al.*, 1998). In the field of innovation, literature is mostly concerned with the evaluation of the performance of a company within a given innovative project. Many authors propose approaches determining the balance between the outcomes and the inputs of innovation. Generally, financial and commercial variables are taken into account (Griffin 2000, Huang 2004).

Moreover, for a better understanding of the global impact of innovation on economy, new statistics and indicators are also developed on a macro-level (European Commission, 1997). Indeed, globalization of economic activities gives criteria for a standardization of the measurement of innovation (Canibano *et al.*, 2000).

We propose a measurement system in order to evaluate the continuous process of innovation of companies. The reference level is not the project but the company itself. Innovation measurement aims at:

- determining the development degree of an organization,
- finding the problematic parts of each studied enterprise,
- comparing innovative companies,
- analyzing and anticipating the temporal behavior of a company.

4. Methodological Proposal: The Potential Innovation Index (PII)

4.1. *Thirteen fundamental pilot practices for company innovation*

The innovation capacity measurement principles depend on the properties (practices) of the innovation process. Boly (Boly, 2004) listed and gathered practices observed in innovating companies in thirteen categories: thirteen fundamental pilot practices for company innovation.

- *Practice 1*: Innovation actors work to develop projects and technology evolution with *design tasks*.
- *Practice 2*: Involves a fundamental *follow-up of each innovative project*.
- *Practice 3*: A global supervision of new innovative projects (budget, deadline…) must be led with an *integration of the strategic dimension* dictated by the management team.
- *Practice 4*: Within *the project's portfolio*, the direction ensures *coherent management* between different initiatives.
- *Practice 5*: The management team and project managers have to *control* and *receive feedback* on innovation processes in order to develop the practices of the actors.
- *Practice 6*: *Suitable context and working conditions* have to be created in order to stimulate innovation.
- *Practice 7*: *Clear steps* toward *necessary competence allocation* to the innovation process.
- *Practice 8*: *Moral support* must be given by the management team and the project managers to the innovation process participants.
- *Practice 9*: *An environment of collective learning* has to exist for the actor, as a project progresses.
- *Practice 10*: An effort of *capitalization of know-how and knowledge acquired* during the former projects must be done, know-how which will be used for forthcoming projects.
- *Practice 11*: *Survey tasks* (technological, competitive, economic, managerial, intelligence) must be organized in order to open-up the company to the environment.
- *Practice 12*: The management team has to *manage the networks* in which the firm is integrated.

- *Practice 13*: *New ideas* from research, marketing or those proposed by the employees must be continuously collected using *creativity*, in order that future projects emerge.

Innovative companies develop these thirteen practical, totally or partially, with more or less relevance, formally organized or not. These elements constitute the basis of the evaluation.

4.2. Formulating PII

The thirteen practices are not directly measurable and as a result if this, each one of them is divided into several sub-characteristics, each of these under-characteristics being observable. Then a mathematical algorithm is proposed. The objective is to provide a unique indicator of the innovation potential of each company.

$$p_i = \frac{\sum_{j=1}^{m} v_j * q_j}{\sum_{j=1}^{m} v_j} \qquad (1)$$

Where:
p_i is the practice development level i,
q_j is the observable variable value j,
v_j is the importance value (weight) of observable variable q_j,
m is the number of observable descriptive variables of practice I,
j is the variable number.

At last, to obtain the global IIP of a company the following formula is proposed:

$$I = \frac{\sum_{I=1}^{n} w_i * p_i}{\sum_{I=1}^{n} w_i} \qquad (2)$$

Where:
I is the potential innovation value of a company,
p_i is the development degree of practice i,
w_i is the importance value (weight) of practice p_i,
N is the innovation fundamental practices number,
i is the number of the practice.

5. Analysis and Results

130 under-characteristics are listed. Table 1 synthesizes the number of observable phenomena by practice as well as the relative weight for each practice of the technological innovation given by the experts.

This observation grid has been experimented within a panel of 20 companies (E01 to E20) to be considered innovative. Table 2 shows these experimental results. The last column presents the average value (Em) of each practice.

Table 1. Under-characteristics and weight by practice.

	Technological Innovation Practices	**Observable Under-Characteristics**	**Weight**
P1	Design	20	38
P2	Project Management	11	3
P3	Integrating Strategic	12	20
P4	Project Portfolio Management	8	1
P5	Organization of Innovation Task	7	10
P6	Feedback on Innovation Process	5	2
P7	Allowance of Necessaries Competences	5	1
P8	Incentive to Innovating	6	2
P9	Memorizing Know-how	3	4
P10	Technological Watch	15	2
P11	Networks	18	2
P12	Collective Training	6	5
P13	Capitalization of Ideas and Concepts	14	10
	Σ	130	100

A Methodology to Measure the Innovation Process Capacity in Enterprises   455

Table 2. Outcomes within the 20 companies panel.

	Innovation Practices	value max (VM)	weigth (VA)	Innovation Process on Enterprise — Data Analysis — value response (VR)																				av
				E01	E02	E03	E04	E05	E06	E07	E08	E09	E10	E11	E12	E13	E14	E15	E16	E17	E18	E19	E20	
1	Conception	20	38	10	16	4	3	7	15	3	16	10	9	4	17	13	15	12	8	9	15	10	8	10,2
2	Project Management	11	3	9	9	0	4	10	7	2	6	7	7	4	11	7	8	5	4	2	6	7	5	6
3	Integrating Strategic	12	20	7	4	0	8	5	6	6	4	4	3	4	4	9	8	3	4	4	9	6	2	5
4	Project Portfolio Management	8	1	4	5	0	3	1	7	4	4	3	3	3	7	6	1	6	4	4	5	6	6	4,1
5	Organization of Innovation Task	7	10	4	6	0	0	5	7	4	4	6	4	3	7	6	7	4	1	4	5	5	3	4,25
6	Feedback on Innovation Process	5	2	2	2	0	0	2	3	1	5	1	1	0	3	4	3	2	2	0	4	2	1	1,9
7	Allowance of Necessaries Competences	5	1	2	5	2	3	4	3	1	3	2	2	2	5	2	4	2	2	4	3	3	2,8
8	Incentive to Innovating	6	2	3	3	0	3	0	6	0	5	0	1	2	6	1	4	0	2	0	5	4	4	2,45
9	Memorizing Know-how	3	4	0	2	0	0	1	2	0	2	1	1	1	2	1	0	0	1	0	2	2	2	1
10	Technological Watch	15	2	11	7	0	6	2	4	3	2	1	5	9	0	8	8	5	6	2	11	8	8	5,3
11	Networks	18	2	7	15	0	4	11	13	8	14	0	9	2	14	12	11	14	10	5	16	13	9	9,35
12	Collective Training	6	5	2	5	0	1	1	3	1	0	1	1	2	4	2	0	0	3	2	4	3	2	1,85
13	Capitalization of Ideas and Concepts	14	10	7	7	3	6	2	8	1	9	2	2	2	3	5	7	4	3	3	10	6	6	4,8
	amount	130	100	68	86	9	41	51	84	34	74	38	48	38	83	76	76	57	50	37	96	75	59	59

Table 3. Pi, Vi and PII values.

Pratiques	E01	E02	E03	E04	E05	E06	E07	E08	E09	E10	E11	E12	E13	E14	E15	E16	E17	E18	E19	E20	moy
1	0,500	0,800	0,200	0,150	0,350	0,750	0,150	0,800	0,500	0,450	0,200	0,850	0,650	0,750	0,600	0,400	0,450	0,750	0,500	0,400	0,510
2	0,818	0,818	0,000	0,364	0,909	0,636	0,182	0,545	0,636	0,636	0,364	1,000	0,636	0,727	0,455	0,364	0,182	0,545	0,636	0,455	0,545
3	0,583	0,333	0,000	0,667	0,417	0,500	0,500	0,333	0,333	0,250	0,333	0,333	0,750	0,667	0,250	0,333	0,333	0,750	0,500	0,167	0,417
4	0,500	0,625	0,000	0,375	0,125	0,875	0,500	0,500	0,375	0,375	0,375	0,875	0,750	0,125	0,750	0,500	0,500	0,625	0,750	0,750	0,513
5	0,571	0,857	0,000	0,000	0,714	1,000	0,571	0,571	0,857	0,571	0,429	1,000	0,857	1,000	0,571	0,143	0,571	0,714	0,714	0,429	0,607
6	0,400	0,400	0,000	0,000	0,400	0,600	0,200	1,000	0,200	0,200	0,000	0,600	0,800	0,600	0,400	0,400	0,000	0,800	0,400	0,200	0,380
7	0,400	1,000	0,400	0,600	0,800	0,600	0,200	0,600	0,400	0,400	0,400	1,000	0,400	0,800	0,400	0,400	0,400	0,800	0,600	0,600	0,560
8	0,500	0,500	0,000	0,500	0,000	1,000	0,000	0,833	0,000	0,167	0,333	1,000	0,167	0,667	0,000	0,333	0,000	0,833	0,667	0,667	0,408
9	0,000	0,667	0,000	0,000	0,333	0,667	0,000	0,667	0,333	0,333	0,333	0,667	0,333	0,000	0,000	0,333	0,000	0,667	0,667	0,667	0,333
10	0,733	0,467	0,000	0,400	0,133	0,267	0,200	0,133	0,067	0,333	0,600	0,000	0,533	0,533	0,333	0,400	0,133	0,733	0,533	0,533	0,353
11	0,389	0,833	0,000	0,222	0,611	0,722	0,444	0,778	0,000	0,500	0,111	0,778	0,667	0,611	0,778	0,556	0,278	0,889	0,722	0,500	0,519
12	0,333	0,833	0,000	0,167	0,167	0,500	0,167	0,000	0,167	0,167	0,333	0,667	0,333	0,000	0,000	0,500	0,333	0,667	0,500	0,333	0,308
13	0,500	0,500	0,214	0,429	0,143	0,571	0,071	0,643	0,143	0,143	0,143	0,214	0,357	0,500	0,286	0,214	0,214	0,714	0,429	0,429	0,343
	6,228	8,634	0,814	3,873	5,102	8,688	3,186	7,404	4,011	4,526	3,955	8,984	7,234	6,980	4,823	4,876	3,395	9,488	7,618	6,128	5,797
Pi	0,479	0,664	0,063	0,298	0,392	0,668	0,245	0,570	0,309	0,348	0,304	0,691	0,556	0,537	0,371	0,375	0,261	0,730	0,586	0,471	0,446
Vi	0,523	0,662	0,069	0,315	0,392	0,646	0,262	0,569	0,292	0,369	0,292	0,638	0,585	0,585	0,438	0,385	0,285	0,738	0,577	0,454	0,454
IIP	0,504	0,660	0,101	0,285	0,383	0,679	0,259	0,601	0,411	0,365	0,269	0,667	0,622	0,648	0,419	0,346	0,356	0,734	0,535	0,380	0,511

In Table 3 partial values are calculated. It includes one line with partial values given by the formula pi = vj/vjmax where vj is the value of the positive answers of the practice J and vjmax is the possible maximum value of the practice j, and another line with the PII value.

To examine the outcomes, two options may be considered:

Firstly, Vi, is the mean value of answers by company. The general mean is 59, it is low in comparison with 65 (this is the half of the higher probable value). In general, only 1 (5%) of all the companies uses less than 25% of the activities suggested (E03), 9 (45%) use more than 50%, and none employs more than 75% of elements.

The company with the lowest performance is E03 (pi = 0.063), followed by E07 (pi = 0.245), E17 (pi = 0.261), and E04 (pi = 0.298).

The companies with the highest performance are E18 (pi = 0.730), E12 (pi = 0.691), E06 (pi = 0.668) and E02 (pi = 0.664). The average value is lower 50% (pim = 0.446).

Secondly, the Potential Innovation Index (PII). The lowest index corresponds to PII(E03) = 0.101 followed by PII(E07) = 0.259, PII(E11) = 0.269 and PII(E04) = 0.285. On the other hand, the most powerful company according to IPI is E18 with PII(E18) = (0.734), behind one finds PII(E06) = 0.679, PII(E12) = 0.667 and PII(E02) = 0.660.

5.1. *Comparison with other methods*

To validate the results obtained by this method, other methods have been tested: The Electre method developed by Bernard Roy (Roy *et al.*, 1993) and AHP method (Analytic Hierarchy Process) by Saaty (Roy *et al.*, 1993; Romero, 1996).

The Electre method is an outranking multiple criteria method. It permits to choose the best action according to a criteria group (Collette *et al.*, 2002). Electre analyzes various solution alternatives in multiple decisions by outranking binary relations. (Sanchez Guerrero, 2003). The method is an efficient process to reduce the solution whole (Romero, 1996). It is based on outranking in a non-compensatory logic with a veto utilizing concordance and discordance concepts (Roy, 1985). Aggregation formulas used by the preference's system in Electra I are among the simplest in comparison with the multiple criteria systems developed later (Roy *et al.*, 1993).

Analytic Hierarchy Process (AHP) supposes that the process to determine the classification of qualification is transitive, i.e., if A is higher than B and B is higher than C then A is higher than C (Sanchez Guerrero, 2003). Analytic Hierarchy Process can be interpreted like a way of implementing an additive aggregation (Roy *et al.*, 1993). AHP is one of the multiple criteria decision aiding techniques most known and most applied (Sugihara *et al.*, 2004; Escobar *et al.*, 2004). The construction of AHP is based on complete aggregation, type additive, characteristic of American school (Macharis, *et al.*, 2004).

The results obtained in application of various methodologies are presented in Table 4, by descending order. Note that the companies classified in first and last position (E18 and E03, respectively) hold their position, just as the companies that are in positions 8, 9, 10, and 16 (E19, E01, E15, and E16, respectively).

Table 4. Enterprise classification.

Method	Classification																			
	1	2	3	4	5	6	7	8	9	10	11	12	13	14	15	16	17	18	19	20
PII	E18	E06	E12	E02	E14	E13	E08	E19	E01	E15	E09	E05	E20	E10	E17	E16	E04	E11	E07	E03
ELECTRE	E18	E14	E13	E06	E02	E08	E12	E19	E01	E15	E09	E17	E05	E20	E10	E16	E07	E04	E11	E03
AHP	E18	E06	E02	E14	E13	E12	E08	E19	E01	E15	E20	E09	E05	E17	E04	E16	E10	E07	E11	E03

Looking at Table 4, four groups of companies may be distinguished:

The behavior of the companies E18, E06, E12, E02, E14, E13 and E08 is not uniform (except for the E18 company) but they all are present of the first at the seventh place (in agreement with distribution given by Figure 1).

Afterwards, and according to the suggested classification method, we note the presence of companies E19 and E01, occupying the 8th and 9th methods for each of the three classifications. E15 company appears in three classification in the site ten and the E09 company is in place 11th in PII and Electra; and 12th in AHP.

The company E16 keeps the same localization in the three methods but the positions of companies E05, E20, E10, and E17 are different in each method, except for E05, which guards the same position in Electra and AHP (13th) but which changes with the PII (12th). In the other cases, the E20, E10, and E17 companies have places 13, 14 15 and 16 by method PII; 14, 15 and 12, by Electre; and in AHP they have places 11, 17 and 14. Finally, the companies E04, E18, E19, and E20; located by PII in 17, 18, 19 and 20; by Electra in 18, 19, 17 and 20; and by AHP in 15, 19, 18 and 20.

In conclusion, even if the classifications order is not exactly identical, the different methods give similar global results.

5.2. Innovation processes classification

Godet (Godet, 1997), suggests the following classification:

1) *Proactive*. Most dynamic and offensive companies, causing the changes desired with a vision in long term and a structured innovation management.
2) *Preactive*. Companies that impose no changes, but anticipate them. These dynamic and offensive companies develop an average term strategic vision.
3) *Reactive*. These companies stay in touch with environmental dynamics. They have short-term vision, and their strategy is defensive.
4) *Passive*. The strategy is defensive; a company clings to the medium to survive.

Godet classification is used to propose a typology of innovative systems. Table 5 shows each practices value. By fixing limits with IIP, the aim is to determine if some practices are class specific or/and to analyze the practice's sub-characteristic spectrum. Proactive companies have an IIP over 0.6. From 0.6 to 0.4 are preactive companies. From 0.4 to 0.3 are reactive companies. Then passive companies have an IIP under 0.3.

Table 5. Values of models by companies with innovative system.

	Innovation Process on Enterprise Data Analysis Value of Model Innovant System						
Innovation Practices	PROACTIVE		PREACTIVE		REACTIVE	PASSIVE	BASICALLY
	Target	Base	Target	Base	Target	Target	Target
1 Conception	0,850	0,750	0,600	0,500	0,450	0,200	0,200
2 Project Management	1,000	0,545	0,818	0,636	0,909	0,364	0,364
3 Integrating Strategic	0,750	0,750	0,583	0,500	0,417	0,667	0,417
4 Project Portfolio Management	0,875	0,625	0,750	0,750	0,750	0,500	0,500
5 Organization of Innovation Task	1,000	0,714	0,857	0,714	0,714	0,571	0,571
6 Feedback on Innovation Process	1,000	0,800	0,400	0,400	0,400	0,200	0,200
7 Allowance of Necessaries Competences	1,000	0,800	0,600	0,600	0,800	0,600	0,600
8 Incentive to Innovating	1,000	0,833	0,667	0,667	0,667	0,500	0,500
9 Memorizing Know-how	0,667	0,667	0,667	0,667	0,667	0,333	0,333
10 Technological Watch	0,733	0,733	0,733	0,533	0,533	0,600	0,533
11 Networks	0,889	0,889	0,778	0,722	0,611	0,444	0,444
12 Collective Training	0,833	0,667	0,500	0,500	0,500	0,333	0,333
13 Capitalization of Ideas and Concepts	0,714	0,714	0,500	0,429	0,429	0,429	0,429
IIP	0,834	0,734	0,622	0,535	0,507	0,396	0,345

Proactive Innovative System. Are considered the companies E02, E06, E08, E12, E13, E14 and E18. The companies of this class type have a PII higher than 60%.

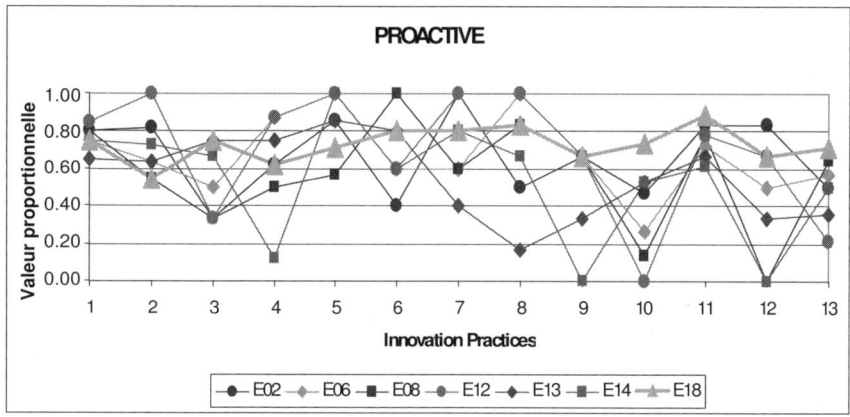

Figure 1. Proactive companies.

Figure 1 helps determine a target model company (ideal) and a basic model company (acceptable minimum) within this class. Target model company results from compilation of maximum values found for each practice, basic model company from the minimum.

Concretely, top management of one company may seek to compete with the best company (E18) or to develop the maximum characteristics held by target model company proactive mention. The PII of proactive target model company is 0.834. We observe that partial values are higher than 0.5.

Preactive Innovative System. This category federates the companies E01, E09, E15, E19. The values of PII are between 41.1% and 53.5%.

In a similar way to the preceding demonstration, a target model and a basic model are defined (see Figure 2).

The PII of a target model company préactive is 0.622. The companies that belong to this category must seek to obtain more than the basic values, in fact E19, and at least the target values (before changing for another class).

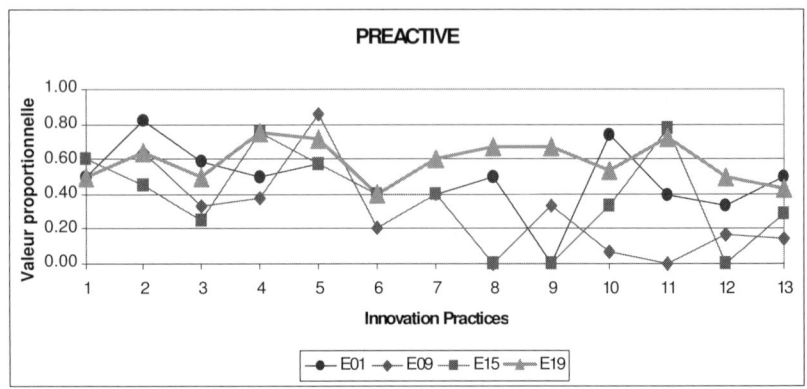

Figure 2. Preactive innovative system.

Reactive Innovative System. The companies E05, E10, E16, E17 and E20 are considered. The PII is between 0.346 and 0.380. The target value of PII is 0.507.

Here, the best-classified company is E05. However, it is necessary to note the even complicated irregular behavior of its partial values, which oscillates between maximum values of obtaining a practice as well as between minimal values.

Does this statement mean that it will not be possible with the companies resting on this category to find parameters to improve? Our answer is no, we think that the companies must obligatorily seek to stick to maximum partial values, without considering basic values.

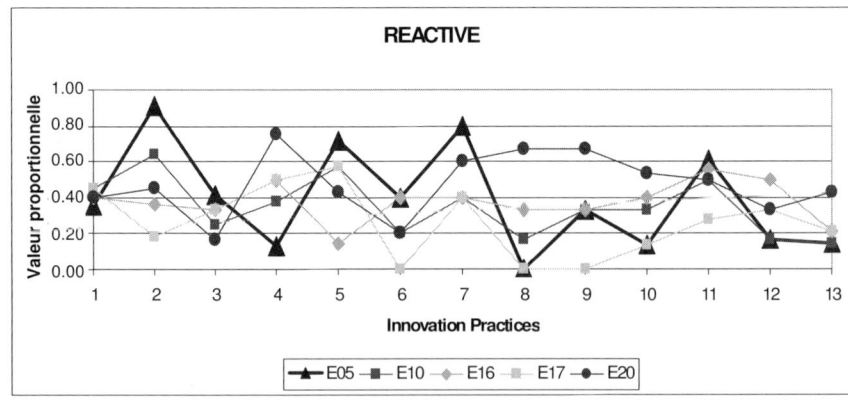

Figure 3. Reactive innovative system.

Passive Innovative System. These are E03, E04, E07 and E11. The value of PII oscillates between 10.1% and 28.5%.

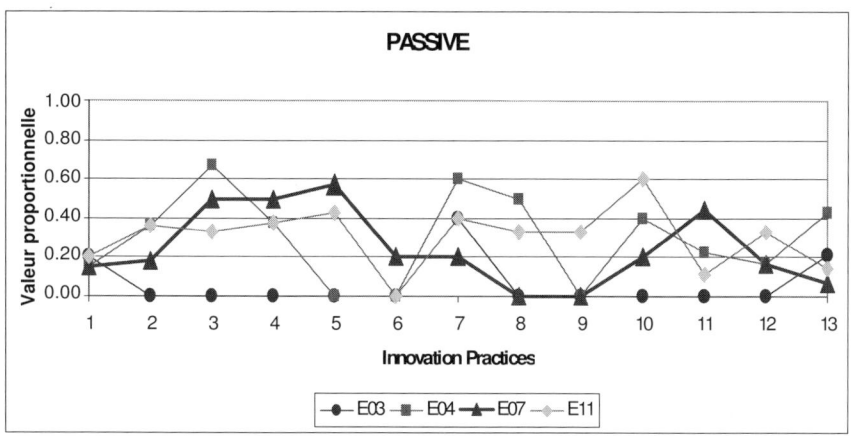

Figure 4. Passive innovative system.

The companies must then develop actions in order to obtain the level of target values, which is then the necessary minimum to obtain the name of innovating system. The construction of target values for this category results from combination of minimal values: we selected the maximum values of these minimal values.

5.3. *Limits and improvement of innovative system*

It appears that the concepts of target and basic models are useful in term of innovation process improvement. They give an operational objective to managers. This is very important considering the fundamental characteristics of innovation: uncertainty and risk. Changing the innovation system will be done according to the needs and resources that the company can invest. The strategy may be first to seek the basic level before evolving to the target. Then going to one class to another one is possible.

6. Conclusions

Innovation is a highly strategic element, and its development becomes basic in all fields: economic, technological, industrial, educational, research, social and governmental (Carlsson, 2002). Thus a theoretical system to measure innovation capacity in a company based on the PII is suggested. This index is:

- an "added" index because the various elements of innovative system take part for whole or part in final note,
- a "static" index, which gives us a value for a precise moment in time. The index obtained is valid in the short term, it should be calculated each time we need to know the situation of innovative system in a company,
- a "cold" index, because it determines potential innovation of company without establishing any other bond with intervening elements in its development,
- an index "of activity", its realization takes into account neither success nor the failure of innovations, only the system to produce new products,
- an "on going index": new practices and under-characteristics may be added to the approach in order to obtain an evaluation system coherent with the state of the art in the field of innovation management.

The outcomes obtained on an industrial sample made up of 20 companies enable a better understanding of the role of PII in term of development and evolution of innovation management. One of the assets of IIP is to give a quantitative tool when observing New Product Development Processes.

Indeed, the measurement of the innovation through PII allows:

- a ranking of companies. The index makes it possible to treat on a hierarchical basis the companies and to establish an order between them.

- a comparison between companies. It is carried out in two ways, direct and indirect. The direct one uses a comparison between the indexes PII of each company, and the indirect one through calculation of distances between the vectors (data of each company).

This evaluation system may be enriched with a larger industrial panel. A database may be established. With the new innovation management practices and sub-characteristics, it constitutes a reference that uses in situ observation for the innovation processes.

References

Afuah A. (1999). La dinámica de la innovación organizacional. Ed. Oxford University Press México, México, Mexico.

Amable B., Barré R., Boyer R. (1997). Les systèmes d'innovation à l'ère de la globalisation. Ed. Economica, Paris, France

Boly V. (2004). Ingénierie de l'innovation organisation et méthodologies des entreprises innovantes. Ed. Hermes Science Publications - Lavoisier, Paris, France.

Cañibano L., García-Ayuso M., Sánchez, M. (2000). Shortcoming in the measurement of innovation: implications for accounting standard setting. *Journal of Management and Governance*, 4, 319-342

Carlsson, Bo; Jacobsson, Staffan; Holmen, Magnus; et Rickne, Annika (2002). Innovation systems: analytical and methodological issues. *Research Policy*, 31, 233-245

Collette Y., Siarry P. (2002). Optimisation Multiobjectif. Ed. Eyrolles, France.

Commission Européenne (1995). Livre vert sur l'innovation.

Dert F. (1997), L'art d'innover ou la conquête de l'incertain. Ed. Maxima, Paris, France

Escobar, M. T.; Aguaron, J.; et J. M. Moreno Jimenez (2004). A note on AHP group consistency for the row geometric mean priorization procedure. *European Journal of Operational Research*, 153, 318-322

Escorsa C.P. et Valls P., Jaume (2001). Tecnología e innovación en la empresa dirección y gestion. Ed. Alfaomega. Bogotá, Colombia.

Godet M. (1997). Manuel de prospective stratégique. Tome 2. L'art et la méthode. Ed. Dunod, Paris, France.

Griffin A.(2000) PDMA Research on new product development practices: updating trends and benchmarking best practices, *Journal of Product Innovation Management*, Vol. 14, p. 429-458.

Himbert, M. (1998). La métrologie : un langage universel pour les sciences et techniques. Récents Progrès en Génie des Procédés 60, no. 12, 15-23

Huang X., Soutar G.N., Brown A., (2004) Measuring new product success: an empirical investigation of Australian SMEs, *Industrial Marketing Management*, 33, p 117-123.

Macharis C., Springael J., De Brucker K., Verbeke A. (2004). PROMETHEE and AHP: The design of operational synergies in multicriteria analysis. Strengthening PROMETHEE with ideas of AHP. *European Journal of Operational Research*, 153, 307-317

Morel L.; Guidat C.; and Rault-Jacquot V. (1998). Nature et questions de métrologie en sciences de l'innovation. Récents Progrès en Génie des Procédés 60, no. 12, 53-62

Romero C. (1996). Analisis de las Decisiones Multicriterio. Ed. ISDEFE, Spain

Roy B., Bouyssou D. (1993). Aide Multicritère à la Decision: Methodes et Cas. Collection Gestion. Série: Production et Techniques quantitatives appliquées à la gestion. Ed. ECONOMICA, Paris, France

Sanchez Guerrero G. (2003). Técnicas participativas para la planeación. Ed. FICA. Mexico.

Sugihari K., Hiroaki I., Tanaka H. (2004). Interval priorities in AHP by interval regression analysis. *European Journal of Operational Research*, 158, 745-754

Turriago H.A. (2002). Gerencia de la innovación tecnológica. Collection Guías Empresariales. Ed. Alfaomega Colombiana. Bogotá, Colombia.

CHAPTER 31

AUDITING TECHNOLOGY DEVELOPMENT PROJECTS

Dominique R. Jolly

CERAM Sophia Antipolis, France

Chief Technology Officers need guidance when they allocate resources to different technology programs. The starting point for making the right choice is to audit all the technology projects. A distinction has to be made between controllable criteria and non-controllable criteria. Controllable criteria refer to the accumulated competencies and assets that depend on a firm's behaviour and decisions. Uncontrollable criteria refer to the intrinsic potential, the appeal or attractiveness, of one given technology that does not depend on the firm's action. A list of 32 criteria for auditing "technological competitiveness" and "technological attractiveness" is discussed and accompanied with appropriate measurement.

1. Introduction

Technology development projects are a core element of corporate renewal, competitiveness and sustainability. Chief Technology Officers (CTOs) role encompasses several issues. They have to define where R&D efforts should be directed in line with the company strategy. Technology strategy is about finding the optimal balance of research versus development, basic research versus applied research, selecting autonomous or cooperative means, choosing between internal development and external acquisition, communicating recommendations for action both vertically and horizontally within the organization, deciding which technologies should be commercialized, etc. R&D managers face a collection of investment opportunities into different technology projects. But, they face as well uncertainty in estimating the future outcomes of these opportunities.

The most sensitive issue where R&D decision markers exert their power is the Go/Kill decisions. From an operational perspective, it means allocating resources (capital, people, physical facilities, equipment, etc.) through an array of significantly different technology development programs. The questions are the following: which programs should be slowed down, scaled back, or even cut off? Which ones should be sustained, expanded, boosted? Which new projects should be launched? This is about screening, prioritizing and selecting projects to be funded and determining the level of funding for each project selected. Screening allows identification of doomed projects. Prioritization provides a rank-ordered list of projects. Starting with the best ones at the top, projects in the list are funded until funded until the budget is exhausted. And, this has to be done according to three constraints: limited funding; reaching an optimal portfolio; alignment with corporate strategy (Sethi et al., 1985; Roussel et al., 1991; Say et al., 2003).

In the process of allocating resources across an array of significantly different technology programs, there is internal competition for limited resources. Unsurprisingly, resources are usually in shorter supply than the number of potential competing projects. The treatment of this question may frequently be the output of a purely intuitive process depending upon unexpected opportunities. But, an un-formalized approach to technology management puts the decision maker under strong pressures from various interest groups. It exacerbates personality's bias as well as individual and emotional preferences.

In order to formalize and systematize this decision process, technology portfolio models were designed in the 1980s to help CTOs tackle this major task. Seminal approaches to technology portfolio modelling should be attributed to A.D.L. (1981), Foster (1981) or Harris et al. (1981). Research conducted by Cooper et al. (2000) shows that businesses implementing a systematic process for managing their project portfolios clearly outperform other businesses.

This is why technology-based companies need a formal and structured audit of their technology portfolio to produce an image of their technological assets. This is precisely the aim of this paper to offer a frame for auditing technology development projects. Section 1 casts a light on the existing models for auditing technology. This shows the limited interest devoted in the literature to the design of auditing criteria and proper measurement techniques. Section 2 suggests that there are two distinct sets of criteria for auditing technology: those that relate to the potential, the appeal or attractiveness of one given technology; and those related to one given firm set of competencies for the technology under study. Section 3 presents two sets of 16 criteria; the main difference between the two sets is that the second set incorporates much more controllable factors than the first one – the last part suggests weighting the different criteria for analysis. Section 4 is the final and conclusive one. It puts the emphasis on practical issues when implementing a technology audit.

2. Existing Models for Auditing Technology

Technology audits are useful for internal purposes, i.e. to define where Research and Development (R&D) efforts should be directed and organized, as well as for external strategies, such as potential licensing, forming alliances, targeting acquisitions, etc.

2.1. *Feeding the decision process*

Once technology development projects have been identified, the next step is to assess the current situation. The action oriented decisions previously mentioned (chiefly, Go/Kill decisions) are served with several inputs: a) Evaluation of the alternative projects; b) Identification of independencies and interrelationships amongst projects.

Different evaluation methods have been developed. Heidenberger & Stummer (1999) and Henriksen & Traynor (1999) have identified different approaches: financial cost-benefit measurement, scoring, mathematical programming, decision and game theory, simulation, artificial intelligence, heuristics, and cognitive. Because of the level of complexity of particular techniques such as mathematical programming, managers might be discouraged to use them (Henriksen & Traynor, 1999). Poh et al. (2001) have compared different methods (based on ranked weightings or on benefit-contribution) with an Analytic Hierarchy Process (AHP). They have shown that the scoring method (i.e. a technique used for R&D project selection since the 1960s!) is the most favourable method for R&D project evaluation.

Evaluation of alternative projects is not enough. Several authors have stressed that most conventional models evaluate individual projects in isolation: they focus on individual opportunities. They don't capture interdependencies when projects are highly coupled (i.e. where the success of A depends on the success of B) and interrelationships between projects such as mutual exclusion or overlap in resources utilization allowing some positive synergies (Stummer & Heidenberger, 2003). Ouellet & Martel (1995) have proposed to measure synergies between projects at three levels: use of resources, technology and payoffs. Dickinson et al. (2001) have suggested drawing a (square) dependency matrix to capture interdependencies between projects.

2.2. *Shortcomings of existing evaluation models*

Limited use might be explained by the drawbacks of these methods. Though much has been written on evaluation models, there is very little work on criteria and measurement to be used for technology auditing.

Many models are derived from financial analysis. They use financial metrics, such as Discounted Cash Flows (DCF), Net Present Value (NPV), Internal Rate of Return (IRR), Return On Investment (ROI) or pay-back period (see e.g. Spradlin & Kutoloski, 1999 or Kirchhoff et al., 2001). More sophisticated versions incorporate probabilities and uncertainties into the financial calculation. The objective is to maximize the potential return per unit of risk (Carter and Edwards, 2001). The immediate limit of these approaches is that by focusing on financial and/or economic returns, these metrics fail to deal with non-monetary criteria. Another serious weakness relates to the input used. Fixed costs, initial investment and variable costs can usually be approximated properly. However, estimating potential future cash (in-) flows and probabilities to compute the expected NPV over a very long planning horizon might be speculative and oversimplified.

Despite financial measurements appear to be very clear-cut and elegant, the data they use are very often based on highly subjective judgements with wide variance (i.e. numbers pulled out of the air).

Some authors (for example Yoon et al., 2002) suggest using patent statistics as a tool for decision making at the micro-level (i.e. not only at the level of the national technological capacity as it has widely used by economists). Bibliometrics are used to identify potential research areas, to assess technological competitiveness and to set up priority in R&D investment. Such techniques are very useful, but unfortunately also very focused in their approach as patent statistics considerably restrict the analysis.

Audit of the technology portfolio cannot be restricted to a single indicator. This is obviously incomplete. Audit needs to rely on several criteria. Most of the models rely on a narrow set of (3 to 6) criteria or pay a limited attention to the justification of the choice of criteria. Neither did the definition of anchored scales for assessment receive much attention.

The model developed by Ringuest et al. (1999) and later by Graves et al. (2000) only takes into account two variables: an estimate of the "success probability" for each project and the "financial return" if the project is successful or unsuccessful. The scoring method developed by Henriksen and Traynor (1999) [for a federal research facility] relies on four criteria: relevance to the organization's strategy, scientific and/or technical risks, reasonableness considering the organization's budget, and perceived return. Dickinson et al. (2001) use five variables: the "NPV", the "overall probability of success", the "level of interdependence" (with other projects), the "capability and process change" and the "alignment with strategic objectives". Linton et al. (2002) use very crude measures of "intellectual property life cycle", "stage of market life cycle", "investment" and "anticipated cash flows". Stummer & Heidenberger (2003) use a limited set of six variables: cash flow, sales, patents, R&D funds, R&D staff and production capacity.

Only a few authors have generated comprehensive lists. Ouellet & Martel (1995) use a more extensive list of 19 different criteria organized into 3 families (interdependencies, intrinsic value and risk). Thanks to an in-depth literature review in a related area (success in R&D projects and new product innovation), Balachandra and Friar (1997) came up with a very extensive list of items but did not match these factors with proper measurements.

The concept under study is not always clear. While many authors tend to focus on "technology attractiveness", "technology merit", "technology value" or "project value", many do not precisely define what they exactly look at. Most of the models mix internal and external criteria and controllable and non-controllable factors. For example, when Linton et al. (2002) suggest the existence of six broad categories – financial, strategic, quality, environment, market, technological – but they fail to consider that some of these categories cover internal (and controllable) and external (and non-controllable) issues. Some comparative models develop a sophisticated mathematical programming, but fail to input reliable data. A survey conducted by Cooper et al. (2000) highlighted a need for better information to feed the process of portfolio management. Because of these limits, an effort was carried to draw an extensive list of criteria for technology auditing. Two sets of 16 criteria will be presented in the next section.

3. Criteria for Analysis

A first paragraph elaborates on the dichotomy of the Greek philosopher Epictete who said: "Amongst things that exist, some depend on us and some do not". The next two paragraphs suggest two sets of criteria for evaluating technology attractiveness and technology competitiveness. These lists originated from literature review, common sense and several workshops conducted with managers involved in executive training.

3.1. *Two distinct sets: Controllable and non-controllable criteria*

Some technology audit models rely on one single dimension (e.g. financial value) and some others use a two-dimensional framework. These tools assume that every technology can be examined and scaled along two dimensions. But these dimensions might vary from one model to another.

It is suggested here that we can differentiate between:

- those that relate to the potential, the appeal or attractiveness of one given technology;
- and those related to one given firm set of competencies for the technology under study.

This dichotomy between controllable and non-controllable criteria is widespread in the management literature. This pattern is verified in many circumstances; it can be found in several domains. It can be found in the field of strategy with the Swot framework. The strengths and weaknesses of the company depend on its internal resources; the firm is free to adopt the behaviour it wants regarding these internal resources that are supposed to be under its control. On the contrary, the opportunities and threats depend mostly on what is happening in the environment. As a matter of fact, the firm has little impact on external elements such as the actions of competitors, suppliers and regulators as well as the choices made by customers – these are mostly non-controllable factors.

Well known portfolio techniques used by strategists are also all based on a two-dimension framework. The BCG matrix combines the market growth and the market share. The General Electric/McKinsey matrix refers to industry attractiveness and business strengths, whilst the ADL matrix uses industry maturity and competitive position. These models are based on the same foundations: they differentiate between the sector attractiveness and the SBU position in its sector. The first dimension is mostly given and not under company's control while the second is supposed to be under the control of the firm.

Now, regarding technology audits, some authors suggesting a two-dimensional matrix are not in line with the controllable/non-controllable dichotomy. For example, Balachandra & Friar (1997) and Balachandra (2001) rely on a three dimension framework: the market (existing, new), the type of innovation (incremental, radical) and the technology (familiar, unfamiliar).

But some other authors are implicitly in line with the dichotomy between controllable and non-controllable variables. Harris et al. (1981)

refer to "relative technology position", which describes clearly controllable factors (such as patent position or key talents), and "technology importance for competitive advantage". The latter is a dimension that can be traced back to factors such as the potential value added or the position in the life cycle (which are mostly specified). Sethi et al. (1985) plot a technology on two dimensions: its "technology importance" (in terms of value added, rate of change, potential market) and the "corporate strengths regarding the technology and relative to its competitors" (in terms of patent, human resource strengths, technology expenditures, etc.). Capon & Glazer (1987) also suggested establishing technology portfolios along two dimensions. First is the "time" – from technology inception to decline. It incorporates both technology and product life cycles by distinguishing pre and post market phases of technology exploitation. Second is the "technology competitive position" of the firm. Brockhoff (1992) and later Ernst (1998) use two dimensions to draw patent applications: "technology attractiveness" (as the growth rate of patent applications) and relative patent position for "company's competitiveness". The approach depicts external and internal features using objective, but narrow measurements. Hsuan (2001) has suggested mapping a given technology along the "benefit provided to customers" and the "competitive advantage"; this allows to portray external vis-à-vis internal features.

In summary, there are things that are mainly under the firm's control, assets that depend on the firm's behaviour and decisions; I will refer to these factors as "**the company's technological competitiveness**". Criteria for auditing a firm's competitive position on a given technology point to internal factors that are within the firm's control. So, on this axis, the position of a given company could be very different from the position of another.

There are also things that do not depend on the firm's actions, which are beyond its control: I will refer here to these elements as the "**the attractiveness of the technology**". Criteria used for the attractiveness of a given technology are important for value creation. These criteria refer mainly to external features that are idiosyncratic to the technology. They are intrinsically related to the technology and are beyond the control of the firm. This means that technological attractiveness is identical for all companies competing in this technology.

3.2. Evaluating technological attractiveness

The attractiveness of a technology appears as a function of various different factors. A list of 16 criteria for depicting "technological attractiveness" is given in Figure 1.

Environmental factors over which the company has a weak control		Weak attractiveness		High attractiveness
Market factors	Market volume opened by technology	low	☐☐☐☐☐☐☐☐	high
	Span of applications opened by technology	narrow	☐☐☐☐☐☐☐☐	wide
	Market sensitivity to technical factors	weak	☐☐☐☐☐☐☐☐	strong
Competition factors	Number of stake-holders	many	☐☐☐☐☐☐☐☐	few
	Competitors' level of involvement	high	☐☐☐☐☐☐☐☐	low
	Competitive intensity	strong	☐☐☐☐☐☐☐☐	weak
	Impact of technology on competitive issues	low	☐☐☐☐☐☐☐☐	high
	Barriers to copy or imitation	low	☐☐☐☐☐☐☐☐	high
	Dominant design	exist	☐☐☐☐☐☐☐☐	non-exist
Technical factors	Position of the technology in its own life-cycle	declining	☐☐☐☐☐☐☐☐	emerging
	Potential for progress	low	☐☐☐☐☐☐☐☐	high
	Performance gap *vis-à-vis* alternative technologies	narrow	☐☐☐☐☐☐☐☐	wide
	Threat of substitution technologies	high	☐☐☐☐☐☐☐☐	low
	Ability to transfer the technology from one unit to another	difficult	☐☐☐☐☐☐☐☐	easy
Other criteria	Societal stakes	threatening	☐☐☐☐☐☐☐☐	supportive
	Public support for development	spartan	☐☐☐☐☐☐☐☐	generous

Source: Jolly (2003)

Figure 1. Evaluating technological attractiveness.

Semantic differential scales are given on the right for each criterion; they were co-constructed in the process of an executive seminar. Semantic differential scales are given on the right for each criterion; they were co-constructed in the process of an executive seminar. It is possible to distinguish between four families: market, competition, technical factors and other criteria.

3.3. Market criteria

Market, demand and customers are key drivers when it comes to making decisions about technology. Bond & Houston (2003) have stressed that it is essential to link technologies to the market – the main difficulty is that the market and technologies are likely to be highly uncertain. Factors in this category should express the expected commercial reward that can be gained from a given technology. Market volume opened by technology, or market sensitivity to technical factors, stem directly from the marketing literature. Technology-driven customers usually represent a limited segment, a narrow sub-section of the market. Most of the time, customers do not look for technical performance in itself. Technology only exists to satisfy their needs and expectations which are expressed through competitive issues.

Relying on the resource-based framework (Prahalad, 1993), technological attractiveness should also emphasize the span of new applications, new functions and/or new customer segments that the technology opens and its impact on the company's performance. Technologies might differ in their ability to reach several markets, to fulfil different expectations, to be embodied in distinct applications, products, processes or services, and to target varied customer segments. The interesting feature of technologies with a large span of applications is that if one of these segments declines, this will not be a disaster. This drop should be compensated by other demand segments. Market attractiveness will be more or less easy to estimate depending on the degree of its newness. Existing markets are relatively easy to estimate. But, it is much more difficult to evaluate potential when the market is entirely new: there is much more uncertainty and many more unknowns about the potential uses or the size of the market.

3.4. *Competitive criteria*

Competition is a strong driver of technological development. Managers must pay close attention to the level of competition in their business when they allocate resources to technology programs. Relying on competition analysis, as depicted by Porter (1980), criteria for evaluating technological attractiveness should emphasize the increase/decrease number of stake-holders, the competitors' level of involvement, the competitive intensity, and the barriers to copying or imitation (i.e. the capacity of the technology to support a resource position barrier - to entry & to imitation). As a matter of fact, investing in a technology would be useless if the company were not able to protect it against competition inclined to copy it. Competition has to be scanned carefully. In particular, the attention paid by competitors to the technology gives an indication of its attractiveness. The more competition is involved, the more likely the technology is to be attractive. As such, this criterion is a sign of attractiveness.

Also very important is the impact of technology on competitive issues (Khalil, 2000), that is the impact of technology on the value and/or cost of the offer in which it is incorporated, as perceived by the client/customer. This refers to the role usually hypothesized for technology: it is a means to create gaps vis-à-vis competition.

The dimensions on which firms are competing are not so important. Firms might be competing on cost, quality, speed of development, speed of delivery, performance, and so on. What is important is the contribution of technology to the building of competitive edge. Technology becomes attractive when it enables a firm to gain competitive advantage.

Finally, as suggested in the work of Abernathy and Utterback (1978), and later of Utterback (1994), the absence or the existence of a dominant design will or will not sustain technological attractiveness.

3.5. *Technical criteria*

These criteria rely on the work of Foster (1986) on the technology life-cycle and the threat of substitution technologies. Whatever the phase of the technology life-cycle, there is always a risk that substitution technologies will emerge. Nevertheless, the threat is probably higher when the technology reaches its mature stage - as the dominant technology has exhibited diminishing marginal returns. In the same vein, the concept of reserve for progress has been studied by many authors (see e.g. Van Wyk et al., 1991).

Technical criteria also include the performance gap vis-à-vis alternative technologies. This gap is limited in the case of an incremental innovation or significant in the case of a radical innovation. Anyway, the technology will have to find its way in a competitive world. A gap must be created to overcome barriers. Otherwise, technological change might be difficult to implement.

Finally, another technical criteria is the ability to implement horizontal transfers. The dominant organizational relationships refer to vertical patterns (between the SBUs and the corporate levels). While technology transfers usually occur at the horizontal level (between SBUs). This criterion might be underestimated because of the way R&D managers are evaluated. Most of the time, a R&D manager is controlled and remunerated according to what he does in his own field. The impact of his actions on other components in the organization and his contribution to other programs are very rarely taken into account by compensation policies or by the hierarchy.

3.6. *Socio-political criteria*

Evaluating technological attractiveness also calls for an examination of the negative by-products and societal pressures that can arise as a consequence of the exploration of entirely new technical fields. The point in question is the societal acceptability of the technology. Some major societal issues can arise from new technologies; the trouble is that societal pressures might impede the development of the business.

The current example of genetically-modified organisms (GMOs) in the pharmaceutical and the seed industry illustrates this point. It means that societal pressures might be an issue for some industries and not for others. On the other hand, we also have to take into account the financial support that is obtainable from public sources for trendy technologies.

3.7. *Evaluating technological competitiveness*

Criteria for evaluating a company's position on one specific technology should be broadly based. A list of 16 criteria for depicting "technological competitiveness" is given in Figure 2.

Internal factors over which the company can exert a strong control		Weak position		Strong position
Technological resources	Origin of the assets	external	☐☐☐☐☐☐☐☐☐	internal
	Relatedness to the core business	unrelated	☐☐☐☐☐☐☐☐☐	related
	Experience accumulated in the field	no experience	☐☐☐☐☐☐☐☐☐	world-class player
	Registered patents	none	☐☐☐☐☐☐☐☐☐	many
	Value of laboratories and equipment	low	☐☐☐☐☐☐☐☐☐	high
	Fundamental research team competencies	low	☐☐☐☐☐☐☐☐☐	high
	Applied research team competencies	low	☐☐☐☐☐☐☐☐☐	high
	Development team competencies	low	☐☐☐☐☐☐☐☐☐	high
	Diffusion in the enterprise	undiffused	☐☐☐☐☐☐☐☐☐	diffused
Complementary resources	Capability to keep up with fundamental S&T knowledge	none	☐☐☐☐☐☐☐☐☐	strong links
	Financing capacity	low	☐☐☐☐☐☐☐☐☐	high
	Quality of relationships between R&D & Production	weak	☐☐☐☐☐☐☐☐☐	strong
	Quality of relationships between R&D & Marketing	weak	☐☐☐☐☐☐☐☐☐	strong
	Capacity to protect against imitation	low	☐☐☐☐☐☐☐☐☐	high
	Market reaction to the company's design	unfavorable	☐☐☐☐☐☐☐☐☐	favorable
	Timetable relative to competition	behind	☐☐☐☐☐☐☐☐☐	ahead

Source: Jolly (2003)

Figure 2. Evaluating technological competitiveness.

Semantic differential scales are given on the right for each criterion. Criteria can be grouped into two families. Some relate to the technical capabilities of the company, i.e. to the technological resources within its control. Others relate to complementary resources which are also within its control.

The evaluation of technological resources should take into account:

- The origin of the assets. This relates to whether there is a dependence of the firm vis-à-vis external suppliers (another company, a public research centre, etc.) or a total independence of the company;
- The relatedness to the core business. That is to say the distance, the alignment, or the potential contribution of the technology to the firm's core competencies. This has been stressed by authors such as Coombs (1996) – relying again on the resource-based framework (Hamel & Prahalad, 1990; Prahalad, 1993). The hypothesis is: the closer the alignment between one technology and one core competence, the higher the R&D support should be. A program which is not in line with the existing technological platform would prevent the company from performing well. This argument

coincides with the current trend of the resource-based theory (Jolly, 2000), which is to refocus on core competencies rather than expand in several unrelated directions;
- The proprietariness. This criterion captures the patents owned by the firm, and the protection issue, as analyzed by Teece (1986) and Ernst (1998);
- The firm's accumulated experience and familiarity in one given technological field, as well as the value of its labs and equipment, the expertise of R&D staff. These criteria have been stressed by authors such as Roussel et al. (1991). Development is known to be the most expensive when compared to "applied research" and "fundamental research". As such, the competencies of development teams are crucial for the success of the program;
- The diffusion of technological knowledge in the company. This stems from the strong emphasis given over the last ten or fifteen years on the value of lateral transfer, sharing knowledge within the group, horizontal development, learning and knowledge management, etc.

Almost as important as technological resources are complementary resources. These include:

- The links established by the firm with the scientific community in order to keep up with the latest developments. CTOs in companies are much more used to applied research and development than to fundamental research. To fill this gap, they have better to launch bridges with providers of fundamental research;
- The ability of the company to finance the development of technology. R&D managers are not morons. They are fully aware of financial issues. They know that they will have to convince their Chief Financial Officer that a given technological program should be capable of attracting financing – either internally or externally. Fighting to attract budgets is a major issue;
- R&D-Marketing and R&D-Production interfaces. These are downstream coupling. R&D no longer lives in an ivory tower. R&D laboratories must not behave as independent units. CTOs have to

recognize the importance of interfaces between functions. Interfaces between R&D and Marketing as well as between R&D and Production are intangible assets that need to be developed. Technological competitiveness depends on the strength of the link between R&D and Marketing. The two functions should establish channels of communication so as to fluidify the transfer of knowledge between them - especially knowledge about consumer behavior on one side, and functionalities offered by the technology on the other. Strong interfaces facilitate implementation whereas weak interfaces handicap business success. In the same vein, CTO's responsibilities encompass the transfer of the knowledge they develop to the forward stages of the value chain. Managers should pay special attention to the quality of the bridges between R&D and Production;

- The capacity to protect against imitation. This criteria is important as any effort to build a technological competitive advantage might be ruined if the technology in question is not protected;
- The impact of the design developed by the company on the market. This refers to the associated probability of the transformation of this design into a dominant one. R&D managers are not disconnected from the reality of demand. They understand that their technical choices have to be accepted by the market;
- The time issue. Timing exemplifies the importance of time in current competitive battles. It is well-known that being late in a technological race creates a competitive disadvantage. How to reduce time to develop, time to industrialize, time to market are very common challenges.

3.8. *Taking into account different weightings*

It can be hypothesized that the 32 criteria presented previously do not have the same importance. Some might be considered as having a greater impact than others. In order to shed light on the relative weight of these criteria, the present section reports the results of research undertook to weight each criterion (Jolly, 2003). Results are presented in Tables 1 and 2.

Table 1. Weightings of attractiveness criteria.

Rank	Criteria for assessing technological attractiveness	Average score	Standard deviation
1	Impact of technology on competitive issues	8.4	0.9
2	Market volume opened by technology	8.3	1.8
3	Span of applications opened by technology	7.7	2.0
4	Performance gap vis-à-vis alternative technologies	7.6	2.0
5	Competitive intensity	7.6	2.3
6	Barriers to copy or imitation	7.3	2.1
7	Threat of substitution technologies	6.8	2.8
8	Competitors' level of involvement	6.7	2.4
9	Position of the technology in its own life-cycle	6.3	2.1
10	Potential for progress	6.3	2.1
11	Dominant design	6.3	2.5
12	Number of competitors	6.2	2.7
13	Market sensitivity to technical factors	6.0	2.3
14	Potential for unit-to-unit transfers	5.7	2.6
15	Societal stakes	5.3	2.6
16	Public support for development	4.4	1.8

Table 2. Weightings of competitiveness criteria.

Rank	Criteria for assessing technological competitiveness	Average score	Standard deviation
1	Development team competencies	8.8	1.2
2	Distance of technology to the company's core business	8.1	1.3
3	Timetable relative to competition	8.0	1.9
4	Financing capacity	7.7	2.0
5	Applied research team competencies	7.2	1.6
6	Market reaction to the design proposed by the company	6.9	2.5
7	Quality of relationships between R&D & Marketing	6.9	3.0
8	Quality of relationships between R&D & Production	6.8	2.0
9	Registered patents	6.7	2.4
10	Experience accumulated in the field	6.4	2.4
11	Capacity to protect against imitation	6.3	2.6
12	Value of laboratories and equipment	6.2	2.5
13	Origin of the assets	6.2	2.0
14	Capability to keep up with fundamental scientific and technical knowledge	5.3	2.8
15	Fundamental research team competencies	4.7	2.5
16	Diffusion in the enterprise	4.4	2.4

Research was carried with a panel of 20 top-managers and experts from high-tech companies or large public laboratories. Each participant was first asked to rate individually the list of 2 x 16 criteria presented above (one list for attractiveness and one list for competitiveness). The scale used started at "one" (not important) and went through to "ten" (very important). Then, a Delphi second round allowed to deal with observable discrepancies amongst the results.

On each of the two dimensions considered in this research, competitiveness and attractiveness, three groups can be distinguished: (1) criteria which were evaluated as very important (average score > 7.0); (2) criteria falling in the average range (average score from 7.0 to 6.0) and (3) criteria exhibiting limited importance (average score lower than 6.0).

4. Conclusion

This section set out to demonstrate the use of a specific method for the evaluation of a company's technological position and attractiveness. The method provides the building blocks for Technology Portfolio Mapping that is used to assist the CTO in his decision-making concerning the allocation of resources (manpower, capital, etc) for the company's technologies. A technology portfolio map of a company, which the group members were familiar with, was built using the method mentioned above. This enabled an analysis of the company's technological situation (attractiveness and position) and subsequent questioning of the company's recent decisions. The company's representative reported that the analysis strengthened the company's recent decisions.

Every company must use some tool, if not several tools, to assist their decision-making process. As was in this case in the allocation of resources on R&D, or on/in any other issues/direction(s), it is a complex ever-changing world that we live in. The forces that shape the market in which a company operates in (forces which will "determine" whether a technology/product/service will succeed) vary in many aspects and represent a complex system that cannot be easily understood, and to an even lesser extent predicted!

References

Abernathy, W. & Utterback, J. (1978). Patterns of industrial innovation, Technology Review.

Arthur D. Little (1981). The Strategic Management of Technology, European Management Forum.

Balachandra, R. (2001). Optimal Portfolio for R&D and NPD projects, Portland International Conference on Management of Engineering & Technology (Picmet' 01), Portland (OR), July 29 – August 2 (cd rom).

Balachandra, R. & Friar, J.H. (1997). Factors for Success in R&D projects and New Products, *IEEE Transactions on Engineering Management*, 44(3): 276-287.

Bond, E.U. & Houston, M.B. (2003). Barriers to Matching New Technologies and Market Opportunities in Established Firms, *Journal of Product Innovation Management*, 20(2): 120-135.

Brockhoff, K.K. (1992). Instruments for Patent Data Analyses in Business Firms, *Technovation*, 12(1): 41-59.

Capon, N. & Glazer, R. (1987). Marketing and Technology: A Strategic Coalignment, *Journal of Marketing*, 51(3): 1-14.

Carter, R. & Edwards, D. (2001). Financial Analysis Extends Management of R&D, *Research Technology Management*, 44(5): 47-57.

Coombs, R. (1996). Core competencies and the strategic management of R&D, *R&D Management*, 26(4): 345-355.

Cooper, R.G., Edgett, S.J. & Kleinschmidt, E.J. (1998). Best Practices for Managing R&D Portfolios, *Research Technology Management*, 41(4): 20-34.

Cooper, R.G., Edgett, S.J. & Kleinschmidt, E.J. (2000). New Problems, New Solutions: Making Portfolio Management More Efficient, *Research Technology Management*, 43(2): 18-33.

Dickinson, M.W., Thornton, A.C. & Graves, S. (2001). Technology Portfolio Management: Optimizing Interdependent Projects over Multiple Time Periods, *IEEE Transactions on Engineering Management*. 48(4): 518-527.

Ernst, H. (1998). Patent portfolios for strategic R&D planning, *Journal of Engineering and Technology Management*, 15: 279-308.

Foster, R.N. (1981). Linking R&D to Strategy, The McKinsey Quarterly, Winter: 35-52.

Foster, R.N. (1986). Innovation, The Attacker's Advantage, Summit Books: New York.

Graves, S.B., Ringuest, J.L. & Case, R.H. (2000). Formulating Optimal R&D Portfolios, *Research Technology Management*. 43(3): 47-51.

Hamel, G. & Prahalad, C.K. (1990). The Core Competence of the Corporation, *Harvard Business Review*, May-June, 79-91.

Harris, J.M., Shaw, R.W., Jr. & Sommers, W.P. (1981). The Strategic Management of Technology, Outlook, 5, Fall/Winter: 20-26.

Heidenberger, K. & Stummer, C. (1999). Research and development project selection and resource allocation: a review of quantitative modelling approaches. International *Journal of Management Reviews*, 1(2): 197-224.

Henriksen, A. & Traynor, A.J. (1999). A Practical R&D Project-Selection Scoring Tool, *IEEE Transactions on Engineering Management*, 46(2): 158-170.

Hsuan Mikkola J. (2001). Portfolio management of R&D projects: implications for innovation management, *Technovation*, 21(7): 423-435.

Jolly, D. (2000). Three generic resource-based strategies, *International Journal of Technology Management*, 19(7/8): 773-787.

Jolly, D. (2003). The issue of weightings in technology portfolio management, *Technovation*, 23(5): 383-391.

Khalil, T. (2000). Management of Technology: The Key to Competitiveness and Wealth Creation, McGraw-Hill International Editions, 483 p.

Kirchhoff, B.A., Merges, M.J. & Morabito, J. (2001). A Value Creation Model for Measuring and Managing the R&D Portfolio. *Engineering Management Journal*, 13(1): 19-22.

Linton, J.D., Walsh, S.T. & Morabito, J. (2002). Analysis, ranking and selection of R&D projects in a portfolio, *R & D Management*, 32(2): 139-148.

Ouellet, F. & Martel, J.F., (1995). Méthode multicritère d'évaluation et de sélection de projets de R&D interdépendants, Revue Canadienne des Sciences de l'Administration/*Canadian Journal of Administrative Sciences*, 12(3): 195-209.

Poh, K.L., Ang, B.W. & Bai, F. (2001). A comparative analysis of R&D project evaluation methods. *R & D Management*, 31(1): 63-75.

Porter, M.E. (1980). Competitive Strategy, New York: The Free Press, Macmillan Pub.

Prahalad, C.K. (1993). The role of core competences in the corporation, *Research-Technology Management*, 36(6): 40-47.

Ringuest, J.F.; Graves, S.B. & Case, R.H. (1999). Formulating R&D Portfolios that Account for Risk, *Research Technology Management*, 42(6): 40-43.

Roussel, P.A.; Saad, K.N. & Erickson, T.J. (1991). Third Generation R&D. Managing the Link to Corporate Strategy, Boston (Massachusetts): Harvard Business School Press.

Sethi, N.K., Movsesian, B. & Hickey, K.D. (1985). Can Technology be Managed Strategically? *Long Range Planning*. 18(4): 89-99.

Spradlin, C.T. & Kutoloski, D.M. (1999). Action-Oriented Portfolio Management. *Research Technology Management*, 42(2): 26-32.

Stummer, C. & Heidenberger, K. (2003). Interactive R&D Portfolio Analysis With Project Interdependencies and Time Profiles of Multiple Objectives, *IEEE Transactions on Engineering Management*, 50(2): 175-183.

Teece, D.J. (1986). Profiting from technological innovation: Implications for integration, collaboration, licensing and public policy, *Research Policy*, 15: 285-305.

Utterback, J.M. (1994). Mastering the Dynamics of Innovation, Harvard Business School Press.

Van Wyk, R.J.; Haour, G. & Japp, S. (1991). Permanent Magnets: A Technological Analysis, *R&D Management,* 21(4), 301-308.

Yoon, B., Yoon, C. & Park, Y. (2002). On the development and application of a self-organizing feature map-based patent map, *R & D Management,* 32(4): 291-300

CHAPTER 32

COMPLEXITY, COST REDUCTION AND PRODUCTIVITY IMPROVEMENT THROUGH HISTORY

Mats Larsson* and Carl-Henric Nilsson**

*School of Economics and Management, Lund University, Sweden;
**Technology Management Centre, Lund University, Sweden

History shows us that in the cost/performance ratio of any given product has decreased continuously. This may wrongly have led us to the conclusion that improvements can continue infinitely at the current pace. Our proposition is that we trust a mental picture that is not consistent with the future level of potential improvements in technology and business, which in turn has repercussions for societal change.

1. Introduction

Anthropologists and historians have increasingly turned to economic explanations behind basic social inventions in human history, such as the transition from hunter-gatherer societies to agriculture, including the domestication of animals and plants. Diamond (1997) argues that agricultural societies created specialised roles in society. The increased productivity in agriculture compared to hunting and gathering made it possible to set aside resources to maintain chiefs and later a growing number of court people, administrators and other officials. Thomas (1982) describes how the invention of better and better ploughs made it possible to plough deeper in the ground, which made the soil more fertile. The result became bigger crops. With agriculture came the domestication of plants, which meant that the cobs of corn became larger (the wild ancestor of corn had cobs that were smaller than a thumbnail)

(Diamond, 1997). This article puts the more recent development in agriculture and business into this historical context.

The industrial revolution, gene technology, biotechnology and other important technological developments are based on the same basic logic as the development of early agriculture. One of the main benefits of these developments is the improvement of productivity in various ways. The cost/performace ratio for a comparable product has continuously been reduced. This has been shown for various products such as crushed and broken limestone as well as integrated circuits and RAM components (Ghemawat, 1985; Hill, 1985). The commonly held perception is that productivity can improve infinitely. Tainter (1988) argues that economic development causes increasing specialisation and, consequently, an increase in complexity in society. He also argues that increasing complexity has been the cause of the collapse of complex societies through history.

We use a theoretical argument taken from agriculture and empirical data from the printing industry in our analysis in order to argue that economic development reduces unit production cost in the direction of zero, but that recent technology developments may reduce, rather than increase, complexity.

2. Proposition

We propose that the picture of unlimited economic growth is not a correct mental picture for productivity improvements. We rather suggest an X-Y diagram with cost and time on the axes indicating the absolute level of cost and time possible is zero cost and zero time hence a finite potential for productivity improvements and economic growth. We also propose that this will have negative implications for the continued development of our society.

3. Theoretical Foundation

Porter (1996) argues that the development of productivity is a trajectory of productivity improvement "outward" with no end to the potential improvement.

Instead we argue that development is clearly moving toward the limit of cost improvement. No product can be produced at a cost that is lower than zero units of input per unit of output. The closer to zero this development moves, the lower the value of a further improvement. This means that in the days when humans were hunters and collectors, the value of being able to set aside one person in the tribe to tend full time to improving society was very high. The archaeologist Tainter (1988) puts this as the ability of a tribe to increase complexity by starting on the path of specialisation, and argues that this increase in complexity brought comparatively large returns to the tribe. Gradually, as the development of agriculture and social complexity in general has developed, the returns to an investment in further complexity decrease and eventually become negative. In a situation today where one farmer is able to produce food for a large number of people, each improvement brings smaller returns. The value of a further investment in complexity (specialisation) becomes smaller and smaller the more complex society grows. Tainter defines complexity as the degree of specialisation in a society. Specialisation is measured as the number of specialised jobs that exist in a society and the number of specialised tools that people use in order to perform various tasks. In the least complex societies there is virtually no specialisation. Tainter mentions the Ik of Africa, as the least complex of all known societies, where everybody cares only for himself. In our society we have several thousand different jobs and tens of thousands of different tools that we use for different purposes (Tainter 1988) and each person is highly dependent on the fact that others perform their jobs in order for society as a whole to function.

Specialisation in society has been highly augmented since the industrial revolution. At this time the industrial production of farming equipment started, the food processing industry and large-scale food distribution into the growing cities grew as well. Farming, food production and distribution had formerly been integrated in the production on farms and were from the industrial revolution and onwards increasingly broken down into more and more specialised businesses, each using more and more efficient organisational models, equipment and marketing tools in order to compete. The specialised functions of suppliers of raw materials machinery and maintenance, agricultural production, refining, wholesale and retail developed. These developments contributed to increased productivity up to the present level where 3 per cent of the total population that is employed in farming in the industrialised world can produce the agricultural raw materials that feed the whole population and also produces a surplus in the developed countries. In total, a larger share of the population is involved in the whole chain of food production and distribution, but the total number has decreased compared to the levels before the industrial revolution.

According to Tainter, complexity could only profitably be increased up to a certain level. In an advanced society, such as the ancient Egypt, the Maya empire and our society, the value of a further increase often becomes negative. The cost of maintaining complexity increases to a level where complexity becomes costly to sustain and society becomes vulnerable to disruptive events that could previously be successfully handled when the level of complexity was lower. The reason for this vulnerability is that a substantial share of the resources in society is used in order to maintain the status quo. If, through some type of shock or crisis, problems appear in any of the supply chains for resources, the society is hurt and run the risk of declining. In early agricultural societies this type of shock may be caused by a prolonged draught or invasion by other clans or nations. In a modern society it can be caused by political upheaval (Tuchman 1984) in a resource rich country or by increased scarcity in oil or other fossil fuels (Heinberg 2003). Tainter (1988) argues that the Mayas and the Romans were able to handle shocks in their growth phases when the returns to increasing complexity were still substantial, that they were less able to handle towards the end of their empires when the returns to increasing complexity were low. Heinberg (2003) argues that the use of fossil fuels has given humanity a brief period with access to cheap extra energy. Each person uses the equivalent of the energy of some fifty slaves that we are able to pay for mainly through the low cost of fossil fuels. We are now approaching the maximum point of oil extraction. The cost of finding new oil resources and exploiting them is now several times higher than they were in 1960. The ability to maintain industrial society based on access to low cost energy will not last. Oil and other existing energy sources will become increasingly scarce and expensive. New sources of energy are not developed rapidly enough in order for production from these sources to grow at the rate that fossil fuels are depleted. Thus, we must expect the cost of energy during the next few decades to increase sharply and our ability to maintain a high level of complexity may thus be reduced (Heinberg 2003).

Larsson (2004) argues that there are definite limits to the reduction of time and cost. Nothing can be done in less than no time and at less than no cost. In many areas companies are coming closer and closer to this limit. This is true especially for information flows, where any piece of information, such as a book or a film can be electronically copied and distributed to any place in the world in close to no time and at close to no cost at all. In the case of material flows we will not be able to produce or distribute physical products at zero cost, but we are gradually coming closer to this limit by systematically reducing the non-value added time and cost in production processes in the direction of zero. Efforts in business, such as Buiness Process Reengineering, Supply Chain Management and the digitalisation of production and administrative processes and work flows inevitably lead to the reduction of time and cost towards zero. This has not previously been noticed by researchers (Larsson 2004).

4. Empirical Evidence

We have applied the theory of increasing complexity to the printing industry in order to study the increasing complexity of printing in terms of increasing specialisation. Our findings are not conclusive, since the development of printing is no longer confined to the printing industry itself, or to the printing industry and the printing machinery industry.

4.1. *Copying by scribes*

Prior to Gutenberg's invention of movable type in the western hemisphere, most duplication of written matter was done by scribes who were employed in the courts of wealthy people (Eisenstein 1983). At this time four or five persons were involved in the transaction of writing, copying and purchasing a book. In case more than one scribe was involved, this does not increase complexity in the sense of Tainter (1988). The persons involved were the author, the scribe, a binder and the buyer. Sometimes, a middleman, such as a book wholesaler (Eisenstein 1983) was also involved.

4.2. *Printing with movable type*

Through the invention of the printing press and movable type, we are able to identify a number of specialised tasks that was significantly larger than the above figure. We identify the following seven tasks:

- writing
- casting of type (Gutenberg devised the original formula for mixing tin, antimony and lead to the material that is still used, nearly unchanged (Adams & Dolin 2002).
- typesetting
- printing
- storing already printed pages
- sorting and assembling a book
- binding

In addition to this a further new role was introduced in the industry, the machine constructor and, the maintenance person. The supply chain, which initially consisted of the suppliers of paper, ink and carpentry (who supplied the table for writing on), was extended by the smith.

Through the mass production of books and other printed matter a system of distribution was gradually developed, which involved:

- sales from the print shop
- wholesalsers
- retailers

Through the "printing revolution" of Gutenberg, it was possible for his team to print and bind 200 copies of the Bible in only three years (Adams & Dolin 2002).

4.3. *Offset printing*

Through the development of later generations of printing technology each of the above specialised tasks has been developed into further specialisation. In a printing company we may have five departments:

- sales
- pre-press
- printing
- post-press
- administration

In each department there are several specialised tasks. In the pre-press area, the material is prepared for printing, formats are chosen and contrasts and colours are set. At the end of pre-press printing plates are made. Similarly, there are a number of tasks in the other departments.

Even if a printing company today may only contain 30 persons or less, this company relies on a large number of suppliers for its efficiency and for the improvement of its productivity. Suppliers of software and hardware supply the computers and the machinery in each department, such as cameras, densitometers, film-processing units and screens in pre-press. Colour printers and black/white printing machines are complemented by computerized control equipment in the printing area and in the post-press area companies have sorting machinery, saddle stitchers and book-binding equioment.

In each department a number of consumables are used. The most obvious would be printing plates, paper and inks, but there will be plastic folios, adhesive tape, glue, cloth and a number of other types of consumables that are used. These consumables are supplied by companies specialised in different areas, such as ink companies and paper companies. Some of the more important suppliers of a modern printing company would be Microsoft, Adobe and IBM. In addition to these companies there would be suppliers of pneumatic valves, rollers for the print engines and other machine parts for printing presses and other equipment that may be purchased via the machine supplier or directly from the subsystems supplier.

Today it takes a few hours to produce 200 bibles. For a very small run, such as 200 copies, the cost of setting up the press and print the make-ready may cost as much as the printing of the actual production run (Adams & Dolin 2002).

4.4. *Digital printing*

"Digital printing has revolutionized the printing industry. It has changed the production cycle and drastically shortened the turnaround time of a job." (Adams & Dolin 2000)

Digital printing again changes the complexity of the printing industry and this time it seems as if it <u>reduces</u> the complexity of the printing process, by compounding a number of tasks in the printing company into the main activity of printing, sometimes with integrated post-press.

A digital printing machine prints all pages in a sequence, similar to an office printer. There is no need for printing plates, storage of already printed pages or sorting in the process of assembling the book or magazine. Instead, a digital printing company may still consist of the same departments as in the offset printing industry, but the work in the printing department may have been "revolutionized", because a number of manual tasks have been taken away. The post-press activities can be organized in-line with the printing, so that the printing and post-press is seamlessly integrated. This gives rise to the term "print-on-demand", because printing of single or a few copies of a book or a magazine becomes possible. Adams & Dolin (2002) not only foresee changes in the printing industry through this development, they foretell the elimination of traditional bookstores.

Through the simplification of the printing process, the printing machinery is also simplified with fewer subsystems and fewer specialized technologies. This development may indicate that there is not necessarily a direct relationship between increased complexity and cost reduction, as has been the case earlier in history (Tainter 1988). In the case of digital printing this technology seems to reduce the complexity in terms of the specialization of tasks. Yet, it decreases the cost of printing. According to our studies, it initially reduces the cost of small runs, up to

100 copies, so that a large publisher of literature for higher education in Sweden produces all runs shorter than 100 copies in a digital press. Adams & Dolin (2002) argue that digital production of colour copies can be done for close to 50 cents a page, which may make the market demand for colour documents to take off. Gradually, the cost of printing each copy, irrespective of the size of print run is being reduced in the direction of zero, through the development of the machinery towards increasingly efficient machines with increasing speed and capacity.

Through the most modern and largest of Xerox digital presses, iGen, (Xerox is the market leader in the market for digital printing equipment), digital printing is competitive against offset printing for increasingly large runs. Runs of 250 or more copies may be comparable to the offset price on a direct price comparison basis. If the cost of storage and waste on offset print runs is included, an offset run of several thousand may be broken down into a number of shorter runs, which offers a number of added values for digital printing, described by Larsson (2004).

The amount of manual labour that is needed in digital printing is very low, compared to offset printing and the only manual tasks in the printing and post-press stages in an on-line post-press situation, is the feeding of paper and the emtying of the tray of finished product. The cost reduction in the digital print industry is expected to continue so that increasingly large runs will be taken over by digital printing (Adams & Dolin 2002).

This study of the printing industry indicates that cost reduction through history has brought the per unit cost of printing increasingly close to zero. The cost is not yet zero and there is further room for cost reduction, even though a substantial share of manual tasks have been taken away since the time of Gutenberg and some more are taken away through the current transition to digital printing. The machine time that remains can be further reduced through increases in speed. However, contrary to the argument of Tainter (1988) as indicated by the case study of digital printing technology and specialisation, cost reduction may not always be dependent on an increase in complexity, even if this has been the case up until very recently, at least in many industries.

5. Conclusion

Only a very small share of technology development propels the economy into entirely novel areas (Larsson 2004). As we are approaching zero cost and time in current processes such as technical operations, commercial operations, financial operations, security operations, accountancy operations and administrative operations, as the key processes in companies were named already in the first decade of the twentieth century by the French organisation theorist Henri Fayol (Sanchez & Heene 1997), we have to expect that the growth in these processes that has historically been caused by cost reduction will decrease as well.

Tainter (1988) forwards the hypothesis that the fall of complex societies through history has been caused by the increasing complexity and the diminishing returns to investing in more complexity at later stages. He uses the example of the cost of maintaining order in the Roman Empire at the cost of maintaining large legions, which had to be paid for through taxes. Tainter also uses the modern example of oil production to illustrate his argument, a theme further developed by Heinberg (2003).

Our study of the printing industry, however inconclusive, indicates that cost reduction may not always be accompanied by increasing complexity. However, in the case of digital printing, the path towards a further reduction in per unit cost seems to be maintained and this, according to Larsson (2004) creates a possible risk of economic collapse for our present society.

With an economy that in many areas is coming closer to the limits of business development and economic growth, by coming closer and closer to zero cost, the value of a further increase in complexity will be reduced further. In many instances of information flow within companies and between companies and public organisations, the cost of creating and distributing information is very close to zero. Ten years ago such flows often involved large numbers of people in a number of different organisations. Today, completely automated and computerised transactions are transmitted inside companies and between companies at the cost of a few cents per transaction (Larsson 2004). In such cases and in many material flows as well, where most of the non-value added cost has been taken away in process improvement projects during the last decades, the value of a further improvement, by definition, becomes smaller and smaller. In the case of material flows in production and distribution the cost of material flows is not coming close to zero, because of the obvious need for resources to produce and distribute material goods. In these cases, development has been focused over the centuries on the reduction of non-value-added cost and this cost has been reduced to very close to zero during the past few decades in particular (Larsson 2004). This has been done through the development of new materials, production technologies and distribution systems, together with management principles that support radical cost reduction in area upon area.

In this sense, as argued for the case of agriculture and printing above, the present may bring us close to the end point of thousands of years of human development at least from the aspect of cost reduction and productivity improvement. Nothing can be done at less than no cost and when we have taken away almost all cost in a number of specialised information flows and almost all non-value added cost in many advanced material flows, we can't continue further on this trajectory. Printing is only used as an example.

Economic growth is based on improvements in productivity (De Long 1991). If the historical improvement in productivity is moving towards an absolute limit in further improvement, since nothing can be done in less than no time or at a cost that is lower than zero, the limit to productivity improvement may prove to be a more difficult obstacle for our society in the near future, than the diminishing returns to increased complexity, as is argued by Tainter (1988).

Acknowledgments

The research behind this article has generously been financed by Sparbanksstiftelsen Skåne.

References

Adams, J.M & Dolin, P.A. (2002) Printing Technology. Delmar, Albany.
Barker, J.A. (1993) Paradigms. The Business of Discovering the Future. Harper Business, New York.
Boudon, R. (1994) The Art of Self-Persuasion. Polity Press, Cambridge.
De Long, J.B. (1991) Productivity Growth and Investment in Equipment: A Very Long Run Look. Harvard University and NBER, Cambridge, Mass.
Diamond, J. (1998). Guns, Germs and Steel. Vintage Random House, London.
Eisenstein, E.L. (1983) The Printing Revolution in Early Modern Europe. Cambridge University Press, Cambridge.
Ghemawat, P. (1985) Building Strategy on the Experience Curve. Harvard Business Review, Mar- Apr 85, 143-149.
Heinberg, R. (2003). The Party's Over. New Society Publishers, Gabriola Island.
Hill, T. (1985). Manufacturing Strategy, Macmillan, London.
Larsson, M. (2004). The Limits of Business Development and Economic Growth. Palgrave Macmillan, London.
Porter, M. E. (1996). What is Strategy? *Harvard Business Review,* Nov-Dec, 61-78.
Sanchez, R. and A. Heene. (1997). Competence Based Strategic Management. John Wiley & Sons, New York.
Tainter, J. (1988). The Collapse of Complex Societies. Cambridge University Press, Cambridge.
Thomas, H. (1982). A History of the World. Macmillan, London.
Tuchman, B.W. (1984). The March of Folly. Ballantine Books, New York.

SECTION V

SUSTAINABILITY

CHAPTER 33

INVOLVEMENT OF SMALL MANUFACTURING FIRMS IN ORGANIC PRODUCTION SYSTEMS

Stéphane Talbot

Université du Québec à Montréal, École des sciences de la gestion, Canada

Mass production systems have been the dominant paradigm in manufacturing industries for the past century. Such systems rely on energy-intensive transformation processes. They require transformed materials which are product specific. They reach high levels of efficiency through centralization and large scale facilities. Considering raises concerns. Such systems have so far proven to be cost-effective but counter-productive from an environmental perspective. The concept of organic production systems has been envisaged by some researchers as a promising alternative. Organic production systems are defined as small-scale and decentralized production facilities, located close to the market they serve. Such systems rely on small and flexible manufacturing equipment and make use of modular materials that can be used, or reused, to make a wide range of products. The present paper explores the applicability of organic production systems to the particular industrial settings of SMEs.

1. Introduction

Mass production systems have been the dominant paradigm in manufacturing industries for the past century. Such systems rely on energy-intensive transformation processes. They require transformed materials which are product specific. They reach high levels of efficiency through centralization and large scale facilities. Considering this traditional grounding, the evolution of industrial manufacturing raises concerns. Such systems have so far proven to be cost-effective but counter-productive from an environmental perspective. The concept of organic production systems has been envisaged by some researchers as a promising alternative. Organic production systems are defined as small-scale and decentralized production facilities, located close to the market they serve. Such systems rely on small and flexible manufacturing equipment and make use of modular materials that can be used, or reused, to make a wide range of products. One of the aims of organic production systems is to improve the environmental performance of manufacturing operations by mimicking biological processes observed in the natural environment (Benyus 1997; Senge, Carstedt, and Porter 2001). In fact, biomimicry would provide a cohesive framework to think about the development of complex control and automation systems (Passino 2005). The organic production system phenomenon is in great part dictated by underlying trends grouped under the following topics: industrial context, manufacturing strategies, and product and process modularity.

1.1. *Industrial context*

The current industrial context prevailing in most developed countries can be qualified as highly turbulent in a majority of markets, giving rise to increased uncertainty as perceived by manufacturing firms (Pine II 1993). This phenomenon can be observed at both ends of the product value chain. For instance, on the demand side, one can notice that the variety of innovating products keeps growing, while demand for them becomes more difficult to predict. On the supply side, manufacturing processes and their supporting technologies are evolving at a steady rate. The consequences of the combination of these factors place increasing pressure on firms involved in the design, manufacturing or support of manufactured products. In addition, market demand is increasingly asking for customized products, forcing companies to develop more complex products faster (Blanchard 1998) and at lower cost (Bernstein and DeCroix 2004).

1.2. Manufacturing strategies

Consequently, manufacturers have no other choice than to adapt to their external environment which is in constant evolution. Some firms are adopting strategies oriented towards the notion of precision. Such strategies enable the enterprise to compete on the basis of the satisfaction of unique customer requirements, while providing high quality, rapid delivery, and competitive prices at the same time (Victor and Boynton 1998). Mass customization is among the strategies addressing the challenges posed by evolving market conditions. Mass customization is defined as a set of processes making it possible to provide customized products and services on a large scale, in an economical manner. This strategy relies on technologies and management practices improving the flexibility of manufacturing operations and cycle times (Pine 1993; Da Silveira, Borenstein, and Fogliatto 2001). Different levels of customization may be achieved in the passage from pure standardization to individualization (Lampel and Mintzberg 1996).

1.3. Product and process modularity

The development of product platforms represents one successful option supporting the implementation of mass customization strategies (Robertson and Ulrich 1998). These authors define the concept of product platform as a collection of assets shared by a family of products. Indeed, such product platforms allow the attainment of high product differentiation levels without soliciting organizational resources in an excessive manner. Product platforms rely on the concept of product architecture, which in turn may be characterized as modular or integral (Ulrich 1995). The principles behind the notion of modularity are central to the reduction of the complexity of large scale systems, where numerous overlapping interrelations are taking place (Ethiraj and Levinthal 2004). Modularity also has the capacity of substantially increasing the number of possible configurations of a given set of components, thus elevating the system's flexibility (Schilling 2000). Finally, through the simplification of operational processes or products, modularity presents a wide ability to save resources such as space, materials, energy, transportation and time (Hawken, Lovins, and Lovins 1999).

The present article argues that the emergence of modular manufacturing practices (Tu et al. 2004) and modular assembly systems (Bernstein and DeCroix 2004) are in part driven by the above trends, and share similarities with the concept of organic production systems described hereafter.

2. Organic Production Systems

The organic production system concept was recently introduced as an alternative to the conventional model for industrial production systems (Demeester, Loch, and Van Wassenhove 2003). According to these authors, classical industrial operations are based on the transformation of specialized materials requiring energy intensive processes implemented in large scale facilities. Such materials are not versatile, since they are designed for very specific functions. At the end of a product's useful life, these specialized materials become waste in most cases because they are not easily convertible or adaptable to other products. Energy intensive processes and end-of-life waste are major sources of pollution. In addition to these charges, these authors argue that such production facilities increase their productivity through high volumes, or mass production, and centralization. These kinds of industrial organizations are thus considered to be insufficiently responsive to local market demand on one hand, and to rely on costly and highly polluting modes of transportation on the other hand.

By way of an alternative, the theoretical concept called organic production system has been developed in order to address some of the environmental issues raised above. As defined, organic production systems are made of many decentralized small-scale production facilities situated close to local markets, while product design and knowledge may still remain centralized. Such organic production systems rely on a small number of modular materials having the potential of being used and reused in different combinations to make a variety of products.

Production processes are conducted through small manufacturing tools, easier to set-up and less expansive than their large scale counterparts. Finally, the advantages stemming from market proximity, increased flexibility, and reduced waste and pollution would be expected to outweigh the lost of efficiency when compared to large scale centralized production systems (Demeester, Loch, and Van Wassenhove 2003).

Considering the above definition, organic production systems require innovation efforts both at the product level (Giudice, La Rosa, and Risitano 2005; Viswanathan and Allada 2006), and at the operational process or manufacturing performance level (Vachon and Klassen 2006). The aim of this article is to contribute to the development of the organic production systems concept, by expanding its conceptualization along these two lines.

3. Research Model

The proposed research model, represented in Figure 1, splits the organic production system concept along two distinctive axes. The horizontal axis, called organic product, measures the intensity of some environmental initiatives applicable to the products manufactured by the firm. The term "intensity" here relates to the level of effort deployed by the firm on each initiative listed on the first half of Table 1. The vertical axis called organic process, measures the intensity of some environmental initiatives related to the firms' operational processes. The term "intensity" here relates the level of effort deployed by the firm on each initiative listed on the second half of Table 1. The combination of both measures provides an overall score for the organic production system as a whole.

The organic production system map model is composed of four quadrants. Quadrant I covers an area gathering firms involved in production systems highly organic on both the organic product and the organic process dimensions. Firms located in this area would be considered as leaders in the implementation of environmental initiatives related to both the product and their processes. At the other end of the diagonal, Quadrant III covers an area gathering firms involved in production systems scoring the lowest on both the organic product and organic process dimensions. This subset of the studied population would gather laggards from an environmental perspective. Quadrant II covers an area gathering product innovators, giving greater attention to the development and implementation of product related environmental initiatives. Finally, Quadrant IV gathers process innovators, focusing their efforts on process related environmental initiatives.

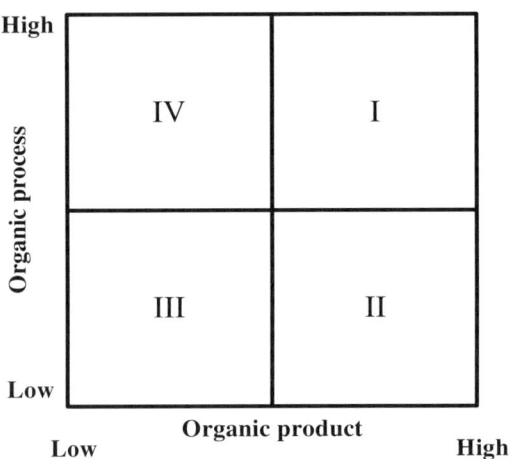

Figure 1. Organic production system map conceptual model.

Table 1 presents the theoretical justifications for the underlying variables, for both the organic product and the organic process dimensions. The reliability coefficient values for the organic product dimension ($\alpha = 0.80$) and for the organic process dimension ($\alpha = 0.85$) are quite satisfactory, since they both exceed the 0.70 lower limit guideline usually recommended (Robinson, Shaver, and Wrightsman 1991). A close look at the underlying variables presented in Table 1 reveals an interesting fact: the organic product construct is mostly composed of variables involving activities conducted internally, while the organic process construct is composed of variables involving inter-organizational collaboration.

Table 1. Organic production system dimensions' theoretical justifications and reliability coefficients.

Variable	Theoretical justification
Organic Product[1]	Cronbach alpha = 0.80
Choose raw materials that can be recycled or that are less harmful to the environment	(Hall 1993; Carlson-Skalak et al. 2000)
Reduce the amount of raw materials	(Lewis et al. 2001)
Design the product to accommodate multiple future users	(Hall 1993; Billatos and Basaly 1997; Mangun and Thurston 2002)
Design the product to be easy to disassemble	(Hall 1993; Lewis et al. 2001)
Design the product to be easy to recycle	(Hall 1993; Sarkis 1995; Lewis et al. 2001)
Design the product to be easy to manufacture	(Hall 1993; Ulrich et al. 1993; Lewis et al. 2001)
Organic Processes[1]	Cronbach alpha = 0.85
Reduce the energy required for manufacturing and assembling the product	(Billatos and Basaly 1997)
Minimize waste	(Sarkis 1995; van Hemel 2001)
Minimize product packaging	(Murphy, Poist, and Braunschweig 1995; Lewis et al. 2001)
Make packaging recyclable	(Murphy, Poist, and Braunschweig 1995; Lewis et al. 2001)
Optimize the distribution network	(Van Woensel, Creten, and Vandaele 2001; Murphy and Poist 2003)
Establish recycling procedures	(Thierry et al. 1995; Klausner, Grimm, and Hendrickson 1998; Krikke, le Blanc, and van de Velde 2004)
Ensure that recuperation infrastructures exist	(Fleischmann et al. 2001; Savaskan, Bhattacharya, and Van Wassenhove 2004)

[1]Measurements based on 7 points Likert scales ranging from 1 "no effort" to 7 "considerable efforts".

4. Empirical Evidences

4.1. *Methodology*

The intrinsic characteristics of SME industrial settings, like their small size, their relatively simple and flexible manufacturing processes, their ability to adapt to changing market conditions, and the fact that they are significant drivers of economic growth and employment (del Brio and Junquera 2003), make them good candidates for the application and study of the organic production system concept. In order to provide solid theoretical bases for the conceptual model studied in this article, a thorough literature review was conducted covering the domain of environmental initiatives enabling the firm to improve the environmental performance of their products and processes.

Then, a survey research was conducted in November 2003. To do so, a pre-tested questionnaire was sent to environmentally responsive SMEs from two distinctive industries, namely, the fabricated metal products industry (SCIAN 332) and the electric/ electronic products industry (SCIAN 335/334). These two industries were selected because of the significant environmental issues to which they are confronted. Environmentally responsive SMEs were voluntarily targeted in order to increase the probability of collecting significant data relative to innovative environmental concepts. A total of 198 valid questionnaires were retained for the purpose of the present study. The final sample is composed of 103 firms from the fabricated metal product industry and of 95 firms from the electric/electronic product industry. Then, a cluster analysis using a hierarchical procedure called the centroid method based on Euclidean distance was performed on the final sample in order to identify significantly distinctive groupings of firms that could be positioned on the organic production system map model. Finally, some characteristics of the products manufactured by the retained firms were analyzed with respect to these groupings.

4.2. *Results*

The results presented in Figure 2 allow the identification of three significantly distinctive clusters, positioned according to their respective average scores with respect to both dimensions of the organic production system map. The characteristics of these three groups are described hereafter.

Group 1 is composed of 122 firms showing an average score of 4.19 on the organic product dimension which, is above the scale's central point of 4 and which is also higher than the average score of 3.50 obtained on the organic process dimension. These average coordinates position Group 1 in Quadrant II, where the environmental efforts along the organic product dimension supersede the efforts made along the organic process dimension. However, it is worth mentioning that Group 1 occupies a rather central position on the map. This group represents the product innovators, and gathers a large proportion of the studied sample (62%).

Group 2 is composed of 31 firms with average coordinates of 1.77 along the organic product axis and 1.49 along the organic process dimension. The firms in this group, which is clearly located in Quadrant III according to its average coordinates, can be qualified as environmental laggards with respect to the proposed organic production system concept. This is the smallest group of the three, with 15% of the sample.

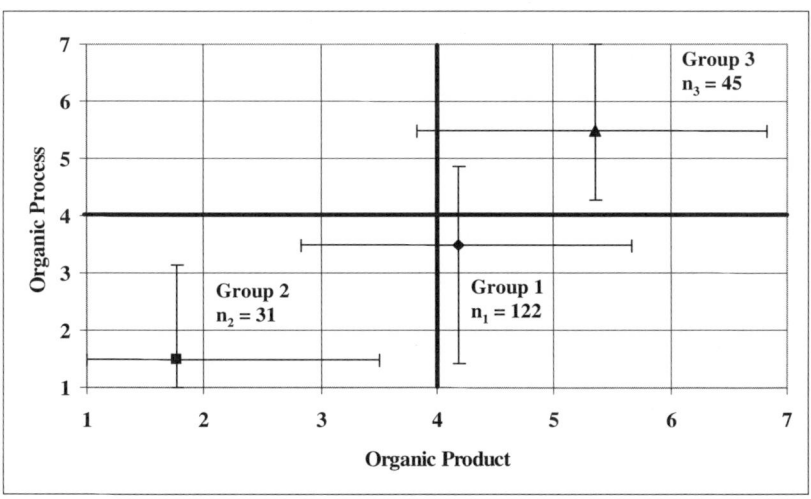

Figure 2. Cluster analysis results.

Group 3 is composed of 45 firms scoring an average of 5.35 along the organic product axis and an average of 5.49 along the organic process dimension. The 45 firms in this group, whose average coordinates are clearly located in Quadrant I, can be associated with environmental leaders in the implementation of organic production systems.

The fact that Quadrant IV is left empty also presents some interest. This might be representative of the fact that SMEs are granted greater autonomy in the environmental initiatives they undertake along the organic product dimension. Indeed, many of the initiatives comprising the organic product dimension may be implemented internally, without affecting suppliers or partners (see Table 1). However, the environmental initiatives in the organic process dimension might involve some important inter-organizational issues, which are supposed to be highly significant in the context of organic production systems relying on many small scale and geographically distributed components. Considering the limited power of SMEs at the collaboration network level, their capacity to implement environmental initiatives reaching beyond the firms' boundaries may be limited. But by taking this reasoning one step further to the firms in Group 3 located in Quadrant I, we may reach the conclusion that these environmental leaders are also demonstrating some leadership in the adoption of inter-organizational collaboration technologies.

Table 2 presents some characteristics of the products manufactured by the studied firms. The theoretical justifications for the selection of those characteristics are provided in Table 3. The products' life span variable is the only one that does not differ significantly among the three groups ($p = 0.343$). It is also interesting to note that Group 3 – the environmental leaders – is almost exclusively composed of exported products (98%), while Group 2 – the environmental laggards – is composed of only 39% of exported products. This could be an indication of the potential positive impact of foreign markets and regulations on SMEs' environmental practices. Group 1 – the product innovators- and Group 3 are both composed in major part of final products (0.74 and 0.67 respectively), while Group 2 is mainly composed of intermediate products (0.39 final products). This might indicate that market proximity, or direct end-user contact through the manufacturing of final products, could entice SMEs to raise their environmental profile.

Table 2. Comparison of product characteristics in the three identified groups.

Product characteristic	Group 1 (n1 = 122)	Group 2 (n2 = 31)	Group 3 (n3 = 45)	p(1)
Products' life span (2)	14.5	11.7	18.3	0.343
Exported products (3)	0.74	0.39	0.98	0.000
Final products (4)	0.74	0.39	0.67	0.001
Niche products (5)	4.1	3.4	4.6	0.000
Standardized products (6)	3.8	3.4	4.5	0.004
Customized products (6)	5.1	4.2	5.7	0.029
Sector (7)	0.38	0.52	0.73	0.000

[1] Kruskall-Wallis test
[2] Measured in years
[3] Dummy variable, 1 = "product is exported" and 0 = "product is not exported"
[4] Dummy variable, 1 = "final product" and 0 = "intermediate product"
[5] Continuous variable based on a 7 points Likert scale, ranging from 1 "lower" to 7 = "higher"
[6] Continuous variable based on a 7 points Likert scale, ranging from 1 "disagree" to 7 = "agree"
[7] Dummy variable, 1 = "electric/electronic product" and 0 = "metal product"

Group 3 scores the highest on the niche products and customized products characteristics (4.6 and 5.7 respectively), while Group 2 scores the lowest on these two characteristics (3.4 and 4.2 respectively). High end or niche products might justify additional environmental efforts while specific environmental requirements might be imposed by customers on customized products.

Table 3. Theoretical justifications of the product characteristics variables.

Product characteristic	Theoretical justification
Products' life span	(Parlar and Weng 1997; Lewis et al. 2001; ISO 2002)
Exported products	(King and Lenox 2001; Toffel 2004)
Final products	(Kahn 2002)
Niche products	(Reinhardt 1998)
Standardized products	(Ulrich 1995; Lampel and Mintzberg 1996; Spring and Dalrymple 2000; Karmarkar 2004)
Customized products	(Lampel and Mintzberg 1996; Spring and Dalrymple 2000)
Sector	(Murphy, Poist, and Braunschweig 1995; Handfield et al. 1997)

Finally, Group 3 – the environmental leaders – is mainly composed of firms from the electric/electronic products industry (73%), while Group 2 – the environmental laggards – is balanced in respect to that variable (52%). The high innovation rate in the electric/electronic products industry, which accelerates the obsolescence of previous generations of products, in combination with increasingly restrictive regulations like the WEEE directive in Europe, might, in part, explain this industry's environmental leadership.

5. Conclusion

The present article intended to explore the applicability of the organic production system concept to the Canadian SME industrial context. In order to do so, the organic production system concept was examined along two dimensions: the organic product and the organic process aspects. Based on information gathered from a sample of 198 SMEs, a multivariate analysis was conducted and allowed the identification of three clusters of firms, namely environmental leaders, product innovators, and environmental laggards. The interpretation of these results, and further analysis of the characteristics of the products manufactured by the SMEs in the studied sample, lead to some interesting findings found to be explainable in the context of the model called organic production system map. The organic production systems concept is a rich and promising research avenue. Considering their small scale, their decentralized nature, and the fact that they use modular materials that can be reused for different types of products, the study of organic production systems could be conducted in conjunction with the study of emerging e-collaboration technologies in order to understand how to best integrate such distributed organizational components along an entire product's life cycle.

References

Benyus, Janine M. (1997). Biomimicry innovation inspired by nature. 1st ed ed. New York: William Morrow.

Bernstein, F., and G. A. DeCroix. (2004). Decentralized pricing and capacity decisions in a multitier system with modular assembly. *Management Science,* 50, No. 9: 1293-308.

Billatos, S., and N. Basaly. (1997). Green Technology and Design for the Environment. London: Taylor and Francis Group.

Blanchard, Benjamin S. (1998). System engineering management. New York: Wiley.

Brio, J. A. del, and Junquera B. (2003). A review of the literature on environmental innovation management in SMEs: Implications for public policies. *Technovation* 23, No. 12: 939-48.

Carlson-Skalak, S., Leschke J., Sondeen M., and Gelardi, P. (2000). E media's global zero: Design for environment in small firm. *Interface,* 30, No. 3: 66-82.

Demeester, L., C. Loch, and L. Van Wassenhove. (2003). Organic production systems. Fontainebleau, France: INSEAD.

Ethiraj, S., Levinthal K., and D. (2004). Modularity and innovation in complex systems. *Management Science* 50, No. 2: 159-73.

Fleischmann, M., Beullens, P., Bloemhof-Ruwaard, J., and Van Wassenhove, L. N. (2001). The impact of product recovery on logistics network design. *Production and Operations Management,* 10, No. 2: 156-73.

Giudice, F., Rosa, G. La, and Risitano. A. (2005). Product design for the environment a life cycle approach. Boca Raton, FL: CRC Press.

Hall, R. (1993). The soul of the enterprise: Creating a dynamic vision for American manufacturing. New York: Harper Business.

Handfield, R. B., Walton, St. V., Seegers, L. K., and Melnyk, St. A. (1997). "Green" value chain practices in the furniture industry. *Journal of Operations Management* 15, No. 4: 293-315.

Hawken, P., Lovins A. B., and Lovins, H. L. (1999). Natural capitalism creating the next industrial revolution. 1st ed ed. Snowmass, CO: Rocky Mountain Institute.

ISO. (2002). ISO/TR 14062:2002. Management environnemental - Intégration des aspects environnementaux dans la conception et le développement de produit. Geneve: International Standard Organisation.

Kahn, K. B. (2002). An exploratory investigation of new product forecasting practices. *Journal of Product Innovation Management,* 19, No. 2: 133-43.

Karmarkar, U. (2004). Will you survive the services revolution? *Harvard Business Review* 82 , No. 6: 100.

King, A. A., and Lenox, M. J. (2001). Lean and green? An empirical examination of the relationship between lean production and environmental performance. *Production and Operations Management,* 10, No. 3: 244-56.

Klausner, M., Grimm, W. M., and Hendrickson, C. (1998). Reuse of electric motors in consumer products: Design and analysis of an electronic data log. *Journal of Industrial Ecology,* 2, No. 2: 89-102.

Krikke, H., Blanc, I. le, and van de Velde, S. (2004). Product modularity and the design of closed-loop supply chains. *California Management Review,* 46, No. 2: 23-39.

Lampel, J. and Mintzberg, H. (1996). Customizing customization. *Sloan Management Review*, 38, No. 1: 21-30.

Lewis, H., Gertsakis, J., Grant, T., Morelli N., and Sweatman A. (2001). Design + environment: A global guide to designing greener goods. Sheffield, UK: Greenleaf Publishing.

Mangun, D., and Thurston, D. L. (2002). Incorporating component reuse, remanufacture, and recycle into product portfolio design. *IEEE Transactions on Engineering Management,* 49, No. 4: 479-90.

Murphy, P. R., and Poist, R. F. (2003). Green perspectives and practices: A comparative logistics study. *Supply Chain Management*, 8, No. 2: 122-31.

Murphy, P. R., Poist, R. F., and Braunschweig, C. D. (1995). Role and relevance of logistics to corporate environmentalism: An empirical assessment. *International Journal of Physical Distribution and Logistics Management*, 25, No. 2: 5-19.

Parlar, M., and Weng, Z. K. (1997). Designing a firm's coordinated manufacturing and supply decisions with short product life cycles. *Management Science,* 43, No. 10: 1329-44.

Passino, K. M. (2005). Biomimicry for optimization, control, and automation. London, New York: Springer.

Pine, B. J. (1993). Mass customization: The new frontier in business competition. Boston, Mass: Harvard Business School Press.

Pine II, B. J. (1993). Making mass-customization happen: Strategies for the new competitive realities. *Planning Review* 21, No. 5: 23-24.

Reinhardt, Forest L. (1998). Environmental product differentiation: Implications for corporate strategy. *California Management Review,* 40, No. 4: 43-73.

Robertson, D., and Ulrich, K. (1998). Planning for product platforms. *Sloan Management Review*, 39, No. 4: 19-31.

Robinson, J. P., Shaver, P. R., and Wrightsman, L. S. (1991). Measures of personality and social psychological attitudes. San Diego, Toronto: Academic Press.

Sarkis, J. (1995). Manufacturing strategy and environmental consciousness. *Technovation,* 15, No. 2: 79-97.

Savaskan, R. C., Bhattacharya, S., and Van Wassenhove L. N. (2004). Closed-loop supply chain models with product remanufacturing. *Management Science*, 50, No. 2: 222-38.

Schilling, M. A. (2000). Toward a general modular systems theory and its application to interfirm product modularity. Academy of Management. *The Academy of Management Review*, 25, No. 2: 312-34.

Senge, P. M., Carstedt, G., and Porter, P. L. (2001). Innovating our way to the next industrial revolution. *MIT Sloan Management Review,* 42, No. 2: 24-37.

Silveira, G. Da, Borenstein D., and Fogliatto F. S. (2001). Mass customization: Literature review and research directions. *International Journal of Production Economics,* 72, No. 1: 1-13.

Spring, M., and Dalrymple, J. F. (2000). Product customisation and manufacturing strategy. *International Journal of Operations and Production Management*, 20, No. 4: 441-67.

Thierry, M., Salomon M., Van Nunen, J., and Van Wassenhove L. (1995). Strategic issues in product recovery management. *California Management Review*; 37, No. 2: 114-35.

Toffel, M. W. (2004). Strategic management of product recovery. *California Management Review*, 46, No. 2: 120-141.

Tu, Q., Vonderembse, M. A., Ragu-Nathan, T. S., and Ragu-Nathan, B. (2004). Measuring modularity-based manufacturing practices and their impact on mass customization capability: A customer-driven perspective. *Decision Sciences*, 35, No. 2: 147-68.

Ulrich, K. (1995). The role of product architecture in the manufacturing firm. *Research Policy*, 24, No. 3: 419-40.

Ulrich, K. T., Sartorius, D., Pearson, S., and Jakiela, M. (1993). Including the value of time in design-for-manufacturing decision-making. *Management Science*, 39, No. 4: 429-47.

Vachon, S., and Klassen, R. D. (2006). Green project partnership in the supply chain: The case of the package printing industry. *Journal of Cleaner Production*, 14, No. 6-7: 661-71.

Van Hemel, C. G. (2001). What sustainable solutions do small and medium-sized enterprises prefer? in Sustainable solutions: Developing products and services for the future. eds. M. Charter, and U. Tischner, 188-202. Sheffield, UK: Greenleaf Publishing.

Van Woensel, T., Creten, R., and Vandaele, N. (2001). Managing the environmental externalities of traffic logistics: The issue of emissions. *Production and Operations Management*, 10, No. 2: 207-23.

Victor, B., and Boynton, A. C. (1998). Invented here: Maximizing your organization's internal growth and profitability. Boston, Mass: Harvard Business School Press.

Viswanathan, S., and Allada, V. (2006). Product configuration optimazition for disassembly planning: A differential approach. *Omega*, 34, No. 6: 599-616.

CHAPTER 34

INNOVATION, COMPETITIVENESS AND SUSTAINABILITY: STUDY ON A PLYWOOD INDUSTRY CLUSTER

Sieglinde Kindl da Cunha* and João Carlos da Cunha**

*Centro Universitário Positivo – UNICENP, Curitiba, Brazil;
**Universidade Federal do Paraná, Curitiba, Brazil

The present article identifies the factors responsible for systemic competitiveness of a plywood industry located in Palmas, State of Paraná – Brazil. Our research is based on the German Development Institute model (GDI), which analyses the systemic competitiveness through the connections and interdependences between the meta, macro, meso and micro economic aspects. The research results identified and classified the key-factors responsible for the systemic competitiveness, those factor associations and the factors that jeopardize such industry growth and increased penetration in the foreign and domestic markets.

1. Introduction

The present article aims to identify the local industrialization strategies based on the group of plywood industries located in Palmas region, in the State of Paraná. The plywood industry was established there primarily because of the comparative advantages of raw-material (natural reserves) availability and inexpensive labor costs. With the depletion of natural wood reserves and the national and international market requirements, the old sawmills turned out to be obsolete and were replaced by the plywood industry owing to its sustainment and growth potential associated with the exploration of competitive system advantages. Development of a favorable environment with competitive clusters is based on system factors. Such factors include: a clear definition of regional development strategies by local agents, stable macro-economy, political and institutional environment, favorable local culture for interaction and cooperation between private and public agents, a company environment that nurtures learning and innovation, micro-economic competitiveness induced by absolute cost advantage, product differentiation, and economies of scale.

The present article is aimed at assessing the competition potential of the plywood cluster in Palmas Region, through identification and classification of the factors (according to importance) responsible for the meta, macro, meso- and micro-economic system competitiveness and by using the main component models.

The identification of responsible factors for the system competitiveness, which is the object of this research, is justified by the importance of such industry gathering within the Palmas region. In particular, the productive structure in terms of job sustainment and increasing participation of the regional plywood industry in the American and European markets.

Firstly, this article reviews the relevant literature on local system competitiveness and development. Then, follows a description of the field research methodology and the main component model used to identify and classify the factors influencing the plywood industry system competitiveness. Section 4 shows and analyses main characteristics of the plywood industry and their importance in terms of job generation and income in the Palmas region. The final sections present the methodology, results, and conclusions.

2. Conceptual Basis

The research development is based on analysis of the system competitiveness as the new industrial pattern support, thus restoring the region role as locus for innovation and competition. (Change this sentence)

Two elements of the Local Development Theory are added to enrich and complement the understanding of empirical potentialities of a cluster system competitiveness.

The first one is related to the globalization phenomenon that, apparently, could lead to the removal of economic barriers thus creating a non-territorialized world without boundaries and dominated by large corporations. In fact, we can see a new phenomenon arising – the possibility of joining local and global environments. Instead of homogenizing the national economic spaces, the globalization process may increase the differences and the competition between regions in the same country.

The second element is related to the fact that, in the knowledge-driven and increased net integration era, the region rises again as a productive organization and innovating locus. To achieve this second element research efforts, success, institutional actions, and learning happen collectively through interaction, cooperation, and complementarities. The agents are immerged in the local cultural environment that is dependent on the historical cultural process. This way, besides these attributes, there is also a continuous process of regional tacit learning. Hence proximity, process flexibility and productive organization are considered very important (Breschi et al., 1997).

The aforementioned elements allow us to restore the region or agglomeration role as basis for innovation and competition. At the same time, to restore and to join the concepts of regional growth or development, industrial district, clusters, productive complexes, industrial agglomeration, external economies, and urban support as organizational forms and conditions for the innovation process and competitiveness gains (Diniz, 2000, p. 4).

The cluster system competitiveness analysis developed in this article is supported by the System Competitiveness and Agglomeration concepts.

The system competitiveness concept is understood as a model where the State and the social actors deliberately create conditions for industrial development to be successful (Altenburg et al., 1998, p. 2).

Among the analytical focuses we highlight the neo-shumpeterian economic focus that explains the system competitiveness in terms of technology, innovation, and knowledge being strategic components of local, regional, national, and sector development (Dost et al., 1988).

According to that focus, the industrial growth new trajectories are based on the new technological paradigm that depends on building new innovation systems. We understand innovation as a process that happens during a trajectory built on continuous learning, such as learning-by-doing, learning-by-using, e learning-by-interacting, learning-by-searching, learning-by-learning intra-companies and between-companies and research and technology institutions. According to the neo-shumpeterianians, not all kinds of knowledge can be codified, that's to say, it is essential to develop the tacit knowledge that is spread through nets of social agents. The institutional structure varies from country to country and region to region according to the different clusters (Lundvall, 1992).

To compete in a global environment, which strengthens the accelerated economic liberalization and brings down the trading and investment barriers, local companies have to adapt to the new international quality standards, answer speed, and flexibility. To overcome common problems, the productive agglomeration has to intensify the cooperative interaction between the component agents.

The advantages associated with cooperation nets are having the possibility to effectively explore the collective efficiencies and/or develop external economies (business cooperation, work productive specialization, collective infrastructure, service specialization, etc.), and having the benefit of increased collective negotiating capacity with input and component suppliers. In addition to these advantages, the cooperation nets make it easy to develop new production models, processes, organizations, technical and market information exchange, consortiums to buy and sell goods and services, and joint campaigns for product marketing and distribution in the domestic and foreign markets.

The productive agglomerate agents move through dense interactions, cooperation, and competitive relations. These include the companies (suppliers, clients and competitors), meso-institutions (public and private institutions, and civil society representatives), macro-institutions (macroeconomic strategies and policies), and socio-cultural structures (social actor ability to formulate sustainable development views and strategies).

The cluster system competitiveness (Altenburg *et al.*, 1998) comprises four net relation levels:

(i) Meta – comprises socio-cultural factors that define the social actor negotiation capacity and ability to formulate strategies and policies of interest to society.
(ii) Macro – macro-economic strategies and structure defined by fiscal, monetary, Exchange, trading and competition policies.
(iii) Meso – supporting structures that facilitate interaction and cooperation between companies (suppliers, clients and competitors), interaction and cooperation with P&D institutions, financing supporting institutions, support and dissemination institutions (marketing, exports, fairs, etc.), labor formation and training institutions, and infrastructure (transportation, communications, power).
(iv) Micro – capacity of a company, or company network, to remain competitive offering goods and services that optimize the cost-efficiency relation, quality, variety, and ability to react upon new opportunities and market changes.

Interaction between companies that are geographically close to each other as well as facilitating larger cooperation nets to reduce growth obstacles for competitiveness between companies and regions.

The interaction, cooperation, and complement concepts restore the local role as the basis for innovation and competition. This brings back the concepts of agglomeration, regional growth, clusters and systems.

Porter (1993, p. 209) defines agglomerations as

"Geographic concentrations of interrelated companies, specialized suppliers, service renders, correlated sector companies and other specific institutions (universities, norm-formulating organs and associations) that compete but also cooperate with each other."

Galvão's (2000, p. 9) concept of clusters is

"All types of agglomeration activities concentrated geographically and sector-specialized, no matter what is their productive unit size

or nature of the economic activity they develop, or if belong to the manufacturing industry, service sector or agriculture."

Brito (2000, p. 8) complements such concept:

"...the industrial clusters should not be considered mere industrial activities spatially agglomerated and present in certain sectors, but local productive organizations where complementary and interdependent relations predominate in the different activities located in the same geographic and economic space. These clusters are conceived as confluence points between the organization of innovating regional-local systems at institutional level and emergent company nets as a standard form of configuration these business systems."

The local productive organizations or clusters comprise companies "networks" and develop integration, cooperation, solidarity and valorization of collective efforts. The result is company and system increased competitiveness, when compared to companies that work on their own.

The local productive organization is associated to a group of companies and institutions concentrated spatially and having a vertical (comprising several stages of a determined chain) and a horizontal (involving factor, competency, and information interchange between similar agents) interrelation. It has an intern configuration including: a) a large company or concentration of similar companies and the identification of their upstream and downstream relationships; b) sectors that have the same suppliers or supply complementary products and services; c) companies and institutions that supply specialized qualifications, technologies, capital, infrastructure and class associations; d) government agencies and other regulatory organs that have influence over an agglomeration (Cassiolato et al., 2000).

There are several typologies to classify the local productive organizations. The typology is chosen according to the research objective. For example, taxonomy can be based on sector segmentation criteria (traditional or hi-tech sectors), agent coordination capacity (diversified or sub-contraction), criteria for the relationship between plywood and its markets (producer-driven or buyer-driven chains), and a type of local productive organizational relationship (horizontal and vertical).

3. Methodology

The research aims at identifying factors that are responsible for the local productive organization system competitiveness of the plywood industry located in the Palmas Region.

The research hypothesis is that the plywood industry located in the Palmas Region has a competitive potential to be developed through institutional innovations strengthened by interaction and cooperation networks.

The field survey was performed in 21 plywood and laminated wood located in Palmas micro-regions.

The region industry records were supplied by Federação das Indústrias do Estado do Paraná (FIEP) - Paraná Industry Federation - and the complementary information was provided by Cadastro da Secretaria de Estado da Fazenda e da Associaçãos Comercial e Industrial do Município (Finances State Departament and Trading Association Records of that region municipalities).

The field research was carried out through interviewing industry owners and directors directly. They answered a questionnaire comprising variables that were potentially responsible for the meta, macro, meso- and micro-economic system competitiveness.

Each variable was assessed pursuant to a 0 to 5 scale, according to the extent of its influence on competitiveness, from the entrepreneur point of view.

In order to select and classify the most influential variables over the system competitiveness at each level (meta, macro, meso- and micro-economic), we used the multi-variable statistical analysis technique called Main Component Analysis.

For this research purpose the variable group used in the n x k matrix was divided into four levels. A meta and macro group (21 x 11), a meso company cooperation group (21 x 14), a meso institution cooperation group (21 x 12), and a micro group (21 x 19).

The main components are computed in such way that the first one represents the largest parcel of the explanatory variable group variance, the second represents the second largest parcel, and so on, with the advantage of the factors being non-correlated variables.

Such technique allows us to reduce and classify variables that are typical of a few explanatory indexes (main components) through linear combinations of original variables.

4. Cluster Organization Characterization

In the last decade, the plywood industry consolidated its position within the Brazilian industry, mainly concerning the international market.

The establishment of new production units, the search for production new technologies and the industrial sector modernization favored the Brazilian plywood growth and exports. On the other hand, the Plano Real and the National Policy to increment exports opened new market opportunities at domestic and foreign levels.

From 1998 on the Brazilian plywood exports shot up due to the floating Exchange regime that increased the Brazilian product competitiveness significantly dropping the price of Brazilian plywood below the international market price. From 1998 to 2000 the Brazilian production increased 184%, and Brazil leaped from the 15th to the 10th place among the world exporters. In Latin America, Brazil holds the 1st place as both producer and exporter.

The lumbering activity that has historically defined Palmas region's productive profile in terms of traditional sawmills, has been changing its profile towards a more advanced wood production chain. Extra stages included a larger production scale and managerial and production process innovations. Thus it increased its participation in the plywood foreign market.

The main characteristics of the traditional lumbering sector, where sawmills predominated, were lack of leadership and less cooperation capacity between companies.

Changes were firstly due to the traditional sawmill's depletion, linked to wood extraction from natural forest reserves and the plywood industry growth bringing expressive gains through segments with greater added value, such as laminated wood and plywood boards. There were technological innovations not only concerning industry, but also in the reforestation activities by introducing pine (pinus), which allows significantly increased productivity by anticipating the cut period and improving the raw-material quality (Cunha *et al.*, 2003).

The field survey results show the characteristics of the Palmas Region cluster.

Concerning size, the companies are small and medium under the leadership of two large companies (Table 1).

Larger companies produce plywood addressed to the international market, one of them being the greater pinus plywood exporter in Latin America. The medium companies address their production mainly to the domestic market and are large suppliers of plywood and laminated wood to building construction. These companies eventually export their production when the larger companies are not able to meet the international market demand. Most of the small companies supply laminated wood to local companies and, in some cases, to the domestic market.

Table 1. Characteristics of Palmas plywood company sizes.

Size	Size According to Number of Employees[1]	Size According to Revenue[2]
Small	15	16
Medium	4	3
Large	2	2
TOTAL	21	21

Source: Field survey.
(1) Classification criteria concerning number of employees: small (0 to 100 employees), medium (101 to 500 employees) and large (over 500 employees).
(2) Classification criteria concerning revenue: small (below R$ 10.000.000), medium (R$ 10.000.000 to R$ 50.000.000) and large (over R$ 50.000.000).

Large companies, and some medium ones, are investing heavily in pinus reforestation in order to overcome the raw-material scarcity issue that is foreseen for the next years. Although they have their own reserves to meet the present raw-material demand, they are supplied by outsourcing companies and need to set reforested areas aside for future use.

The small and medium companies, which have less capital, depend exclusively on raw-material supplied by other companies. It is foreseen that such companies tend to meet bottlenecks that will hinder their growth concerning a raw-material supply.

The increasing importance of large companies in the international market, both for job and income generation as well as control of raw-material sources explains the expansion of their leadership.

The plywood industry products are classified as low differentiation level commodities. So, it is the production scale, not the differences, that determines the local productive organization competition potential. The technological pattern added to machinery and equipment is easily purchase in the domestic market. Therefore, the largely widespread technology is not an obstacle to the entry of new companies. The barriers to such entry are established by the amount of capital needed to invest in plants that maximize economies of scale and by investments in forest reserves in order to maintain the control over the raw-material sources.

5. System Competitiveness Analysis

According to the methodology stated above, the analysis of the plywood local productive organization system competitiveness was divided into the meta, macro, meso- and microeconomic levels. The meso-level analysis was divided into meso-cooperation between companies and meso-interaction with institutions because of the number of variables and different relationship characteristics between companies, universities, and research institutions.

According to the tabulations, the variables that most affect the sector competitiveness are those classified in the meat and macroeconomic levels with an interference degree of 3.0 in a 0 to 5 scale. The micro level variables and the ones that define the cooperation between companies have an interference slightly over 2.5. Finally, although the meso-level variables - local productive organization partnerships with supporting institutions – affect the system competitiveness just a little, they are variables that can potentially differentiate the local productive organization in competitive terms (Figure 1).

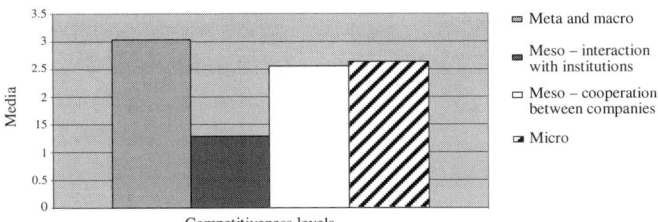

Figure 1. Competitiveness mean intensity (Source: Field survey).

To select and classify the variables with higher influence degree over system competitiveness we used the technique of multi-varied statistic analysis, also called Main Component Analysis (MCA). This technique allows us to reduce and classify the variables on a small number of explanatory indexes (main components) through linear combinations of the original variables.

5.1. *Meta and macroeconomic competitiveness*

The variables observed in the meta and macroeconomic dimensions suggest assessment of the external factors relevant to the plywood industry's competitiveness. These indicators refer to cultural factors such as capability of the agents that define the strategy development and political-economic decisions that affect company competitiveness as: fiscal, exchange, monetary, trading and industrial policies.

Table 2 classifies the factors by the importance of their influence over the plywood industry competitiveness at meta and macroeconomic levels. Four components that account for 78% of the competitiveness sector were selected through MCA.

Factor 1 explains 42% of the macroeconomic level competitiveness and highlights the fiscal and environmental policies as being the most representative factor.

Concerning state and federal taxes, the fiscal policy has burdened significantly the final price of products directed to domestic market, thus reducing the small and medium company competitiveness. Such variable is not significant to exporting companies that are not burdened by the Kadir Law.

Table 2. Main component matrix – Meta and macroeconomic levels.

Variables	Factor 1	Factor 2	Factor 3	Factor 4
Public sector capability to coordinate the sector development	0.010	0.340	0.973	0.112
Actor capability to lead and coordinate the sector development	0.527	0.372	0.298	0.521
Influence of class representations over government decisions	0.535	0.571	0.389	0.031
Exchange rate influence	0.042	0.941	0.002	0.017
Trading policy influence	0.060	0.623	0.517	0.504
Interest rate influence	0.276	0.168	0.157	0.700
Financing policy influence	0.693	0.269	0.181	0.176
Environmental policy	0.903	0.095	0.005	0.094
Federal fiscal policy influence	0.889	0.182	0.071	0.162
State fiscal policy influence	0.948	0.126	0.057	0.036
Infrastructure influence	0.768	0.062	0.401	0.126
Variance percentage	**41.68**	**17.34**	**10.22**	**8.90**
Accrued variance percentage	**41.68**	**59.02**	**69.25**	**78.15**

Source: Field survey.

The importance of the "Environmental Policy" variable is totally justified since by the time this field survey was carried out the Environment Ministry was publishing the Portaria 507 (Administrative rule 507) establishing environmental preservation areas in several municipalities, specially affecting Palmas municipality in Paraná. Such administrative rule limits pinus reforestation in new areas within that region and this restriction brings social and economic negative impacts, thus imposing new challenges on the communities affected by it. This policy affects larger companies in that region and requires them to invest heavily in reforestation activities in order to meet the future raw-material demand. If the Portaria 507 text is not changed in the medium term, the local companies' trend will be towards moving to other regions that do not impose restrictions concerning raw-material supply.

Regarding factor 2, exchange policy is an element that has high impact on the sector competitiveness, mainly on the larger exporting companies that between 1998 and 2002 were highly benefited from the Exchange policy and from the increased foreign demand. Therefore, the exporters improved their profitability, investments in modernization and productive capacity regarding the industry itself and pinus reforestation. The increased international prices of plywood also favored the companies that supply the domestic market, seeing that plywood boards are commodities that have their intern prices linked to the foreign market.

This research highlights as the main factor 3 variable the public sector capability to coordinate local development strategies. Such result can be explained by the fact that the organization of local companies still does not have an institutional leadership (associations, federations), thus counting on the local or regional public sector for developing organization strategies.

Concerning factor 4, explanations on interest rate are not deepened. In spite of the financial system high interest rates, they have not been a limiting factor to the sector competitiveness. There are two main reasons why the local productive organization is independent of the financial system: firstly, company profit growth allows plough back; second, companies do not borrow from the traditional financial system when market interest rates are high.

5.2. Meso-competitiveness – Cooperation between companies

Four factors can explain the meso-competitiveness between companies. The first one comprises the cooperation and interaction activities addressed to improved penetration into the market and reduced production costs, with a variance percentage of 44%. The second factor refers to competitiveness induced by associative activities for developing tacit knowledge technology and promoting innovations, with an explanation degree of 16%. The third factor highlights the cooperation in the production chain, with an explanation degree of 11%. Finally, the fourth and last factor, which has relatively less influence, highlights cooperation to borrow financial resources.

These four factors explain 84% of the competitiveness at meso-level – cooperation between companies, according to Table 3.

Factor 1 highlights four cooperation variables that are relevant to the organization competitiveness. Two of them valorize cooperation addressed to creating market opportunities: local brand development and

Table 3. Main component matrix – Meso-competitiveness – Cooperation between companies.

Variables	Factor 1	Factor 2	Factor 3	Factor 4
Geographic nearness	0.193	0.292	0.269	0.678
Relationship with the production chain links	0.103	0.010	0.921	0.148
Cooperative activities addressed to labor formation and training	0.817	0.158	0.422	0.160
Information flow and tacit knowledge interchange	0.224	0.859	0.036	0.127
Cooperative projects addressed to technological activities and innovation promotion	0.346	0.733	0.071	0.074
Cooperation to develop products and processes	0.239	0.546	0.449	0.361
Cooperation to ease the access to financing sources	0.255	0.741	0.122	0.705
Influence over the sector policies	0.758	0.218	0.016	0.339
Local brand development	0.875	0.018	0.259	0.237
Cooperation to buy inputs	0.557	0.641	0.427	0.173
Cooperation for selling and exporting activities	0.640	0.359	0.265	0.495
Joint fairs and events	0.845	0.191	0.076	0.246
Power to negotiate with supplier and customers	0.732	0.423	0.291	0.101
Supplier development	0.883	0.065	0.072	0.44
Variance percentage	**44.06**	**15.75**	**11.40**	**7.41**
Accrued variance percentage	**44.06**	**59.82**	**71.22**	**84.07**

Source: Field survey.

joint activities to plan fairs and events. The two remaining activities show businessmen are concerned with cost and quality: cooperation to labor formation and training, and cooperation for development of suppliers.

Factor 2 highlights the need of developing cooperative activities addressed to technological development and organization innovation, and to improved information and tacit knowledge flow between companies. The international competitive pattern of the large exporting companies is promoting changes in the technological profile of the lumbering industry in that region. From the interviews, we can notice that the businessmen are clearly concerned about carrying out joint activities when seeking for better quality and cost reduction. Such concerns affect the decision making process concerning the production capacity improvement through machinery and equipment modernization, raw-material improved quality, increased production scale, certification, investing in labor formation and qualification and improved export logistics. The solution to meet the international market new demands is to improve tacit knowledge spreading and develop new technologies and innovations for the whole sector.

The concern about strengthening the production chain links is highlighted in factor 3. Nowadays, a significant part of the demand for raw-material in plywood industry in Palmas Region comes from suppliers external to the region. Mainly due to the restrictions imposed by Portaria 507, there is a trend towards an increased dependence on external suppliers. Concerns about this issue are highly justified, since spatial distance between industries and suppliers will probably increase in the next years.

The factor 4 variable – "Cooperation to ease the access to financing sources" has little influence over the meso-competitiveness. In fact, this is a concern shared by the small and medium companies that are outsourced and have less capital.

Although the competition and fragmentation of the plywood industry production structure has historically characterized the agent culture in the region, the relevant variables in the main component analysis indicate that this trend is changing due to the strengthened interaction between companies. Specifically, this interaction seeks to accomplish a kind of synergy in order to strengthen the local productive organization as a whole.

5.3. *Meso-competitiveness – Company interaction with institutions*

As shown in Figure 1, the relationship between companies and supporting institutions are weak and not capable of generating synergies within the cluster. In the main component analysis there are three institutions in factor 1 which maintain some partnerships with the local organization companies: universities, Banco Regional de Desenvolvimento do Extremo Sul (BRDE) (Regional Bank for the South Region Development) and Instituto de Tecnologia do Paraná (Tecpar) (Paraná Technological Institute) - Table 4. But the businessmen interviewed clearly stated that such relationships are specific, non-continuous, informal and have barely contributed to improvement in the sector competitive performance.

The research results clearly show that the plywood cluster has not explored the system competitiveness potential through partnerships and interaction with research institutions and universities, in order to exchange information and develop technologies and innovations for the sector. It hasn't also explored its main organization and representation channels provided by associations, federations and labor unions.

Table 4. Main component matrix – Meso-competitiveness – Company interaction with supporting institutions.

Variables	Factor 1	Factor 2	Factor 3
Senai	0.604	0.447	0.554
Sebrae	0.297	0.382	0.65
Senac	0.056	0.899	0.262
Universidades	0.815	0.2	0.13
BRDE	0.928	0.076	0.182
Tecpar	0.899	0.273	0.014
Research Institutes	0.499	0.48	0.469
Federations	0.071	0.956	0.081
Associação Brasileira da Indústria da Madeira (Abim)(*Brazilian Wood Industry Association*)	0.855	0.148	0.14
Labor unions	0.108	0.096	0.77
Associação da Indústria da Madeira (AIM) (*Wood Industry Association*)	0.532	0.463	0.233
Variance percentage	**51.22**	**16.00**	**10.66**
Accrued percentage	**51.22**	**67.22**	**77.88**

Source: Field survey.

In the factor analysis we noticed there is a lack of class representation leaderships (federations, associations, etc.) when the sector negotiates and represents its interests before the government, and actors have low capacity for leading and coordinating the sectors development. In that region, the traditional lumbering cultural inheritance (extraction, sawmills) is still part of behavior for older businessmen, who are relegated to second place in the local agent potential for organizing, cooperating, and defining the local organization growth strategies. However, such culture is being transformed by strengthening new leaderships and changing the traditional leadership behavior.

5.4. Micro competitiveness

Micro competitiveness comprises four factors: (1) scale and quality; (2) raw-material suppliers; (3) finished product prices; (4) technology.

Amongst explanatory variables for factor 1, the raw-material quality and qualified labor are fundamental factors for the finished product. Such changes result from the pressure of plywood importers, which are increasingly demanding concerning wood product quality.

Production scale is another variable adding much weight to competitiveness of micro level in order to improve the fixed cost scale and reduction. Last years' investments allowed for industry enlargement and modernization. There is a trend towards an industry concentration process with the incorporation or bankruptcy of small companies with less capital, which face difficulties concerning their insertion in the foreign market and access to raw-material sources.

Table 5. Main component matrix – Microeconomic competitiveness.

Variables	Factor 1	Factor 2	Factor 3	Factor 4
Access to technology	0.227	0.630	0.236	0.427
Investments in the company technological capacitating	0.479	0.196	0.542	0.335
Low production cost	0.514	0.685	0.010	0.162
Scale	0.781	0.211	0.313	0.116
Product quality	0.889	0.095	0.017	0.097
Continuous production innovations	0.366	0.530	0.258	0.538
Production process improvement	0.665	0.258	0.123	0.481
Product price	0.475	0.384	0.761	0.028
Delivery time	0.700	0.191	0.337	0.126
Relationship with suppliers	0.775	0.500	0.042	0.028
Raw-material quality	0.784	0.049	0.386	0.248
Raw-material price	0.580	0.394	0.599	0.011
Relationship with costumers	0.399	0.785	0.054	0.061
Product diversity	0.620	0.076	0.508	0.174
Technical specification adequacy	0.650	0.272	0.510	0.219
RH capacitating	0.788	0.326	0.236	0.292
Company infrastructure	0.476	0.109	0.224	0.603
Environmental issues	0.721	0.327	0.138	0.107
Variance percentage	39.57	15.65	13.21	8.25
Accrued percentage	39.57	55.22	68.43	76.68

Source: Field survey.

Finally, but not less important, is the "production process improvement" variable that is responsible for the cluster microeconomic competitiveness.

6. Result Synthesis

The following table shows a synthesis of the variables with negative and positive impacts according to competitiveness levels. The variables consolidate the local productive organization of plywood in Palmas.

Table 6. Weak, strong and in-consolidation aspects by competitiveness levels within the plywood cluster in Palmas Region.

	Meta and Macroeconomic Levels	Meso Level – Cooperation between Companies	Meso Level – Interaction with Supporting Institutions	Microeconomic Level
Strong Aspects	Exchange policy. International market potential growth. Intern and external market prices.			Scale. Raw-material quality. Raw-material price. Finished product price.
Weak Aspects	Environmental policy – Portaria 507. Lack of cooperation to define and apply local development strategies. Fiscal policy.	Competitive relations. Local leadership fragility. Fragility of P&D cooperation activities.	Fragility, informality and discontinuity of the relationship with Universities and research institutions. Fragility of representative leadership associations and federations.	Production diversification. Company technological capacitating. Plants.
Aspects Being Consolidated		Cooperative relationship with customers. Cooperation in human resource formation activities. Joint activity to improve raw-material quality, finished product and certification.	Relationship with consumers. Spreading information and tacit knowledge.	Human resources capacitating.

Source: The authors.

7. Conclusion

The Main Component Analysis shows the following local productive organization competitiveness aspects as the strongest ones at meta and macro levels: exchange policy, demand growth and international price growth. The competitiveness supported by such variables is extremely fragile, since they go beyond company and local organization grasp.

At meso-level, the analysis does not mention any strong aspect, what shows lack or fragility concerning interaction and cooperation among the local organization agents. Te local agents should plan active attitudes to define sustainable growth strategies for the whole local organization, thus intensifying partnerships through cooperative and less competitive relationships. There is a gap in the relationship between companies regarding information and tacit knowledge spreading, as well as adequate technological development for the local organization. Companies could increase their cooperative actions when seeking for market increased penetration through partnerships and development of trading logistics that could benefit the local organization as a whole.

The Cluster most fragile aspect is lack of cooperation with universities, research institutions, and supporting and extension entities. Such relations are fundamental for local sustainable development in the long term, since they generate synergies concerning growth, tacit knowledge spreading, adequate technological development, negotiation capacity, etc.

The competitiveness of the plywood local productive organization in Palmas Region is sustained by microeconomic variables such as production scale, raw-material cost and quality, and finished product quality and price. These competitive advantages are due to company individual efforts and not to local productive organization collective efforts.

Some variables show a trend towards consolidating cooperative activities, mainly concerning the improvement of interaction with suppliers seeking better prices and quality of raw-material. Some measures addressed labor formation and qualification, as well as strengthening both relationships with clients and fair/event organization and participation.

The research results primarily show how the Plywood Local Cluster in Palmas Region is going through a period of transition from an informal to an organized Cluster.

References

Albuquerque, E. M. (2000). Análise da performance produtiva e tecnológica dos clusters industriais na economia brasileira. Rio de Janeiro: IE/UFRJ, (Série Nota Técnica, 28/00).

Altemburg, T.; Hillegard, W.; Stamer, J. M. (1988). Building System Competitiveness. Berlin: German Development Institue (GDI). .

Breschi, S.; Malerba, F. (1997). Sectorial Innovation Systems: Technological regimes, shumpeterian dynamics and spacial boundaries. In: EDQUIST, C. (Ed.). System of innovation: technologies, instituitions and organizations. London: Pinter.

Brito, J. (2000). Características estruturais dos clusters industriais na economia brasileira. Rio de Janeiro: IE/UFRJ. (Série Nota Técnica, 29/00).
Brito, A. F.; Bonelli, R. (1997). Políticas industriais descentralizadas: as experiências européias e as iniciativas subnacionais no Brasil. Brasília: IPEA. (Série Textos para Discussão n. 492).
Cassiolato, J. E.; Lastres, H. M. M. (2000) Sistema de inovação: políticas e perspectivas. *Parcerias Estratégicas,* n. 8, p. 237-255.
CNI – Confederacao Nacional da Industria. (1988). Agrupamentos (clusters) de pequenas e médias empresas. Brasília: CNI.
Cunha, S. K.; Oliveira, M. A. (2003). Arranjo Produtivo Local e o Novo Padrão de Especialização Regional da Indústria Paranaense na Década de 90. Ed. IPARDES, Curitiba.
Dinez, C. C. (2000). Global- Local: Interdependências e Desigualdades ou Notas para uma Política Tecnológica e Industrial Regionalizadas no Brasil. Technical Note n.9 IE/UFRJ, Rio de Janeiro.
Dosi, G., Freeman, C., Nelson, R. and Soete, L. (1988). Technical Change and Economic Theory. Pinter, London.
Galvao, O. J. A. (2000). "Clusters" e distritos industriais: um estudo de caso em países selecionados e implicações de políticas. *Planejamento e Políticas Públicas,* n.21, p. 3-50.
Galvao, A. C.; Vasconcelos, R. R(1999). Política regional à escala sub-regional: uma tipologia territorial com base para um fundo de apoio ao desenvolvimento regional. Brasília: IPEA. (Série Texto para Discussão n 665).
Garcia, R. (2001). A Importância da dimensão local da inovação e a formação de clusters em setores de alta tecnologia. *Ensaios FEE,* ano 22, n.1, p. 115-142.
Lundvall, B. (1992). National Systems of Innovation: Towards a Theory of Innovation and Interactive Learning, Pinter Publishers, London.
Porter, M. E. (1999). Competição: Estratégias Competitivas Essenciais.. Rio de Janeiro: Campus.

CHAPTER 35

SUSTAINABLE DEVELOPMENT IN COMPANIES: AN INTERNATIONAL SURVEY

Daniela Ebner and Rupert J. Baumgartner

*Institute for Economic and Business Management,
Montanuniversität Leoben, Austria*

In this article, the results of an international survey on Sustainable Development (SD) as an aspect of business management are presented. 95 companies, listed in internationally known indices, are evaluated based on homepages and sustainability reports concerning their integration of ecological and social aspects as well as a super-ordinate "generic" dimension. The analysis has been made to find out specific regional trends and variations between industries on the implementation of SD aspects. Distinct regional differences could be distinguished concerning emphasis being put on the SD dimensions. Apart from regional differences also (enormous) industry-typical variations became obvious concerning SD as aspect of the management.

1. Introduction

Discussions about SD and related terms such as Corporate Social Responsibility (CSR), Corporate Sustainability or Corporate Citizenship have increased in recent years (Robin and Reidenbach, 1987; Zambon and Del Bello, 2005; Labuschagne *et al.*, 2005; Salzmann *et al.*, 2005). They have become key words of modern economics (Baumgartner, 2004; Castka *et al.*, 2004). The notion of SD by Brundtland has become the common basis for sustainability (Gauthier, 2005; Korhonen, 2003). Nevertheless, advocates of SD and CSR have continued to produce definitions on social responsibility and its related terms (van Marrewijk,

2003; Quazi and O'Brien, 2000; Zambon and Del Bello, 2005; Fergus and Rowney, 2005). In this paper, the term SD is based on Brundtland's definition (World Commission on Environment and World Commission on Environment and Development, 1987) and the concept of Triple-Bottom-Line (Elkington, 1998). In recent years, companies have tended to focus more on sustainability (Moir, 2001). Empirical studies (e.g. Knox *et al.*, 2005; Quazi and O'Brien, 2000; Jenkins, 2004; Whitehouse, 2006; Welford, 2005; Snider *et al.*, 2003) concerning the integration of SD into companies already exist, however, the main focus is often on specific sustainability dimensions, industries or countries rather than offering an insight into SD as a whole.

2. Research Purpose

This paper deals with the question how the sustainability aspect is implemented and performed at a micro-economic level. Furthermore, it analyses whether special regional trends like the focus of one of the three dimensions are recognizable and how Austrian companies operate concerning the implementation of SD in comparison to the global trend.

3. Methodology

3.1. *Method*

In this extensive study, 95 companies were examined for the implementation of SD and its integration as an aspect of the management. All companies are quoted on stock exchanges and listed in the indices (as of 31st July 2004)

- ATX (Austrian Traded Exchange): all 20 companies
- Dow Jones Industrials Average: all 30 companies
- Nikkei Stock 225 Average: the 20 companies with the highest market capitalization
- Dow Jones Eurostoxx50: the 25 companies with the highest market capitalization

Data was obtained from analysing annual and sustainability reports, homepages and press texts of the respective companies. Distinct regional differences could be subdivided by the emphasis being put on the three SD dimensions.

3.2. Framework for survey

In order to analyse the companies, a framework was developed, covering the ecological and social dimension of SD as well as a super-ordinate "generic" dimension. These three dimensions were divided into nine aspects, from which 46 criteria were derived (see the detailed list of criteria in Appendix A). They were transformed into questions which had to be answered by the analyst with "Yes" or "No" (or "not apparent" if no related information was available) according to information from the homepage or the official company report. Moreover, open questions were included in the framework which provided the opportunity to present the actions taken by the company more clearly.

Generic Dimension: The aim of the generic analysis is to give an overview of the company's vision and strategy. Although the generic dimension is not part of the general sustainability model, we can gain a broad picture about the integration of SD into the company and its culture by studying the company, its vision and aims in general.

Ecological Dimension: When talking about SD, the ecological dimension has been mostly emphasized. Its aim is to integrate environmental management systems into the organization. Also, to raise awareness of certain consequences which result from the company's behaviour towards the environment. Possibilities to support life sustaining systems and to prevent environmental destruction are discussed within the ecological dimension. In this survey, the ecological dimension is divided into four categories. At first, information about the company's environmental policy and strategy is analysed, followed by the question about the existence of environmental management systems (EMS). Whether the enterprise behaves ecologically well is part of the third category. It also tackles questions concerning the production process in the company. Criteria concerning the reporting of the ecological dimension are also part of the second sustainability pillar.

Some aspects are similar to the indicators shown above in the generic dimension but are adapted to the needs of analysing the ecological pillar.

Social Dimension: The main focus in the social dimension is the human being. In sustainable companies, programs are started to guarantee that the stakeholders' needs – those of the employees and the society – meet. Similar to the ecological dimension, the categories are politics, management systems and reporting.

4. Findings

In this chapter, the results of the analysis are discussed to give an overview of the relevance and implementation of SD in companies quoted on stock exchanges. The findings for each single index, for each industry and each dimension are presented in the appendices A-C.

4.1. *Findings for each index*

The analysis of the generic dimension for each single index shows an interesting result. Whereas only a third of the American companies implements SD into their strategy, half or more of the Japanese and European companies set SD goals. It is remarkable that a significant number of corporations indicate on their homepages to believe in SD and that it was part of the organizational culture, although it is not articulated in their business strategy. Considering stakeholders, participating in a Corporate Governance Codex and integrating SD into annual reports or own SD reports are practiced by the majority of companies. Corporations listed in the Dow Jones index show a slightly better result than Austrian companies. All of the analysed Japanese companies publish a SD report. The best results concerning generic dimension present corporations listed in the Eurostoxx50. Also, European companies are most active in the UN Global Compact.

The findings of the ecological dimension can be attributed to the very positive commitment of Japanese companies to environmental protection. Only a few criteria are significantly below 50 percent, such as responsibility at the top management, eco-marketing and the external check of environmental reports. In comparison, the other indices show poorer results, except the criteria "external check of the reports" and "eco-marketing". Some indicators seem to be already fulfilled by the firms when analysing the companies' homepages, whereas others seem to be not important at the moment. However, about 85 percent of the corporations (excluding Austrian companies with 55 percent) indicate to follow a proper environmental strategy. On average, two thirds of the companies have implemented an EMS which means that a considerable knowledge of ecological related topics can be assumed. Concerning product development and production process, a lot of actions and measures seem to have been already set by the companies. The efficient use of materials and other resources is an important issue, whereas eco-marketing has not been known or implemented into practice yet. American, Japanese and European companies present quite positive results, Austria seems to be in its early stages concerning environmental aspects or communicating their environmental commitment to the public: not much information can be found on the homepage or in official reports.

Some concurrent results occur when analysing the social dimension: On average, two thirds of the companies demonstrate to have a social plan and a social strategy. Only a third of the companies mention to have implemented a social management system (SMS) to guarantee social commitment. From analysing homepages and official reports, on can see that US homepages are often geared towards CSR. In CSR reports all dimensions are described, however, a special emphasis is often put on the social aspect. The Dow Jones index provides the best results regarding the implementation of a SMS or programs to reduce accidents at work. The gap between a social plan and the implementation of a SMS is, except for the Dow Jones, rather large. None of the analysed Japanese companies investigate the employees' satisfaction and only a small number describes aims and achievements concerning the social aspect. Europe presents as good results as in the other dimensions; especially the number of companies describing relevant social activities and following external standards for reporting should be underlined. The results of Austrian companies are quite positive as well. The high percentage of evaluating employees' satisfaction (35 percent), which shows the respect to the employees, has to be particularly stressed.

4.2. Findings for the industries

In a second step, the results were divided into industries in order to give a worldwide overview about the willingness of industries to accept and integrate SD as aspects of business management. There are several approaches to divide the economy into different branches of industry and to assign the subcategories to the main industries. The classification according to the MSCI is used to identify the specific industries (MSCI, 2005).

Approximately 50 percent of companies in industries like Consumer Discretionary, Energy, Telecommunication Services and Health Care have implemented SD into their vision. However, sustainability in the vision seems not to be common in the Information Technology (IT). The best three industries, which have integrated the sustainability aspect into the organizational culture, reach more than 70 percent. The criterion whether the responsibility of SD lies in the hand of a top manager varies extremely in its evaluation. Especially the question whether the company is listed in a sustainability index (for this survey the Dow Jones Sustainability Index and the FTSE4Good have been chosen) turns out to be interesting. Energy is one of the industries which shows a great commitment to SD in general, but the result of this sector concerning the admission in sustainability indices is not better than average. Being listed in a sustainability index such as FTSE 4 Good or Dow Jones Sustainability Index reflects the real commitment to sustainability of the company. Why Energy is not often listed is discussed in the interpretation.

The majority of industries have established a specific environmental strategy and in the industries of Consumer Discretionary and Health Care, half of the companies have a top manager who is responsible for the environmental sustainability. The implementation of an EMS is an important step towards sustainability. On average, two thirds of companies of all industries indicate to use an EMS to achieve environmental goals and standards. The poorest result is to be found in the sector of Financials where SD does not seem to be an important issue. Half of the companies of Consumer Staples, Health Care and the IT sector educate their employees on environmental issues. When talking about the purchase of "green products", companies from IT and Consumer Staples declare to be seriously concerned about the quality and the trader they buy the materials from. However, no corporation of the Health Care sector shows a similar concern. Ecological optimization of the products and the development of environmentally friendly products play a main role in IT, Energy and Telecommunication Services. Environmental Reporting varies between 88 percent (Telecommunication Services) and 25 percent (Financials).

A social vision in which the human being is focused on is created at least in half of the companies of each industry but especially in Health Care and Consumer Staples companies. Although a social vision seems to be unusual, the implementation of a social management system is not common. Only the sectors of Industrials, Energy and IT show a result higher than 50 percent. The amount of companies which evaluates the employees' satisfaction is not high at all, only some industries (Energy, Materials) take this into account. Whereas some industries seem to integrate SD in their daily business and follow the above mentioned social sustainability standards by publishing quantitative data and programs to reduce the number of industrials accidents, other sectors such as Financials, Industrials and Materials do not make SD a very important issue of the company.

4.3. *Findings for each dimension*

The international survey discussed in this paper includes 95 companies from different countries and continents. By putting all companies together, an overall view about the dimensions can be achieved. This overall view highlights the progress of SD as aspect of business management in general.

The result of the generic dimension shows that 44% of the 95 companies indicate to have integrated SD into their vision, whereas slightly more than the majority has not done this so far. By analysing the homepages and official reports of the companies, it seems that 60% have implemented a mentality of sustainability into the organizational culture, a third takes part in the UN Global Compact and more than three quarters consider their stakeholder as an important part of their business.

Although more than half of the companies have implemented systems and guidelines for a better SD into their organization, only 27% of the companies appoint a top manager of the firm to be responsible for the company's sustainability. Nevertheless, more than 60% are listed in a sustainability index.

The result of all companies concerning ecological aspects is also worth having a closer look at. Many indicators are fulfilled satisfyingly by the companies, in two thirds of the companies an EMS guarantee a focused view on environmental aspects. However, more training on environmental issues for the employees is generally required. In the process of purchasing and in marketing more could be done, yet nearly all of the companies indicate to have implemented waste management which monitors the use and recycling of resources. Eco-Reporting seems to be common (60%), although only a minority of the companies use external standards or mandate an external auditor for a validity check of the quantitative data.

Some indicators reflect that social sustainability is of importance to the companies, other factors such as the implementation and integration of a SMS or the evaluation of the employee's satisfaction show potential areas for improvement. The publication of accidents at work as well as programs to reduce risks should be supported more.

5. Interpretation

The survey concerning SD in business management covers 95 companies situated in industrialized countries which are listed in the specific regional indices such as Dow Jones Industrials Average, Nikkei 225, Eurostoxx50 and ATX. These companies present the economy, its performance and development of the countries. Their commitment to sustainability reflects how the model of SD is developed in the specific geographical regions. Moreover, it shows the deficiencies the companies have concerning SD and presents several differences and methods how companies implement or live SD in the specific countries. Therefore, it is possible to draw conclusions about the attitudes towards SD in the countries, about the pressure which is put on the companies by the government and the pressure which is exerted by other stakeholders.

Due to the analysis, American companies show a positive result of integrating sustainability into their business. A high number of indicators of the ecological and social dimension are fulfilled by the American companies. Nevertheless, important indicators from the generic dimension, such as how SD is implemented into the vision or whether SD is an integral part of organizational culture, do not seem to be as important. Their attitude towards SD is to promote everything they do that protects the environment as commitment to CSR and SD, however, factual information is missing sometimes. It seems that in the United States pressure is especially exerted by the shareholders.

The analysed Japanese companies are the leader in environmental sustainability and they show the best results of all indices within the ecological dimension. All of them have special environmental policies and work to decrease their environmental footprint to maintain and protect the environment. However, results of the social dimension are disappointing in comparison with the very positive ecological results. In Japan, environmental issues play an important role because of the geographical and demographical situation.

The commitment to SD in Europe is quite high, in every dimension the European companies which are listed in the Eurostoxx50 present very positive results. It seems that the attitude towards SD has been elevated and that a generic model which values all three SD dimension equally is preferred. It seems that there is pressure coming from governments to implement and integrate SD into daily business. Also, companies are put under pressure to work in a sustainable way by the stakeholders. The lobbies in Europe, such as shareholders, environmental organizations and social partnerships have a lot of power and are able to influence the strategy and behaviour of companies. Although Austria is part of Europe, its commitment to SD is not as high as estimated at the beginning. Concerning the generic dimension, Austrian companies show similar results as the US companies, even so, the results regarding ecological and social dimension are not as positive. Companies which have implemented SD into business management behave in a correct sustainable way, however, SD is not as much promoted on the homepage and in official reports as for instance in the United States. Nevertheless, there is potential to increase the acceptance of sustainability and to profit from the positive trend which exists in Euroland.

Industries such as Telecommunication Services, Consumer discretionary, IT and Energy do a lot for implementing and integrating sustainable aspects into their business, whereas other industries such as Financials and Materials are not really committed to SD so far. The percentage of companies listed in a sustainability index show how effective and sincere their commitment to sustainability is in praxis – Telecommunication Services, IT and Consumer Discretionary show the best results. Especially the Energy sector has to be highlighted. In this analysis, the Energy sector achieves very positive results in all dimensions and a lot of data concerning SD and its implementation into the company is published. Nevertheless, only half of these companies are actually listed in Sustainability Indices. Although they convey to do a lot for SD, their engagement might be not enough, or criteria such as the operation of a nuclear power station or the extraction or processing of uranium exclude the company automatically of being listed in sustainability indices (FTSE, 2005).

When putting together the data of all 95 companies, all dimensions show positive results. There are only a few indicators which are not fulfilled by the majority of the companies. Potential exists for indicators such as the implementation of SD into the vision - the crucial one - or the responsibility of SD in the hands of a top manager for the generic dimension. Ecological aspects, such as the education of employees concerning environmental sustainability, eco-marketing or the use of external standards and auditors for the reporting should be discussed further. The social dimension can be ameliorated by increasing the implementation of SMS into the companies, by evaluating the employee's satisfaction and by inventing programs to reduce the number of accidents at work.

The results are quite satisfying, however, they show that there is still more to do for a better sustainable world and business.

Further research has to be carried out for an in-depth analysis regarding the economic dimension of SD and the appreciation of employee's satisfaction in Japan. Additionally, this survey based on public available data will be combined with internal data obtained from questionnaires and interviews analysing the same group of companies.

References

Baumgartner, R. J. (2004). Sustainable Business Management: Conceptual Framework and Application. In: *IAMOT (International Conference on Management of Technology)*. Ed: Y. A. Hosni, R. Smith and T. Khalil, Washington.

Castka, P. et al. (2004). Integrating corporate social responsibility (CSR) into ISO management systems - in search of a feasible CSR management system framework. In: *The TQM Magazine*, Vol. 16, *3*, p. 216-224.

Elkington, J. (1998). Cannibals with Forks. The Triple Bottom Line of the 21st Century. Capstone Publishing, Oxford

Fergus, A. H. T. and Rowney, J. I. A. (2005). Sustainable Development: Lost Meaning and Opportunity? In: *Journal of Business Ethics*, Vol. 60, *1*, p. 17-27.

FTSE (2005). FTSE 4 Good Index Series - Inclusion Criteria. URL: http://www.ftse.com/ftse4good/FTSE4GoodCriteria.pdf

Gauthier, C. (2005). Measuring Corporate Social and Environmental Performance: The Extended Life-Cycle Assessment. In: *Journal of Business Ethics*, Vol. 59, *1-2*, p. 199-206.

Jenkins, H. M. (2004). Corporate Social Responsibility and the Mining Industry: Conflicts and Constructs. In: *Corporate Social Responsibility and Environmental Management*, Vol. 11, *1*, p. 23-34.

Knox, S. et al. (2005). Corporate Social Responsibility: Exploring Stakeholder Relationships and Programme Reporting across Leading FTSE Companies. In: *Journal of Business Ethics*, Vol. 61, *1*, p. 7-28.

Korhonen, J. (2003). On the Ethics of Corporate Social Responsibility - Considering the Paradigm of Industrial Metabolism. In: *Journal of Business Ethics*, Vol. 48, *4*, p. 301-315.

Labuschagne, C. et al. (2005). Assessing the sustainablity performances of industries. In: *Journal of Cleaner Production*, Vol. 13, p. 373-385.

Moir, L. (2001). What do we mean by Corporate Social Responsibility? In: *Corporate Governance*, Vol. 1, 2, p. 16-22.

MSCI (2005). Global Industry Classification Standard (GICS). URL: http://www.msci.com/equity/GICS_Sector_Definitions_2005.pdf

Quazi, A. M. and O'Brien, D. (2000). An Empirical Test of a Cross-National Model of Corporate Social Responsibility. In: *Journal of Business Ethics*, Vol. 25, *1*, p. 33-51.

Robin, D. P. and Reidenbach, R. E. (1987). Social Responsibility, Ethics and Marketing Strategy: Closing the Gap Between Concept and Application. In: *Journal of Marketing*, Vol. 51, *January*, p. 44-58.

Salzmann, O. et al. (2005). The Business Case for Corporate Sustainability: Literature Review and Research Options. In: *European Management Journal*, Vol. 23, *1*, p. 27-36.

Snider, J. et al. (2003). Corporate Social Responsibility in the 21st Century: A View from the World's Most Successful Firms. In: *Journal of Business Ethics*, Vol. 48, *2*, p. 175-187.

van Marrewijk, M. (2003). Concepts and Definitions of CSR and Corporate Sustainability: Between Agency and Communion. In: *Journal of Business Ethics*, Vol. 44, *2-3*, p. 95-105.

Welford, R. (2005). Corporate Social Responsibility in Europe, North America and Asia. 2004 Survey Results. In: *The Journal of Corporate Citizenship*, *Spring*, p. 33-52.

Whitehouse, L. (2006). Corporate Social Responsibility: Views from the Frontline. In: *Journal of Business Ethics*, Vol. 63, *3*, p. 279-296.

World Commission on Environment and Development (1987). Our Common Future. The Oxford University Press, Oxford

Zambon, S. and Del Bello, A. (2005). Towards a stakeholder responsible approach: the constructive role of reporting. In: *Corporate Governance*, Vol. 5, *2*, p. 130-141.

Appendix

Appendix A - Findings for each index

FINDINGS OF EACH INDEX	Dow Jones		Nikkei		Eurostoxx		ATX	
Number of companies	30		20		25		20	
	Yes	No	Yes	No	Yes	No	Yes	No
GENERIC DIMENSION								
SD part of the vision?	37%	63%	50%	50%	56%	44%	35%	55%
SD part of the organizational culture?	50%	50%	70%	25%	76%	24%	45%	40%
Participation in the UN Global Compact?	7%	93%	10%	90%	96%	4%	15%	85%
Consideration of stakeholders?	77%	17%	85%	10%	92%	8%	50%	35%
Guidelines to implement SD?	50%	50%	50%	40%	84%	16%	40%	55%
SD-Report or own chapter in annual report?	63%	30%	100%	0%	88%	12%	55%	40%
Participation in a corporate governance codex?	97%	0%	65%	0%	96%	0%	85%	0%
Responsibility for SD at the top management?	30%	57%	35%	40%	20%	76%	25%	55%
Listed in sustainability indices?	63%	37%	65%	35%	88%	12%	30%	70%
ECOLOGICAL DIMENSION								
Environmental strategy?	83%	17%	100%	0%	84%	16%	55%	35%
Responsibility at top management?	33%	57%	40%	45%	20%	76%	15%	55%
Consideration of stakeholders?	63%	30%	80%	15%	80%	16%	35%	40%
Consideration of ecological aspects?	80%	20%	95%	5%	96%	4%	50%	35%
Standards valid for all affiliations?	80%	20%	95%	5%	84%	4%	60%	35%
Implementation of an EMS?	70%	30%	85%	15%	60%	24%	40%	50%
Training on environmental issues for the employees?	47%	23%	50%	10%	36%	40%	15%	30%
Purchase?	50%	17%	60%	10%	24%	40%	15%	35%
Ecological optimization of production?	73%	20%	85%	5%	56%	36%	45%	35%
Environmental orientation during product development?	63%	23%	75%	15%	72%	20%	50%	30%
Eco-marketing?	10%	30%	15%	30%	4%	68%	25%	35%
Use and recycling of resources?	87%	13%	100%	0%	96%	0%	60%	30%
Environmental reporting?	57%	43%	85%	15%	68%	32%	30%	60%
Description of relevant aspects of the activities?	73%	23%	100%	0%	72%	28%	45%	45%
Presentation of quantitative data?	57%	40%	90%	10%	80%	20%	30%	60%
Description of aims and its achievements?	53%	43%	85%	15%	56%	44%	35%	55%
External standards for reporting?	27%	60%	50%	20%	52%	40%	20%	70%
External check of the reports?	13%	63%	15%	35%	28%	64%	20%	70%

Appendix A (Continued)

FINDINGS OF EACH INDEX	Dow Jones		Nikkei		Eurostoxx		ATX	
Number of companies	30		20		25		20	
	Yes	No	Yes	No	Yes	No	Yes	No
SOCIAL DIMENSION								
Social plan/strategy?	77%	20%	70%	5%	88%	8%	50%	30%
Consideration of the stakeholders?	73%	23%	60%	5%	92%	8%	50%	25%
Standards valid for all affiliations?	87%	13%	65%	5%	88%	8%	60%	15%
Implementation of an SMS?	53%	43%	15%	30%	32%	28%	20%	55%
Evaluation of employee's satisfaction?	7%	13%	0%	30%	16%	80%	35%	15%
Internal ethical and social standards?	80%	3%	50%	10%	88%	0%	70%	15%
Publication of the number of accidents at work?	37%	40%	30%	35%	36%	56%	30%	35%
Programs to reduce industrial injuries?	47%	30%	25%	35%	32%	60%	30%	30%
Activities in developing countries?	30%	17%	5%	20%	32%	24%	15%	35%
Social sponsoring?	67%	10%	80%	0%	88%	4%	50%	30%
Social responsibility reporting?	50%	47%	45%	35%	68%	32%	20%	65%
Description of relevant aspects of the activities?	63%	33%	40%	35%	76%	24%	20%	65%
Presentation of quantitative data?	37%	60%	25%	45%	60%	36%	20%	65%
Description of aims and its achievements?	47%	50%	20%	50%	52%	40%	25%	60%
External standards for reporting?	23%	63%	25%	35%	52%	40%	20%	65%
External check of the reports?	7%	67%	5%	55%	32%	60%	15%	70%

Sustainable Development in Companies: An International Survey 549

Appendix B - Findings for the industries

FINDINGS FOR EACH INDUSTRY (only "yes"-answers)	Consumer Discretionary	Consumer staples	Energy	Financials	Health Care	Industrials	Information Technology	Materials	Telecommunication Services	Utilities
Number of companies	13	10	11	20	4	8	11	7	8	2
GENERIC DIMENSION										
SD part of the vision?	54%	30%	55%	35%	50%	38%	36%	43%	50%	100%
SD part of the organizational culture?	62%	70%	73%	50%	50%	50%	55%	71%	63%	50%
Participation in the UN Global Compact?	8%	10%	36%	15%	25%	13%	18%	0%	38%	0%
Consideration of stakeholders?	69%	90%	82%	60%	100%	88%	91%	57%	88%	50%
Guidelines to implement SD?	69%	60%	73%	40%	50%	38%	64%	57%	63%	50%
SD-Report or own chapter in annual report?	85%	80%	73%	65%	50%	50%	100%	71%	88%	100%
Participation in a corporate governance codex?	77%	100%	91%	85%	75%	100%	100%	86%	75%	100%
Responsibility for SD at the top management?	46%	10%	36%	10%	25%	13%	27%	57%	25%	50%
Listed in sustainability indices?	62%	50%	55%	65%	50%	50%	91%	29%	100%	50%
ECOLOGICAL DIMENSION										
Environmental strategy?	77%	100%	100%	65%	100%	63%	91%	57%	88%	100%
Responsibility at top management?	46%	10%	27%	15%	50%	13%	36%	43%	25%	0%
Consideration of stakeholders?	69%	100%	82%	45%	75%	38%	64%	43%	88%	50%
Consideration of ecological aspects?	85%	100%	100%	70%	100%	50%	82%	57%	88%	100%
Standards valid for all affiliations?	85%	100%	100%	55%	100%	63%	91%	71%	75%	100%
Implementation of an EMS?	62%	70%	82%	30%	75%	63%	73%	71%	88%	100%
Training on environmental issues for the employees?	38%	50%	45%	25%	50%	38%	45%	43%	25%	0%
Purchase?	46%	60%	45%	15%	0%	25%	73%	14%	50%	50%
Ecological optimization of production?	77%	70%	100%	15%	75%	63%	82%	71%	88%	50%
Environmental orientation during product development?	69%	70%	91%	40%	50%	75%	82%	43%	75%	50%
Eco-marketing?	15%	20%	27%	0%	0%	13%	0%	14%	13%	100%
Use and recycling of resources?	85%	100%	100%	70%	100%	75%	100%	71%	88%	100%
Environmental reporting?	69%	80%	82%	40%		38%	73%	29%	88%	50%

Appendix B (Continued)

FINDINGS FOR EACH INDUSTRY (only "yes"-answers)	Consumer Discretionary	Consumer staples	Energy	Financials	Health Care	Industrials	Information Technology	Materials	Telecommunication Services	Utilities
Number of companies	13	10	11	20	4	8	11	7	8	2
Description of relevant aspects of the activities?	77%	90%	91%	55%	100%	50%	73%	57%	75%	100%
Presentation of quantitative data?	85%	80%	91%	45%	75%	38%	64%	29%	75%	50%
Description of aims and its achievements?	77%	80%	82%	25%	75%	38%	64%	43%	50%	50%
External standards for reporting?	31%	70%	45%	30%	0%	25%	36%	29%	38%	50%
External check of the reports?	8%	40%	64%	5%	0%	13%	9%	0%	38%	0%
SOCIAL DIMENSION										
Consideration of the stakeholders?	69%	90%	82%	55%	100%	63%	73%	71%	88%	0%
Standards valid for all affiliations?	77%	90%	82%	65%	100%	75%	91%	86%	75%	0%
Implementation of an SMS?	31%	40%	55%	5%	50%	63%	55%	29%	13%	0%
Evaluation of employee's satisfaction?	15%	10%	27%	15%	0%	0%	9%	29%	13%	0%
Internal ethical and social standards?	77%	60%	82%	70%	75%	75%	73%	86%	88%	0%
Publication of the number of accidents at work?	62%	10%	73%	5%	0%	13%	55%	43%	38%	0%
Programs to reduce industrial injuries?	62%	20%	55%	10%	50%	25%	55%	43%	25%	0%
Activities in developing countries?	15%	30%	18%	5%	25%	38%	36%	29%	38%	0%
Social sponsoring?	62%	100%	73%	80%	100%	38%	73%	43%	88%	0%
Social responsibility reporting?	54%	50%	64%	40%	25%	38%	64%	29%	50%	0%
Description of relevant aspects of the activities?	62%	50%	73%	40%	75%	38%	73%	29%	50%	0%
Presentation of quantitative data?	62%	50%	55%	10%	25%	13%	55%	14%	50%	0%
Description of aims and its achievements?	46%	50%	64%	15%	25%	25%	64%	14%	38%	0%
External standards for reporting?	31%	40%	45%	30%	0%	13%	27%	29%	38%	0%
External check of the reports?	8%	20%	55%	5%	25%	13%	0%	0%	25%	0%

Appendix C - Findings for each dimension

FINDINGS FOR EACH DIMENSION	Yes	No
GENERIC DIMENSION		
SD part of the vision?	50%	50%
SD part of the organizational culture?	70%	25%
Participation in the UN Global Compact?	10%	90%
Consideration of stakeholders?	85%	10%
Guidelines to implement SD?	50%	40%
SD-Report or own chapter in annual report?	100%	0%
Participation in a corporate governance codex?	65%	0%
Responsibility for SD at the top management?	35%	40%
Listed in sustainability indices?	65%	35%
ECOLOGICAL DIMENSION		
Environmental strategy?	100%	0%
Responsibility at top management?	40%	45%
Consideration of stakeholders?	80%	15%
Consideration of ecological aspects?	95%	5%
Standards valid for all affiliations?	95%	5%
Implementation of an EMS?	85%	15%
Training on environmental issues for the employees?	50%	10%
Purchase?	60%	10%
Ecological optimization of production?	85%	5%
Environmental orientation during product development?	75%	15%
Eco-marketing?	15%	30%
Use and recycling of resources?	100%	0%
Environmental reporting?	85%	15%
Description of relevant aspects of the activities?	100%	0%
Presentation of quantitative data?	90%	10%
Description of aims and its achievements?	85%	15%
External standards for reporting?	50%	20%
External check of the reports?	15%	35%

Appendix C (Continued)

FINDINGS FOR EACH DIMENSION	Yes	No
SOCIAL DIMENSION		
Social plan/strategy?	**60%**	5%
Consideration of the stakeholders?	**65%**	5%
Standards valid for all affiliations?	**15%**	30%
Implementation of an SMS?	**0%**	30%
Evaluation of employee's satisfaction?	**50%**	10%
Internal ethical and social standards?	**30%**	35%
Publication of the number of accidents at work?	**25%**	35%
Programs to reduce industrial injuries?	**5%**	20%
Activities in developing countries?	**80%**	0%
Social sponsoring?	**45%**	35%
Social responsibility reporting?	**40%**	35%
Description of relevant aspects of the activities?	**25%**	45%
Presentation of quantitative data?	**20%**	50%
Description of aims and its achievements?	**25%**	35%
External standards for reporting?	**5%**	55%

CHAPTER 36

ENHANCEMENT OF ENVIRONMENTAL PERFORMANCE THROUGH TOTAL PRODUCTIVE MAINTENANCE

Rupert J. Baumgartner

Institute for Economic and Business Management,
Montanuniversität Leoben, Austria

Requirements of different stakeholders force companies to pay attention to environmental aspects of their business and to improve corporate environmental performance. In addition, every organization has to develop its own competitive edge. Due to its influence on economic success, mainly through the impact on invested capital and product quality, the efficient and effective use of plants and production equipment is an important aspect of corporate success. The contribution of maintenance management for increasing corporate environmental performance is discussed in this article. Based on the concept of Total Productive Maintenance (TPM) a framework for environmental consideration is proposed. As example of maintenance prevention an integrated assessment tool, which combines simultaneous economic, maintenance and environmental aspects, is presented.

1. TPM and Environmental Performance

According to ISO 14031 (ISO, 1999) environmental performance is defined as "results of an organization's management of its environmental aspects". Environmental aspects are caused by activities, products or services of an organization, which can interact with the environment: a direct environmental aspect means, that it was caused by the organization itself and that it can be influenced by the organization. An indirect aspect means that it was caused by the organization but cannot be influenced by the organization any more. Stahlmann extends this definition and defines environmental performance as reduction of environmental burdens regarding global, international, national, regional and company specific targets, which should be orientated on the concept of Sustainable Development (Stahlmann and Clausen, 2000, p. 31). The basic definition is: "Sustainable development meets the needs of the present without compromising the ability of future generations to meet their own needs." (World Commision on Environment and Development, 1987, p. 43). It includes ecological, economic and social aspects (Gladwin *et al.*, 1995). The basic aim of Sustainable Development is Sustainability, which means a status, where the functions of the biosphere are permanently secured und where social and economic requirements can be satisfied. This aim can be reached by regarding the following four principles (Robèrt *et al.*, 2002):

1. Principle 1: Eliminate contribution to systematic increases in concentrations of substances from Earth's crust. This means substituting certain minerals that are scarce in nature with others that are more abundant, using all mined materials efficiently, and systematically reducing dependence on fossil fuels.
2. Principle 2: Eliminate contribution to systematic increases in concentrations of substances produced by society. This means systematically substituting certain persistent and unnatural compounds with ones that are normally abundant or break down more easily in nature, and using these ones in an efficient and effective way.
3. Principle 3: Eliminate contribution to the systematic physical degradation of nature through over-harvesting, introductions and other forms of modification. This means drawing resources only from well-managed eco-systems, systematically pursuing the most productive and efficient use both of those resources and land, and exercising caution in all kinds of modification of nature.
4. Principle 4: Contribute as much as to the meeting of human needs in our society and worldwide, over and above all the substitution and dematerialization measures taken in meeting the first three objectives. This means using all our resources efficiently, fair and

responsibly so that the needs of all people whom we have an impact on, and the future needs of people who are not yet born, stand the best chance of being met.

Based on the definition of Stahlmann and ISO, environmental performance is defined as the performance of an organization in an ecological sense.

TPM is characterized trough a close cooperation of production and maintenance supported by performance measurement (Hartmann, 1995, Nakajima, 1995). According to Nakaiima, TPM consists of five key elements (Nakaiima, 1988):

1. TPM aims to maximize equipment effectiveness (overall efficiency).
2. TPM establishes a thorough system of preventive maintenance for the equipment's entire life span.
3. TPM is implemented by various departments in a company.
4. TPM involves every single employee, from top management to the workers on the factory floor.
5. TPM is based on the promotion of preventive maintenance (PM) through "motivation management" involving small-grouping activities.

2. Potentials of TPM for Improving Environmental Performance

The implementation of TPM requires activities in several corporate action fields like target systems, maintenance strategies, organizational development, training of employees, maintenance prevention, reward systems and controlling systems (Biedermann, 2002, p. 11). These action fields are the starting point of TPM for improving environmental performance.

The contribution of each action field for improving environmental performance can be assessed regarding the influence on direct or indirect environmental aspects (see Table 1).

Table 1. Direct and indirect environmental aspects.

Direct Environmental Aspects	Indirect Environmental Aspects
Emission to the atmosphere	Product-related effects
Discharge and drain into waters	Administration and planning decisions
Avoidance, utilization, reutilization and disposal of (dangerous) trashes	Assortment and composition of services
Utilization and contamination of soils	New markets
Utilization of natural resources and raw materials including energies	Equity investment, credit allocation, assurance services
Local phenomena (noise, smell, dust, agitation,...)	Composition of the product range
Traffic (in view of goods, services and employees)	Environmental effort and behaviour of contractor and furnisher
Environmental mischance and environmental effects	
Effects on biodiversity	

Source: EU (2001, p. 27).

Target systems can support the improvement of the environmental performance, if the consumption of resources and energy and the release of emissions and waste are considered. These effect mainly the direct environmental aspects, but in the scope of maintenance prevention also the indirect ones (e.g. environmental performance of supplier). The targets have to be included in **controlling** systems.

In the focus of **maintenance strategy** is the bundle of measurements of maintenance (Biedermann, 2002, p. 12). Here, an optimal frequency of maintenance has to be chosen, because a too low adjusted equipment leads to a higher demand of material and energy. A too high chosen frequency leads to an increasing demand of maintenance materials (Wiethoff, 1996, p. 104).

In the field of **organizational structure** there is no direct reference to environmental aspects.

Environmental aspects like efficient use of resources and handling of waste can be part of **training activities**.

Maintenance prevention can provide a high contribution to an increase of the environmental performance. Following phases of maintenance prevention can be divided: product design, facility layout, creation, installation, start-up and operation (Wildemann, 1997, p. 65). This field offers the strongest influence on direct environmental aspects (necessary input of resources and energy, emissions and other consequences) and indirect environmental aspects (for instance environmental performance of suppliers and customers).

Regarding **reward systems and motivation** bonuses can be awarded for improvement activities.

This presented framework is a general discussion of the relationship between TPM and corporate environmental performance. The potential of TPM to improve the corporate environmental performance has to be assessed individually for each organization.

3. Maintenance Prevention and Environmental Performance: Integrated Assessment

In this section the role of maintenance prevention regarding environmental performance is discussed and an example of integrated assessment of economic and environmental aspects as decision support tool is presented.

Maintenance prevention is defined as the bundle of measures, which anticipates the drop out of equipment and therefore decreases the frequency of maintenance measures or even avoid them. Hereby the spectrum ranges from no prevention to the integration of maintenance in the planning and acquisition of the equipment (Biedermann, 2001, p. 18).

The decision about purchasing a new equipment is based on multiple objectives and represents a classical decision problem. Usually the objectives have competing target correlations. They have to be balanced during the evaluation of alternative equipments. For this task the concept of integrated assessment can be used in an adapted form (Baumgartner and Zielowski, 2003, Baumgartner, 2004).

The decision for choosing a facility, by paying attention to environmental aspects, includes an economic, an environmental and a technological/facility dimension. These dimensions have to be measured with specific criteria. The following criteria can be taken:

- net present value for the economic dimension (Hawawini and Viallet, 2002, p. 6)
- eco-points (Meier *et al.*, 1998, p. 527) as a value for environmental performance
- maintainability (Wildemann, 1997, p. 68)

With this criteria three individual results are obtained, which can be used for making a decision. But this is only possible for these cases, where a hierarchy can be made for each alternative or if the decision maker can balance the economic, environmental and technological/facility dimension against each other. Otherwise the individual results, which are based on different units, have to be combined to an overall result. Furthermore the concrete circumstances of the decision have to be considered in the evaluation. The relative meaning of the dimensions depends on various, time variable factors. Therefore an attempt, based on the Fuzzy Logic, can be used for making a specific evaluation (see Figure 1).

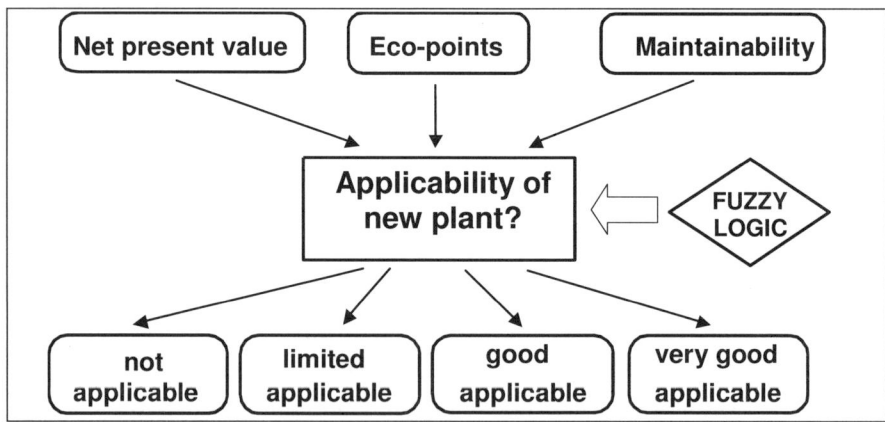

Figure 1. Integrated assessment as decision.

Fuzzy Logic has been developed by Lofti A. Zadeh. (Zadeh, 1965, Zadeh, 1975). This method enables specific weighting of different aspects and translates blurred input signals into stable global results. The fuzzy-based scoring-model contains the following steps (Mechler *et al.*, 1993, p. 69):

- Definition of linguistic variables and rule set
- Fuzzification of the linguistic variables
- Inference: identifying the conclusion through combination of linguistic variables with the rule set
- Defuzzification of the conclusion

Rules are defined as "if-then" sentences. An example could be:
If the improvement of "environmental performance" is only marginal and "net present value" degrades strongly, then "applicability of equipment" is unacceptable.

Rules consist of a condition and a consequence part. In the condition section several premises (here "environmental performance" and "net present value") are combined with logical relations to conclusions. Premises and consequences are defined as linguistic variables, the values (linguistic terms) of the variables are words or sentences.

The criteria (net present value, eco-points, maintainability) and the conclusion (applicability of equipment), which represents the result of the assessment, are expressed as linguistic variables defined on a scale from 0 to 100 (see Table 2). Based on these variables the rules for the decision are verbalized. Alternatives leading to a decreasing environmental performance or showing a negative net present value are dropped out.

Table 2. Linguistic variable with terms for premises and conclusion.

	Net present value	Environmental performance	Maintainability	Applicability of equipment
Range	0-100	0-100	0-100	0-100
Linguistic Term	cero	cero	poorly	not applicable
	marginal	marginal	satisfactory	limited applicable
	high	high	good	good applicable
	very high	very high	very high	very good applicable

The membership functions of the terms of the linguistic variable have to be determined, which is called fuzzification. In the next step, the inference, conclusions from the premises to the consequences have to be made. For instance, the membership function for *environmental performance* (marginal) is $\mu = 0.3$, *net present value* (high) is $\mu = 0.7$ and for *maintainability* (good applicable) is $\mu = 0.8$, the inference defines the membership function for *applicability of equipment* (limited applicable). Consists the condition part of the rule of one variable the conclusion has the same value for μ. For two or more variables, it depends on the logical operator used in the condition part. Basis operators are "and", which represents a minimum operator, and "or", which is a maximum operator. In our example, "and" is used, therefore the conclusion is defined with $\mu = 0.3$ (Figge, 2000, p. 123).

Via the defuzzification a sharp value is calculated out of the affiliation levels of the conclusion. For this purpose several methods can be used. An adequate one is the so called Singleton-method (Figge, 2000, p. 124). Out of the defuzzification a value between 0 and 100 is gained, according to Table 2. This value corresponds with the terms "Applicability of the facility". With this evaluation the alternative equipments can be ranked regarding economic, environmental and maintainability criteria.

4. Summary and Perspective

In this paper the contribution of TPM for improving corporate environmental performance is discussed. Regarding environmental aspects in the action fields of TPM can support the improvement of corporate environmental performance; especially target systems, training activities, maintenance strategies and maintenance prevention have to be named. Exemplary as instrument of maintenance prevention an adopted version of integrated assessment has been presented. This instrument enables the simultaneous consideration of economic, maintainability and environmental aspects during purchasing decisions of new equipment.

References

Baumgartner, R. J. (2004). Sustainability Assessment - Einsatz der Fuzzy Logic zur integrierten ökologischen und ökonomischen Bewertung von Dienstleistungen, Produkten und Technologien.In: Techno-ökonomische Forschung und Praxis (U. Bauer, H. Biedermann and J. W. Wohinz. DUV, Wiesbaden

Baumgartner, R. and Zielowski, C. (2003). Integrated Product and Technology Assessment reflecting Sustainable Development and Organisational Learning. In: From Information to Knowledge to Competencies: Key Success Factors for Innovation and Sustainable Development: 12th International Conference on Management of Technologies. Ed: Y. Hosni, T. Khalil and L. Morel-Guimaraes, Nancy.

Biedermann, H. (2001). Knowledge Based Maintenance. In: Knowledge Based Maintenance: Strategien, Konzepte und Lösungen für eine wissensbasierte Instandhaltung (H. Biedermann, ed.). TÜV-Verlag, Köln.

Biedermann, H. (2002). Instandhaltung und Wissensmanagement: Die Entwicklung zur lernenden Instandhaltungsorganisation. In: *Industrie Management,* Vol. 18, 2, p. 9-12.

EU (2001). Regulation (EC) No 761/2001 of the European parliament and of the council of 19 March 2001 allowing voluntary participation by organisations in a Community eco-management and audit scheme (EMAS)

Figge, F. (2000). Öko-Rating: ökologieorientierte Bewertung von Unternehmen. Springer, Berlin

Gladwin, T. N. et al. (1995). Shifting Paradigms for Sustainable Development: Implications for Management Theory and Research. In: *Academy of Management Review,* Vol. 20, p. 874-907.

Hartmann, E. H. (1995). Erfolgreiche Einführung von TPM. Verlag Moderne Industrie, Landsberg

Hawawini, G. and Viallet, C. (2002). Finance for executives: Managing for Value Creation. South-Western/Thomson Learning, Cincinnati

ISO (1999). Environmental management -- Environmental performance evaluation -- Guidelines. ISO (International Organization for Standardization), Geneva

Mechler, B. et al. (1993). Fuzzy Logic: Einführung und Leitfaden zur praktischen Anwendung ed), Vol. 1, 289. Addisn-Wesley, Bonn; Paris; Reading, Massachusetts. ISBN 3-89319-443-6

Meier, M. A. et al. (1998). Bewertungsmethoden in der Ökobilanz-Rückblick und Perspektiven. In: gwa, 7/98.

Nakaiima, S. (1988). Introduction to Total Productive Maintenance (TPM). Productivity Press, Cambridge, MA

Nakajima, S. (1995). Management der Produktionseinrichtungen. Campus Verlag, Frankfurt/Main

Robèrt, K.-H. et al. (2002). Strategic sustainable development - selection, design and synergies of applied tools. In: *Journal of Cleaner Production,* Vol. 10, 3, p. 197-214.

Stahlmann, V. and Clausen, J. (2000). Umweltleistung von Unternehmen ed), Vol. 1, 331. Gabler, Wiesbaden. ISBN 3-409-11723-7

Wiethoff, Z. (1996). Öko-Controlling in der Instandhaltung: Schaffung einer Informationsbasis für umweltgerechtes Handeln durch Umweltkennzahlen. In: Controlling, Vol. 8, 2, p. 102-109.

Wildemann, H. (1997). Total Productive Maintenance: Leitfaden für ein integriertes Instandhaltungsmanagement. TCW Transfer-Centrum, München

World Commision on Environment and Development, W. (1987). Our Common Future. Oxford University Press, Oxford

Zadeh, L. A. (1965). Fuzzy Sets. In: *Information and Control,* 8, p. 338-353.

Zadeh, L. A. (1975). The Concept of a Linguistic Variable and Its Application to Approximate Reasoning-II. In: *Information Sciences,* 8, p. 301-357.

CHAPTER 37

INTEGRATING SUSTAINABLE BUSINESS MANAGEMENT INTO DAILY BUSINESS VIA GENERIC MANAGEMENT

Rupert J. Baumgartner

Institute for Economic and Business Management,
Montanuniversität Leoben, Austria

Bringing sustainability issues into the boardroom of companies is a central and important issue for a successful development of sustainable societies. The concept of Sustainable Business Management allows the integration of sustainability aspects into corporate activities; but it has to be embedded in a management system. Basically there is no special management system required, but it is advantageous to take an integrated and holistic management approach like Generic Management as basis.

1. Sustainable Business Management

Sustainable Business Management emphasizes on innovation, stakeholder requirements and efficient as well as effective business processes. It is supported with tools for identification, implementation and controlling of sustainability aspects in companies and provides a general framework for the integration of sustainability aspects in corporate structures and processes (Baumgartner, 2004, p. 2).

Sustainable Development has been defined by the Brundtland report as ethical concept (World Commission on Environment and Development 1987). Robèrt defines four principles, which concretize the Brundtland definition (Robèrt et al., 2002):

1. Eliminate contribution to systematic increases in concentrations of substances from the earth's crust.

2. Eliminate contribution to systematic increases in concentrations of substances produced by society.
3. Eliminate contribution to the systematic physical degradation of nature through over-harvesting, introductions and other forms of modification.
4. Contribute as much as we can to the meeting of human needs in our society and worldwide, over and above all measures taken in meeting the first three objectives.

Müller-Christ and Hülsmann worked out the basic interpretations of Sustainable Development in management literature (Müller-Christ and Hülsmann 2003, p. 266). Three main interpretations can be distinguished: an innovation-based, a normative and a rational interpretation. All of them are important for the concept of Sustainable Business Management. The innovation based interpretation is an extension of the (eco-) efficiency concept and points out the importance of resource efficient processes and technologies. Innovation will be an important driver for further improvements in this field. The normative interpretation of Sustainable Development focuses on justness and equity. From this point of view, business management has to be aware of the impact it has on society and nature. Usually stakeholder will pay attention to these impacts and force corporations to avoid them; therefore the stakeholder-concept, introduced by Freeman, is an important aspect of Sustainable Business Management [Freeman, 1984, p. 269]. The rational interpretation of Sustainable Development has the focus on the sustainable use of resources. Organizational activities must secure the availability of all types of resources – therefore sustainability can be defined as extended economic principle (Müller-Christ and Hülsmann, 2003) and is an advancement of the resource based view of strategic management.

Summarizing these interpretations, three points are essential for sustainable corporations:

1. Innovation is recognized as an essential element for corporate sustainability.

2. Stakeholder requirements and demands have to be detected and actively managed.
3. The rational approach focuses on the effectiveness of business activities.

The combination of these interpretations form the basis for the framework of Sustainable Business Management (Figure 1).

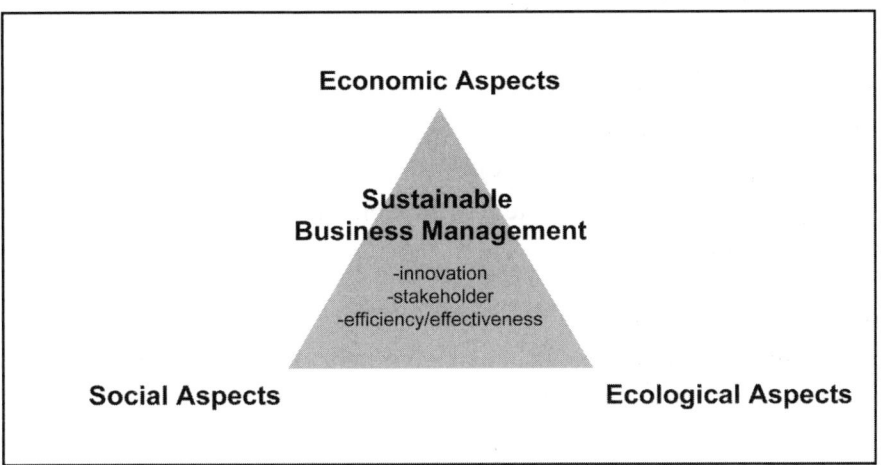

Figure 1. Model of sustainable business management.

The individual business case for Sustainable Development depends on the specific situation of a company, but according to the four general principles of sustainability and the interpretations of Sustainable Development, the following general statements describe aspects and possible measures (Baumgartner 2002):

- Companies have to manage flows of material and energy. The objective has to be an absolute and relative reduction of them.
- Companies have to develop and construct their products and services in a way, which allows an eco-efficient process of usage and disposal. Products have to be energy and material extensive, and easily be reused or recycled within the economy.

- Companies have to redefine their business – the focus has to be on the solution for the customer, not the product or the technical characteristics of it. And the solutions provided by the company have to be sustainable.
- Companies have to respect social principles within the company, the society and the world.
- Companies have to be competitive and secure/increase their corporative value.

2. Generic Management

To integrate the main aspects of Sustainable Business Management in daily business it is necessary to rely on a corporate management system. Ideally this management system is flexible and holistic and focuses on following aspects:

- Regarding stakeholders
- Consideration of organizational environment including nature
- Focus on long-term corporate success and value generation

Generic Management is a flexible and holistic management concept and is defined as:

Generic Management is an integrated model for managing internal and external demands regarding dynamic and complex processes in order to guarantee a successful and sustainable corporate development.

This definition is illustrated by the philosophy of Generic Management; with the structural model the organization-specific potentials and necessities can be identified.

2.1. *Generic management – Basic philosophy*

The basic philosophy of Generic Management consists of three basic elements: stakeholder orientation, flexibility and value generation (Figure 2). Each aspect is essential for business success.

In organizational studies, mainly in new institutionalism, stakeholders play a central role. New institutionalism assumes that organizational structures and expiries are forced by stakeholders (Meyer and Rowan, 1977, Tolbert and Zucker, 1996). In daily business, organizations are tending to orientate themselves on single stakeholder requirements and take these as a base for organizational activities which influence structures and processes. In consequence specialized management systems for quality issues like ISO 9000 series or environmental issues like ISO 14001 have been developed and implemented. This leads to sub optimal solutions due to parallelism of these systems and limited view on requirements of selected stakeholders. In this case, it is not possible to manage stakeholder issues integrated and proactive. In a Generic Management system the separation of these subsystems has to be overcome.

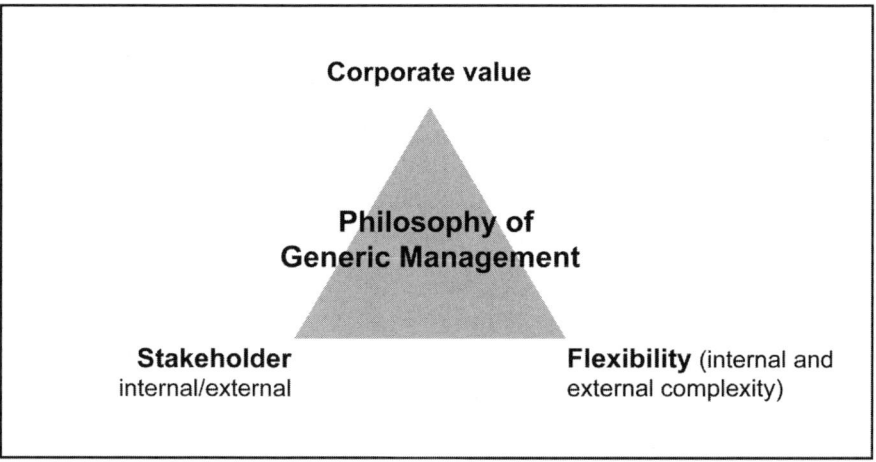

Figure 2. Generic management philosophy.

Hereby, stakeholder requirements are managed in an integrated way and not only by separately formulated programs. This is also a connection to the concept of Sustainable Development as sustainability issues are often formulated by stakeholders and pressure groups.

Flexibility enables an organization to manage the dynamism and complexity of the environment and the organization itself. Managing flexibility has to be based on the basic principles of cybernetics and systems engineering. Outgoing from Ashby's law – complexity can be mastered only by complexity – it is crucial to enable an organization to recognize and manage internal and external dynamic and complexity (Ashby, 1970). Relevant issues are organizational learning, the principle of self-organization and agility management (Pieler, 2003, Sarkis, 2001, Schreyögg and Noss, 1995). Therefore formal as well as informal structures and processes have to be examined, adapted and improved constantly.

The third aspect deals with the corporate value. Every organization has its specific agenda. This agenda describes together with the vision the basic orientation and the leading parameters of an organization (Haberfellner, 1975). In case of profit organizations earning adequate profit is part of the agenda. But profit as only measure of success ignores long-term effects and is no sufficient basis for strategic decisions. Performance measurement instruments like Balanced Scorecard (BSC) (for BSC Kaplan and Norton, 1992, Kaplan and Norton, 1993) show the importance of non-financial aspects. Corporate value has to be regarded in an integrated way. Several value management concepts have been developed during the last years, like shareholder value (Rappaport, 1986, Rappaport, 1995) or Economic Value Added (EVA) (Stewart, 1991, Stewart, 1994). They are focusing only on monetarily measurable factors, too, which can lead together with extensive reimbursement regulations for the management board to visible wrong developments (Enron, Worldcom). Nevertheless, it may not be overlooked that these attempts with the consideration of future monetary flows and the integration of market prices show an advantage compared with classical profit-loss concepts. So corporate value has to be seen as integrated measure, which includes financial figures and additional worth components. Examples for additional components are corporate flexibility, intellectual capital or stakeholder value as extension of Shareholder value. It is essential that every organization has to decide which additional worth components are used for complementing the corporate value. This has to be based on the basic orientation and vision of the company.

The philosophy allows an integrated view on three central aspects for corporations. The following structural model concretizes this normative view.

2.2. Generic management – Structural model

With the aid of the structural model, the individual situation of an organization related to the philosophy of Generic Management can be evaluated. It is a combination of the three aspects of the philosophy with an input – process – outcome model of an organization (Figure 3). The nine identified fields are used to assess the organization in the light of Generic Management.

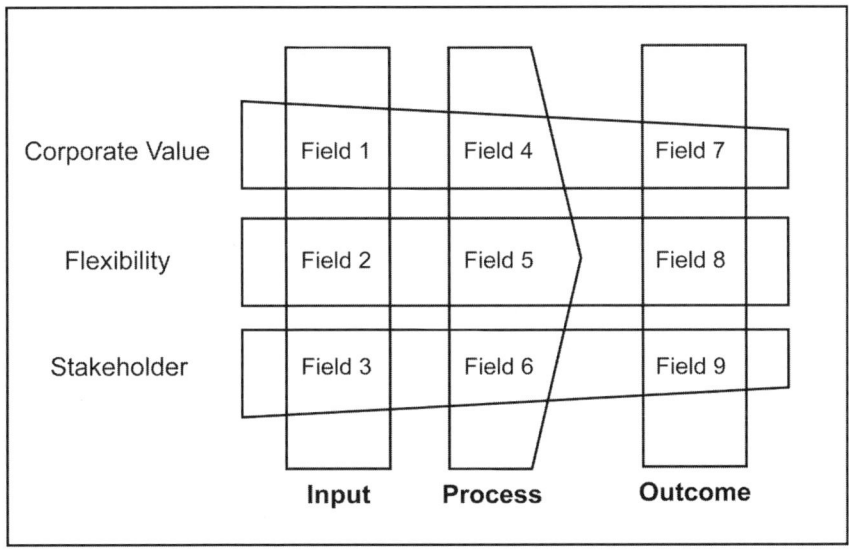

Figure 3. Generic management: Structural model.

Input (resources and potentials) enables value generation and is structured into three potentials: the human, the structural and relationship potential (following to Biedermann, 2003, for human potential also Roos and Roos, 1997). The human potential deals with selection, training and qualifying of employees. The structural potential consists of material and energy flows and the infrastructure, facilities and equipments. Formal and informal organizational structure, networks with suppliers and cooperation's with stakeholders regarding inputs are forming the relationship potential.

Column two represents the processes of value generation; potentials and inputs are used to create saleable products and services. Emissions, waste and other undesired outputs have to be regarded, too.

In column three the outcome is presented, which indicates the effects of organizational activities. Examples therefore are sold products and services, revenues, profit, image, brand value and value generation as well as reactions from stakeholder.

For assessment, following questions in the nine fields are used:

- Field 1: How can the corporate value be influenced by potentials and inputs? This regards human, structural and relationship potential.
- Field 2: How does the potentials influence organizational flexibility and the possibility to adapt changing internal and external circumstances?
- Field 3: Which requirements have internal stakeholders to the input side, for instance quantities and quality of raw materials? Which requirements have external stakeholders to the input side?
- Field 4: How influence processes the corporate value?
- Field 5: How influence processes adaptability and flexibility? How can flexibility be implemented in processes?
- Field 6: Which requirements have internal stakeholders to the processes? Which requirements have external stakeholders to the processes?
- Field 7: Which effects arise through the activities (sold products, brand value…) on the corporate value?
- Field 8: Which effects arise by the corporate activities on the flexibility? Which effects show the organizational flexibility?
- Field 9: Which effects arise by the corporate activities on the internal and external stakeholders?

As a result, an individual picture of strengths and weaknesses of the analyzed organization is obtained. Usually it is impossible for an organization to improve and redesign procedures in all nine fields; so this result is the basis for prioritization of all activities in Generic Management. The identified activities have to be implemented consequently.

3. Connections between Sustainable Business Management and Generic Management

Generic Management is an integrated and holistic management approach and can be used for implementation of Sustainable Business Management. Figure 4 shows the relationship between them. Sustainable Business Management with the economic, ecological and social aspects, concretized through the principles developed by Robèrt, allows the formulation of specific corporate sustainability strategies and goals. It can also contribute to the competitiveness and is forced by rising societal interests in sustainability issues.

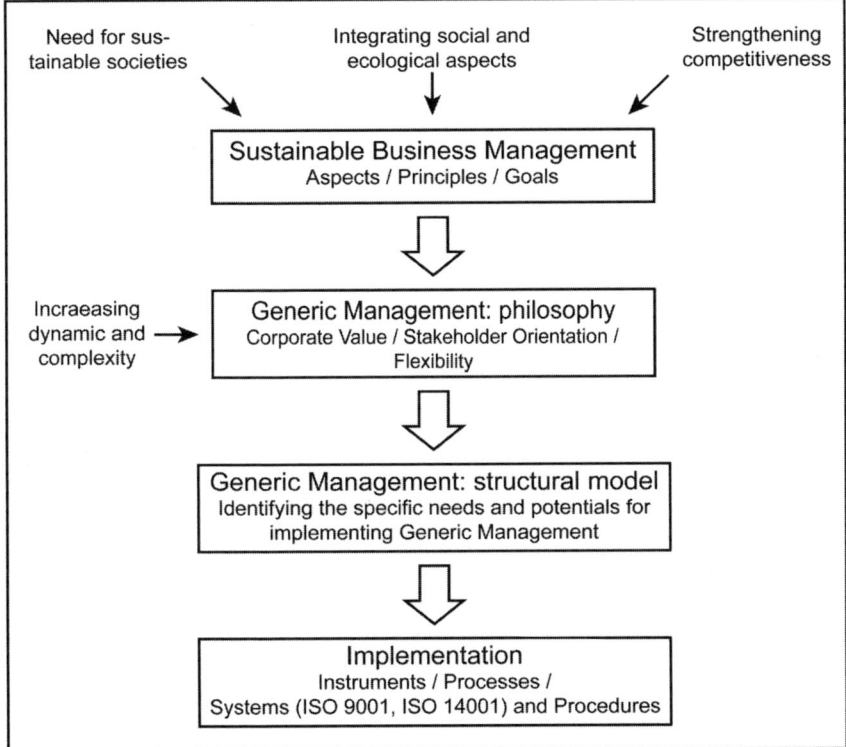

Figure 4. Interrelation between sustainable business management and generic management.

Beside the challenge of Sustainable Development, increasing complexity and rising intensity of competition are evident for corporations. The Generic Management approach allows combining these two aspects in an integrated concept. It enables the identification of needs and potentials; with this instrument at least Sustainable Business Management can be implemented.

4. Conclusion

This paper gives a brief overview about Sustainable Business Management, which allows the integration of sustainability aspects into corporate activities, and presents with Generic Management a management approach for the implementation of Sustainable Business Management. Generic Management is an integrated management system, which combines Sustainable Business Management with central challenges of companies regarding complexity and competitiveness. It consists of two elements - the philosophy and the structural model. The first builds the normative basis; the second allows the identification of needs and potentials.

References

Ashby, W. R. (1970). An Introduction to Cybernetics. Chapman & Hall, London.

Baumgartner, R. J. (2004). Sustainable Business Management: Conceptual Framework and Application. In: *New Directions in Technology Management: Changing Collaboration Between Government, Industry and University: 13th International Conference on Management of Technologies.* Ed: Y. Hosni and T. Khalil, Washington.

Biedermann, H. (2003). Wissensbilanz als Strategie- und Steuerungselement. In: *Werte schaffen - Perspektiven einer stakeholderorientierten Unternehmensführung* (K. Matzler, H. Pechlaner and B. Renzl, ed.). Gabler, Wiesbaden.

Freeman, R. E. (1984). Strategic Management. A Stakeholder-Approach. Pitman Publishing, Boston.

Haberfellner, R. (1975). Die Unternehmung als dynamisches System: Der Prozesscharakter der Unternehmensaktivitäten.In: *Forschungsberichte für die Unternehmenspraxis* (W. Daenzer. Verlag Industrielle Organisation, Zürich. ISBN 3 85743 8320.

Kaplan, R. and Norton, P. (1992). The Balance Scorecard - Measures That Drive Performance. In: *Harvard Business Review*, Vol. Jan.-Feb. 1992, p. 71-79.

Kaplan, R. and Norton, P. (1993). Putting The Balanced Scorecard to Work. In: *Harvard Business Review*, Vol. Sep.-Oct. 1993, p. 134-147.

Meyer, J. W. and Rowan, B. (1977). Institutionalized organizations. In: *American Journal of Sociology*, Vol. 83, p. 340 - 363.

Müller-Christ, G. and Hülsmann, M. (2003). Quo Vadis Umweltmanagement? Entwicklungsperspektiven einer nachhaltigkeitsorientierten Managementlehre. In: *DBW*, Vol. 63, *3*, p. 257-277.

Pieler, D. (2003). Neue Wege zur lernenden Organisation. Gabler, Wiesbaden. ISBN 3-409-21888-2.

Rappaport, A. (1986). Creating Shareholder Value: The New Standard for Business Performance. New York.

Rappaport, A. (1995). Shareholder-Value: Wertsteigerung als Maß für die Unternehmensführung. Stuttgart.

Robèrt, K.-H. et al. (2002). Strategic sustainable development - selection, design and synergies of applied tools. In: *Journal of Cleaner Production*, Vol. 10, *3*, p. 197-214.

Roos, G. and Roos, J. (1997). Measuring your Company´s Intellectual Performance. In: *Longe Range Planning*, Vol. 30, *3*, p. 413 - 426.

Sarkis, J. (2001). Benchmarking for agility. In: *Benchmarking*, Vol. 8, *2*, p. 88-107.

Schreyögg, G. and Noss, C. (1995). Organisatorischer Wandel: Von der Organisationsentwicklung zur lernenden Organisation. In: *DBW*, Vol. 55, *2*.

Stewart, G. B. (1991). The Quest for Value - The EVA Management Guide. New York.

Stewart, G. B. (1994). EVA: Facts and Fantasy. In: *Journal of Applied Corporate Finance*, *2*.

Tolbert, P. S. and Zucker, L. G. (1996). The Institutionalization of Institutional Theory. In: *Handbook of organization studies* ed.), p.175-190. London.

SECTION VI

SOCIAL AND EDUCATIONAL ASPECTS IN MOT

CHAPTER 38

THE NEW COMPANY-SCHOOL RELATIONSHIP IN THE KNOWLEDGE AGE

Luís Henrique Piovezan and Afonso Carlos Corrêa Fleury
Escola Politécnica da USP, São Paulo, Brazil

This paper studies changes in the Company-School relationship given the business and social paradigm changes from the Industrial Age to the Knowledge Age. It considers and evaluates a new model for the Company-School relationship. The proposed model is based in the concept of Business and Vocational School as processes. The proposed model considers tree drive factors for an effective relationship between Business and Vocational School. These drive factor influence the process of knowledge absorption, courses planning, courses evaluation, structures organization and joint technology development. A case study is planned to validate this model.

1. Introduction

The globalization, automation, new information, communication technologies, and the more exigent requirements for quality and productivity all point to radical changes in the leadership and management of companies. In this new paradigm, knowledge is becoming the more important input in the enterprise activity and needs adequate management in a new turbulent environment. In the same way, optimization of productive processes and the easiness of process automation are modifying the requirements of vocational formation.

These new requirements occur because, according to Ernst and Kim (2001), the capacity of absorption of Knowledge by the company is a more important factor for the company success in the Knowledge Age than raw material, financial resources or energy availability. Although the internal factors can be important in the determination of knowledge absorption capacity, much of this capacity is conditioned by external factors, mainly related to the culture and the adequacy of the available school formation. The increase of educational system importance becomes, therefore, evident in the part linked to operational work force formation and direct supervision formation. Thus, new requirements emerge on the Vocational School. One of these new requirements is a new relationship model between Business and Vocational School.In this direction, this paper objective to study how the relationship between Business and Vocational School in this new paradigm is occurring, and which factors of success of this relation are significant.

2. The New Paradigm of Business

Authors such as Galbraith (1986), Toffler (1995) and Drucker (1999) detail the changes that are occurring in the social and economic structures in the second half of the century. These changes have driven business to a new paradigm. There are diverse models that describe how companies must act in this new paradigm; Bartlett *et al.* (2002) describe one of these models. According to these authors, this new paradigm characterizes the change of basic business resource strategy from financial capital and organizational capacity to human and intellectual capital. This new paradigm demands a change in focus to people ability. According Bartlett *et al.* (2000), "in the traditional model, the company depends on knowledge and ability of high administration and on the effective systems and processes power for exploitation and transference of this knowledge and experience in daily activities. (...) In contrast, the Individualized Company is based on the high quality and ability of all the employees."

This new relationship between Businesses and Individuals generates the necessity for different formations of people. Especially, it demands a new relationship between Business and the Vocational School. In this new relationship, the operational staff develops aspects such as competence and confidence. A new configuration for the management of people must be developed.

3. People Management: Competence of People in the Company

The presence of intangible knowledge makes knowledge acquisition and development by the company more important than the abilities currently developed in the workers. They have also to develop values and attitudes (and, therefore, competences) to transform data and information into usable knowledge by the Company. However, the competence concept is complex and demands the definition of its several aspects. Bogner *et al.* (1999) indicate that the concept of business competence is both dynamic and complex, and related to the competitive business strategy. However, according to authors, nor always this complexity is considered. In the other hand, Bogner *et al.* (1999) had only analyzed the competence model concept under the perspective of the Business. Although this perspective already presents an ample complexity degree, Bogner *et al.* (1999) do not considered the individual role in this model. In this direction, Lastres *et al.* (2002) report that "the complex and dynamic character of the new knowledge requires special emphasis in the permanent and interactive learning, as form of individuals, companies and other institutions to become apt to face new challenges and to be enabled to a more positive insertion in the new scene".

Thus, the relationships between the diverse institutions (Business, Vocational School, etc.) must be in form of partnerships. In this form, the new paradigm seems to demand a new action form of the Vocational School, mainly focused in the relationship with the Company. These new requirements are one of the causes that have demand the changes in educational system and, in particular, in the Vocational School. Together with other causes as, for example, the new technologies adoption by the school, the politics changes, the new scientific knowledge, the Vocational School is also subjects to a new paradigm.

In line with this view, SENAI-DN (2003) writes "the emergence of new trends in terms of productive paradigms in an economic globalization context as well as the revalorization of the human contribution in the labor face. A revitalization of pedagogic structures and practices so that they meet professional formation necessities. In particular, with superior qualifications, an ample understanding of the productive process, with a higher adaptation capacity, flexibility and versatility, conditions to deal with unusual situations, decision taking, problem solving, team work, to evaluate consequences, and operate with criteria of quality and performance statistics, to cite some attributes."

Otherwise, it is necessary to isolate the driver factors that directly influence these reforms in Vocational School and, with the indication of the main driver factors, actions must be recommended to the effectiveness of the relationship between Business and Vocational School.

According to SENAI-DN (2002), these drive factors are:

- New systems and methods of production and work - technological innovations,
- New means of production: machines and equipment, tools and instruments and material,
- New quality control techniques and analysis,
- New procedures of maintenance and repairing,
- Changes in the organization of the work,
- Changes on environmental legislation and on security normalization.

From the study of international cases (not detailed in this text), the main drive factors to changes in the Vocational School had been established. These factors are:

- New systems and methods of production and work - technological innovations,
- New means of production: machines and equipment, tools and instruments and material,
- Changes in the organization of the work.

4. The Relationship Model

From the competence concept and from the changes in the relationship between Business, Vocational School and Individual, a model for the relation between Business and Vocational School is constructed. This model comprehends this relationship as a process.

4.1. *The competence concept*

The competence concept is a consequence of Vocational School paradigm change. The previous paradigm was based on the principle that a Vocational School pupil would have to develop abilities. In the new paradigm, the Vocational School also needs to develop knowledge on pupils because of the increase of complexity and changes in Business processes. In the same way, the Vocational School remarks that the pupil must have their attitudes developed mainly to mobilize their knowledge and abilities.

Thus, the Vocational School changes from a paradigm where the school has to develop only abilities to a paradigm where the school has to develop knowledge, abilities and attitudes. The composition of knowledge, abilities and attitudes to be developed on pupils is called competence. Consequently, the competence concept can be defined as the set of knowledge, abilities and attitudes that must be developed in the pupils so that they effectively and efficiently act in the Company. This concept is developed mainly by Zarifian (1999).

4.2. *The relation between company, school and individual*

In the previous paradigm, the Vocational School trained the individual that applied its abilities in the Company. The strongest relations were between the School and the Individual and between the Individual and the Company. The relations between School and Business were limited to the survey of training necessities, that is, of the abilities to be developed.

Given the stability of the abilities, these surveys could be carried through a very low a frequency. In other words, the understanding of the production methods of for the determination of the training necessities was enough. Figure 1 illustrates this relationship. The continuous lines indicate the strong relationships and the dotted lines indicate the weak relations. This relationship can be considered linear therefore the ability is developed in the Vocational School and applied in the Company.

In the new paradigm, the productive processes are flexible and unpredictable. Information and knowledge creation is large. Therefore, the learned tasks and abilities in a course or training lose their usefulness quickly. The Business demands people who have, beyond abilities, competences that allow to continuous learning and the constant adaptation of the individuals.

A stronger relationship must exist between Business and Vocational School in way to consider these factors. Figure 2 illustrates this new relationship.

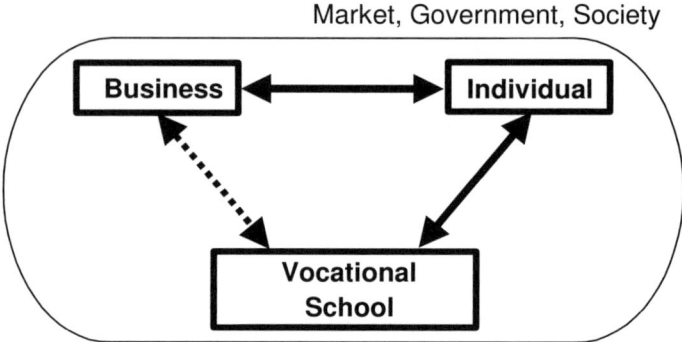

Figure 1. The relation between business, school and individual in the previous paradigm.

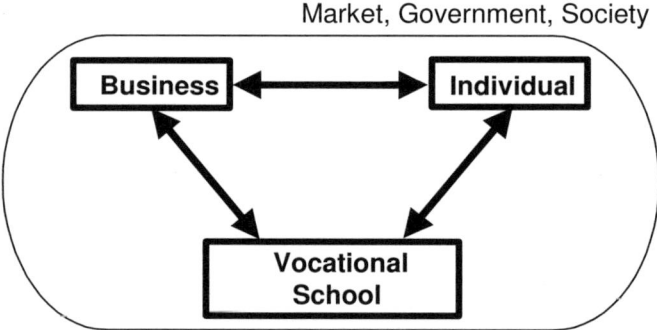

Figure 2. The relation between company, school and individual in the new paradigm.

Thus, the requirements on Vocational School are not only to transmit how to work in parts of a management or production system, but they must to make possible to the individual learn new abilities and new knowledge and develop adequate values.

4.3. *The relationship between business and vocational school*

This item presents a proposal of model of the relationship between Business and Vocational School. Piovezan (2004) and Piovezan & Fleury (2004) show the evolution of proposed model.

The proposed model is based on the necessity of consideration of the three driver factors for an efficient relationship between Business and Vocational School (item 1.3). This model considers the educational system as a process (Coombs, 1976). In the new paradigm, the modified relationship between Business and the Vocational School gains importance. This bigger participation also does not mean subordination of the Vocational School purposes to the Business purposes, but the joint achievement. This joint achievement start on the adequate and unified agreement of how the three specified driver factors acts in the business structure.

To reach this joint achievement, the Vocational School must concentrate on some conditions. These conditions are:

a) Hierarchy reduction and enlargement of Vocational School autonomy
b) Predisposition to partnerships
c) Less directive vision and bigger pupil autonomy
d) Best quality educator practice
e) Learning vision based on complexity

In the same way as Vocational School, Business must concentrate on some conditions so that this relationship occurs efficiently. These characteristics can be indicated by the attendance to the model presented by Bartlett *et al.* (1994, 1995a and 1995b). According to these authors, the business focus changes from strategy to purpose, structure to process, and systems to people characterize the new business paradigm. From these Business and Vocational School conditions and considering the drive factors in the relationship, one can define the relationship model.

The relationship model indicates that five processes occur in the relationship between Business and Vocational School. These processes are:

a) Process of technological and organizational knowledge absorption by the Vocational School on new systems and methods of production and work (technological innovations), new means of production and changes in the job organization.
b) Process of courses planning based on new systems and methods of production and work (technological innovations), new means of production and changes in the job organization.
c) Process of evaluation of courses in function of new systems and methods of production and work (technological innovations), new means of production and changes in the job organization.
d) Process of improvement of organizational and educational structures in function of new systems and methods of production and work (technological innovations), new means of production and changes in the job organization.
e) Process of joint development of technology motivated by new systems and methods of production and work (technological innovations), new means of production and changes in the job organization.

Figure 3 illustrates this model.

According to model, process characteristics depend on driver factors. Thus, the relationship efficiency will depend on the understanding of these factors by Business and by Vocational School.

In accordance with the model, the process of technical and organizational knowledge absorption occurs by:

a) Vocational School technical personnel visitation in Business with technological and organizational innovations.

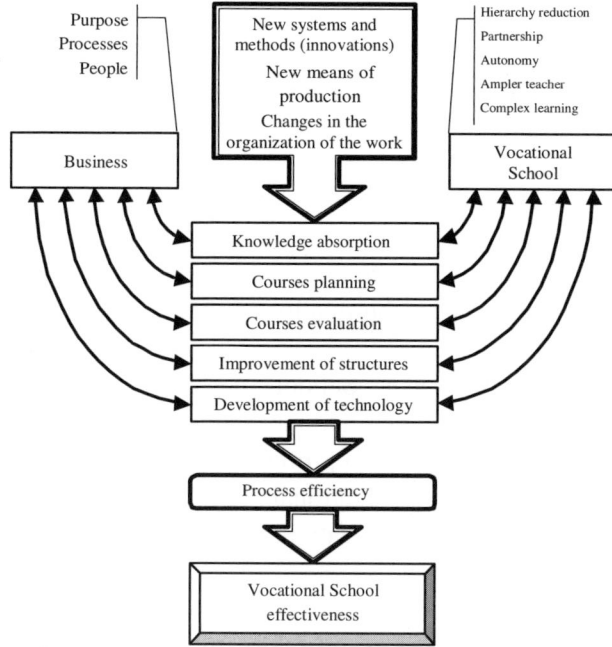

Figure 3. Model of the relation between business and vocational school.

b) Diagnostics and case studies carried through for the Vocational School in Business with technological and organizational innovations.
c) Surveying of University research and monitoring of new technologies developments and new organizational shapes (in fairs, congresses, etc).
d) Creation of scholar structure for the innovations adoption by Vocational School, mainly by debates, prizes and spreading of technologies.
e) Creation of in-house organization for new knowledge incorporation based mainly in debates and technological diffusion.

The process of courses planning must be based on:

a) Analysis and improvement of new technologies development scenarios and perspectives.

b) Participation of companies, labor unions, business associations, and representatives for citizens in this process.
c) Analysis of the tasks and activities in the job places.
d) Use of courses planning methods based on competences.

The process of evaluation of courses must include:

- Evaluation of egresses performance by interviews and by post, including information on the quality of their performance and their market absorption.
- Meetings for Vocational School performance evaluation by companies, business societies and labor unions.
- The participation of pupils and professors in prizes and competitions related with technology or vocational education.
- The external evaluation of Vocational School by external systems as ISO 9000, among others.

The process of improvement of organizational and educational structures must include:

- Courses and training revisions based on evaluations of previous courses.
- Adoption of new pedagogic and administrative practices to function with course evaluations and course results.
- Resource appropriateness based on the results of course evaluations.

The process of joint development of technology must include:

- Technological information systems structuring directed for companies and people.
- Implementation of structure for applied research.
- Partnerships with technological companies and other institutions for new technologies development.

5. Research Agenda

For confirming this model, it chooses a lone case study method (Yin, 2005). This choice was made because it is a representative vocational school. The main research question is how the relationship between Business and Vocational School occurs in the new paradigm.

The chosen vocational school is part of the vocational school net of Sao Paulo Regional Department of SENAI (National Service of Industrial Learning). The choice of this regional department is in function of the projection of this entity and the good results gotten in the 2006 SENAI Knowledge Olympiads.

Amongst the diverse schools of this net, it was selected Lencois Paulista Training Center, located in the Lencois Paulista city, in function to have gotten 22.2% of the medals of SENAI-SP (6 in 27) and 33.3% of the gold medals (5 in 15). All the six pupils sent by the Lencois Paulista Training Center had gotten medals.

The chosen research instruments are structured interview, informal dialogue and personal observation. The structured interviews will be conducted with the school principal, a school instructor, a factory owner who uses the Lencois Paulista Training Center services, and pupils. The interviews must be carried out in the proper school. This research will be finished in November of 2006.

References

Bartlett, Ch. A., Ghoshal, S. (1994). Changing the Role of Top Management: Beyond Strategy to Purpose. *Harvard Business Review*, nov-dec.

Bartlett, Ch. A., Ghoshal, S. (1995a). Changing the Role of Top Management: Beyond Structure to Process. *Harvard Business Review*, jan-feb,

Bartlett, Ch. A., Ghoshal, S. (1995b). Changing the Role of Top Management: Beyond Systems to People. *Harvard Business Review*, may-jun.

Bartlett, Ch. A., Ghoshal, S. (2002). Building Competitive Advantage Through People. *MIT Sloan Management Review,* Winter.

Bogner, W. C.; Thomas, H., McGee, J. (1999). Competence and Competitive Advantage: Towards a Dynamic Model. British Journal of Management, v.10.

Coombs, Ph. H. (1976). A Crise Mundial da Educação. São Paulo, Ed. Perspectiva, Debates, 112.

Drucker, P. F. (1999). Sociedade Pós-Capitalista. São Paulo, Pioneira. São Paulo, Publifolha.

Ernst, D., Kim, L. (2001). Global Production Networks, Knowledge Diffusion, and Local Capability Formation: A Conceptual Framework. Aalborg, Denmark. DRUID, Nelson & Winter Conference.

Galbraith, J., Kenneth. A. (1986). Era da Incerteza. São Paulo, Pioneira.

Lastres, H. M. M., Albagli, S.; Lemos, C. e, Legey, L.-R. (2002). Desafios e Oportunidades da Era do Conhecimento. São Paulo em Perspectiva, v. 16(3), p. 60-66,

Piovezan, L. H. (2004). Mudanças no Mundo do Trabalho da Construção Civil e Conseqüências para a Educação Profissional. Estudos Econômicos da Construção, v.6, n.1(9).

Piovezan, L. H., Fleury, A. C. C. (2004). A Formação da Mão-de-Obra Operacional na Era do Conhecimento. Curitiba, XXIII Simpósio de Gestão da Inovação Tecnológica.

SENAI-DN. (2002). Metodologia [para] elaboração de perfis profissionais. 2. ed. Brasília, 61 p. (Certificação Profissional Baseada em Competências, fase 2).

SENAI-DN. (2003). Metodologias para desenvolvimento e avaliação de competências: formação e certificação profissional.Brasília, 35 p.

Toffler, A. (1995). A Terceira Onda. Rio de Janeiro, Record.

Yin, R. K. (2005). Estudo de Caso: Planejamento e Métodos. 3ª ed. Porto Alegre, Bookman,

Zarifian, Philippe. El Modelo de Competencia y los Sistemas Productivos.

CHAPTER 39

ALWAYS CONNECTED: HOW YOUNG BRAZILIANS USE SHORT MESSAGE SERVICES (SMS)

Marie Agnes Chauvel

Ibmec-RJ, Brazil

This paper aims to study the role played by short messages services (SMS) in the reinforcement of social and emotional links in teenagers and young adults communities. The paper is based on a qualitative research carried out during the year of 2004 in Rio de Janeiro with twenty teenagers and young adults, between 16 and 25 years old, users of SMS, who were submitted to in-depth interviews. Results suggest that short messages play a major role in the teenagers and young adults identity building process. Faced with an uncertain future and a high level of urban violence, these consumers use short messages to keep permanent contact with other young people, usually a small number of correspondents with whom they exchange a large number of messages daily, most of them composed with very few words (like: "are you there?"). This permanent contact with others seems to be used as a tool to increase self-confidence and emotional comfort.

1. Introduction

Text messages via mobile phone are today a very widespread means of communication, mainly among teenagers. It is estimated that, each month, more than 100 billion SMS (Short Message Services) messages are exchanged worldwide (Rheingold, 2003). Currently, more than 460 million SMS messages are exchanged per month in Brazil (Paiva, 2004).

The use of SMS has become an important economic and social movement and contributed to changes in cultural aspects of several countries. In Japan, the *Keitai* (mobile phone) and SMS messages created an alternative space for youth, far away from the watchful eyes and authority of their parents (Ito, 2001). In the Philippines, a group of teenagers organized, through SMS messages, public manifestations which led to the deposition of the country's president, Joseph Estrada, in 2001 (Goodman, 2003). In Finland, SMS generated new terminology, habits and social norms, significantly altering the behavior of the local youth (Kasasniemi and Rautianen *apud* Rheingold, 2003).

In Brazil, studies about this service are still rare. The objective of the research described in this paper was to fill this gap, seeking to understand the dynamics of the use of SMS in Brazil, from the point of view of its users.

2. Literature Review

The literature review for this study touched on three themes: social links and emotional communities, adolescence and identity building, and the specificity of social context in Brazil.

2.1. *Social links and emotional communities*

The spread of SMS messages among teenagers seems to be connected to a large extent with both the need to share feelings with someone else and to have their own space – even if only virtual. Clearly having an objective that is more than the exchange of information, the messages serve as elements of aggregation, emotional safety, and entertainment. Through them, it is possible to connect to partners, feel valued as a person and member of a group, and additionally fill empty spaces – temporal and emotional (Rheingold, 2003; Ito, 2001; Johnsen *apud* Garfalk, 2001; Mäenpää and Kopomaa *apud* Rheingold, 2003; Taylor and Harper, 2001).

What seems to stimulate the use of the service among teenagers - in a general sense - is the search for an emotional connection with others rather than a value of use in itself. This fact contrasts with a sociological current that sees individualism as the highest manifestation of today's society (Lipovetsky, 2002). At least in terms of the use of SMS by teenagers, what can be observed is much more a gregarious impulse than a process of individual affirmation.

For some authors (Cova, 1997; Maffesolli, 2002), this could be the sign of a new type of consumer: post-modern by definition and gregarious by nature. For this type of consumer, the individualist project of affirmation and differentiation would lose space to a new type of experience, more connected to the sharing of emotions and group participation.

2.2. *Adolescence and identity building*

As stated by Giddens (2003), today's family is much more democratic, equalitarian, and plural than before. Further it is significantly less directive in relation to the destiny of its younger members. Since the family is, especially the parents, the main influencing element in the formation of identity (Alberti, 2004; Erikson, 1987), it is reasonable to suppose that changes in its composition and functioning have significant consequences in the formation of the individual.

According to Erikson (as stated by Shultz and Shultz, 2002), "adolescence, between 12 and 18, is the phase when we have to face and resolve the crisis of the ego's basic identity", i.e. to build our self-image and integrate ideas about what we are and what we want to be. In that phase, the adolescent feels compelled to look for answers to two fundamental questions: "Who am I?" and "What do I want?" (Alberti, 2004).

Kimmel and Weiner establish a relationship between the feeling of identity and the demand for approval of others. The less developed the identity is, the more the need for support from external opinions for self-evaluation and the less the understanding of the other as a distinct entity (Kimmel and Weiner *apud* Schoen-Ferreira et al, 2003).

The emotional neediness that characterizes adolescence to a large extent should be reduced significantly at the end of this period. In many cases, however, this does not happen. Even though the identity crisis is faced as a transition process between childhood and adulthood, its duration could vary significantly from culture to culture and be affected by diverse factors, such as late entrance into the job market (Erikson, 1987).

2.3. Specificities of the Brazilian social-cultural context

Two aspects of the Brazilian context are important in order to understand both the dynamics of social relations among youths and their families, and the role of the mobile phone (and text messages) in their daily lives. Firstly, the relational character of Brazilian culture. Secondly, the role played by urban violence and its representations.

According to DaMatta (1986), Brazilian society is characterized by two conciliating antagonistic logics: that of the *individual*, seen as a "subject of the universal laws that modernize society", and that of the *person*, anchored in the "webs of social relationships created in the family, neighborhood, god parenting, the name and, above all, kinship." (DaMatta, 1986, p. 95 to 97).

Further, according to DaMatta (1983), for those two facets of society correspond two distinct spaces: the "home", place for the *person*, a space that is "loving, where harmony must reign over confusion, competition and disorder" (p. 27) and the "street", place for the *individual*, "a dangerous place", "ruled by anonymous laws", where there is "no love, consideration, respect or friendship" (p. 29). Within that context, becoming an adult also means leaving the safe world of the "home", in order to face the somewhat threatening universe of the "street".

In Rio de Janeiro, this passage from the home universe to the street universe is even more threatening due to the high level of urban violence the city faces. According to Costa (2004), the violence promotes a general feeling of "unprotection, fragility and distrust in relation to the effective capacity of the authorities". Nicolaci-da-Costa (2004) observed that protection against urban violence in Rio de Janeiro is, among adolescents, an important reason for using mobile phones.

3. Objectives and Method

Given the lack of previous research about the subject in Brazil, it was opted to undertake an exploratory study, of a qualitative nature. The research was based on in-depth personal interviews. The gathering of the data followed the "saturation" principle, being interrupted only when the undertaking of new interviews was not able to generate additional data that addressed the object of the study (Bertaux, 1980). The data was collected in Rio de Janeiro. The interviewees were users of mobile phones and SMS services, of both sexes, between 16 and 25 years old, and belonging to classes A and B. An additional selection criterion was that interviewees should send at least ten SMS messages per week in order to guarantee a minimum level of knowledge and intimacy with the service. Twenty interviewees were interviewed.

4. Results

The description and analysis of the data was structured according to the following points:

- The insertion of the mobile phone into the daily routines of the interviewees
- The reasons for using SMS
- The process of adhering to the service
- Virtual relationships

4.1. *The insertion of the mobile phone into the daily routine of the interviewees*

Mobiles are an inseparable component of the routine of the interviewees, who in the absolute majority affirm to take them everywhere they go. Most of them expect their interlocutors to always use mobiles and to be able to be reached at any time and any place.

> "*How often do I talk? Every day, every hour, every minute (...). Like, a woman doesn't go out without earring; I don't go out without earrings or without my mobile*" (Gabriela).

More than a simple instrument of communication, the mobile seems to be a fundamental element of emotional safety. For some of the interviewees it is even explicit when they talk about their dependence on phones, associating the lack of them to a feeling of distress.

> *"I think we end up becoming dependent (on mobiles). How I would react if I was forced to be without it for a week? I would feel a bit isolated. I prefer not to even think about that possibility" (Eliza).*

It seems that the urban violence in Rio de Janeiro increases the importance of the mobile phone as an element of emotional safety of the interviewees. Not only to deal with the real violence, but also with the imaginary one. In a certain sense, the mobiles are related to a feeling of protection.

> *"When my car got stolen, I was left in the street alone. Desperate, I picked up my mobile and asked my dad to pick me up. I cried. My legs trembled. If I didn't have my mobile, I don't know..." (Joana).*

Mobiles, by making possible continuous contact with others, end up transforming that contact into a kind of permanent need, which contributes to the control of the anxiety and the emotional comfort of users. That feeling seems to be so strong that several of the interviewees never turn off their mobiles, even when they to go to sleep.

> *"There's some times that I have attacks, I want to talk to so and so, like at three in the morning: hi so and so, what's up? (...) I call (at that time) because I get called too. So I call, to those that call me, I call" (Gabriela).*

4.2. Reasons for using SMS

The results of the research indicate that the main need fulfilled by the service is an affective one. The users seek through SMS the maintenance and reinforcement of friendship, connections, and affection with their partners. Frequently they use it as a way to reinforce their self-esteem, fight loneliness, or even as an element of emotional safety in the midst of an environment that is sometimes aggressive. The importance of SMS messages from the affective point of view is revealed through four main factors: interlocutors of the messages; types of message sent; feelings associated with the messages; and expected response time for the messages. From the rational point of view, the messages tend to be valued for aspects related to their cost, practicality, and ability to tear down communication barriers.

- The interlocutors of the messages: Even though the interviewees in general had a large number of contacts stored in their mobile's phonebook, most of the messages are exchanged with a very small number of interlocutors: intimate friends, boy/girlfriend and, eventually, family. Even the few that send messages to a larger number of people state that they would not send messages to strangers or to people they do not know very well. In a general sense, the research data suggests that a certain level of intimacy is necessary to exchange messages. It seems that, the higher the level of intimacy, the higher the number of messages sent. Between boyfriends and girlfriends, messages are usually very frequent.

"I think I send messages to five or six people: my friends, my brothers and my boyfriend" (Joana).

- Type of message sent: The strong association between affection and motivation for the use of the service becomes explicit when analyzing the content of the messages described by the interviewees. Love declarations, manifestations of missing someone, consolation for friends during difficult times are amongst the messages most commonly sent.

"Notifications, arrange dates, confirm dates, romantic. (...) In reality, what I exchange most is pretty things with my boyfriend" (Aline).

"Sometimes I send one to cheer up a friend who is sad" (Elaine).

That type of message reinforces the hypothesis that the mobile phone and – by extension – SMS messages are contributing towards the redefinition of the concept of "being together". Not only because one can anticipate the moment of a physical encounter, but principally by allowing the continuous sharing of information and feelings. In practice, through SMS messages, it is possible to be "together" even when you are far apart. Even though some of the interviewees refer to part of the messages sent as "silly stuff", "unimportant things", in reality they seem to play an extremely important role in the psychological comfort of senders and receivers. One of the interviewees was very explicit about this when saying: "I don't spend much time alone, I don't like being alone". She informed us that she also uses the messages as a way to find and relate to others.

- Feelings associated to the messages: In general, the act of receiving a message is associated to positive feelings of happiness and reinforcement of self-esteem. The more intimate the person that sends the message, the more chances that it will be perceived as important and will awake a feeling of happiness.

"Lord, it makes my day better, I get happier. (...) I get happy with all of them, but the main and unforgettable one is from my boyfriend, who said he loved me" (Joana).

If receiving messages means being remembered, being important to someone, then not receiving them means exactly the opposite: *"I would feel like I was forgotten"* (Elaine).

- Expected response time for the messages: Most of the time, sending a message presupposes immediate reception of an answer. Especially when the interlocutor is intimate or when a question that is considered important is made. In those cases, it is common that the sender of the message can get relatively anxious in the expectation of an answer. In some cases, the lack of a response is taken as "bad will" on the receptors part and as proof of the carelessness of the other side. This may lead to an angry reaction of demanding an answer (through sending repetitive messages or through a phone call).

"It's uncomfortable (the delay in receiving responses). It bothers you a bit. Because, sometimes, you even think the person has bad will" (André).

- Rational justifications for sending messages: From the rational point of view sending messages tends to be justified through four main factors:

practicality, lower cost when compared to other communication alternatives, speed of sending, and the possibility to overcome barriers (psychological, situational, and relative to the location of use). Additionally, value is given by the fact that messages can be received by a recipient under any circumstances.

4.3. Process of acquiring the service

Friends and boy/girlfriends are the biggest influences in the process of acquiring the service, a finding that reinforces the highly relational and affective components of SMS. The first experiences in the use of SMS are related to receiving messages. Initially, as responses to phone calls, the messages end up stimulating the receiver into testing the service.

> *"I started to use it for speaking with my boyfriend. Silly messages just to say "hi" or that I arrived at home etc. Afterwards I started to exchange more with friends" (Danielle).*

4.4. Virtual relationships

The results of this research reveal that the interviewees avoid relating with strangers through SMS. Simply mentioning exchanging messages with strangers caused very emphatic reactions from a large number of interviewees. Some of them reacted in the following way: "people they don't know who they are", relating that type of behavior to people without friends.

> *"I don't like it because they are people that you don't know (...). The world is really dangerous. Who knows if it´s not a psycho... God forgive me" (Gabriela).*

5. Conclusion

More than simple instruments of communication, mobiles are important elements of emotional safety for the interviewees in this study. Many of them revealed a relationship of dependence with their handsets. In a certain way, losing their mobile would be like losing contact with the world and be unprotected.

Since, adolescence is a period in which the search for others is extremely intense, it seems that the experience of immediate communication, made possible by the mobile, has become a part of the repertoire of the expectations of the interviewees. This expectation probably contributed significantly to altering their habits as compared with previous generations. It is as if each search (here represented by the phone call) should correspond to an immediate meeting (receiving the message and establishing communication). If this meeting is frustrated it would become a reason for anguish and, in a certain way, a feeling of abandonment.

The way in which the interviewees perceive urban violence and relate it to their day to day lives is another element that reinforces the need to use the mobile for their emotional safety. After all, faced with a picture where, at least from an imaginary point of view, "everything is possible", the possibility of immediate contact serves as a means to reduce distress. SMS messages also seem to be a way to keep constant contact with the safe world of "the home" (DaMatta, 1983), which helps to face the universe of the "street" and its threats.

References

Alberti, S. (2004). O Adolescente e o Outro. Rio de Janeiro. Jorge Zahar Editores

Bertaux, D., (1980). L'approche biographique, sa validité méthodologique, ses potentialités, Cahiers Internationaux de Sociologie, vol XIX, juillet/décembre. PUF, Paris.

Costa, J.F. (2003). O mal-estar da nossa civilização. Continente Multicultural, São Paulo, N° 32, Agosto.

Cova, B. (1997). Community and Consumption. Towards a definition of the 'linking value' of products or services. *European Journal of Marketing,* Vol. 31, no. 3/4, pp. 297 a 316.

Da Matta, R. (1983). Carnavais, Malandros e Heróis, 4a. Ed., Rio de Janeiro. Jorge Zahar Editores.

Da Matta, R. (1986). O modo de navegação social: a malandragem e o "jeitinho". In: O que faz o Brasil, Brasil? 2a. Ed. Rio de Janeiro: Rocco. Cap. 7, pp. 93-105.

Erikson, E. H. (1987). Identidade, juventude e crise. Tradução de Álvaro Cabral. Segunda edição. Rio de Janeiro. Editora Guanabara.

Garfalk, C. (2001). Kids on the Move. Telenor Xpress, 1;
http://www.telenor.com/xpress/2001/1/kids_move.shtml.

Giddens, A. (2003). Mundo em descontrole: o que a globalização está fazendo de nós. Tradução de Maria Luiza Borges. Terceira Edição. Rio de Janeiro. Record..

Goodmann, J. (2003) Enviroment and Society. Receiver, 8;
http://www.receiver.vodafone.com/articles/index00.

Ito, M. (2001). Mobile Phones, Japanese Youth, and the Re-placement of Social Contact. In: Society for the Social Studies of Science Meetings. Boston;
http://www.itofisher.com/PEOPLE/mito/Ito.4S2001.mobile.pdf.

Ito, M. (2003). Mobiles and the Appropriation of Place. Receiver, 8;
http://www.receiver.vodafone.com/articles/index07.

Lipovetsky, G. (2002). O Império do Efêmero. A moda e o seu destino nas sociedades modernas. Tradução: Maria Lúcia Machado. São Paulo: Companhia das Letras.

Maffesoli, M. (2002). O Tempo das Tribos. O declínio do individualismo nas sociedades de massa. Tradução: Maria de Lourdes Menezes. 3ª Edição. Rio de Janeiro: Forense Universitária.

Nicolaci-da-Costa, A. M. (2004). Impactos Psicológicos do Uso de Celulares: Uma Pesquisa Exploratória com Jovens Brasileiros. Psicologia: Teoria e Pesquisa, Vol. 20, no. 2, pp. 165-175.

Nisbet, R. A. (1973). Comunidade. In: Foracchi, M. M.; Martins, J. S. Sociologia e Sociedade. Leituras de Introdução à Sociologia. Rio de Janeiro: livros Técnicos e Científicos Editora S.A.

Paiva, F. (2004). "Resultados anima operadoras". Teletime. São Paulo. Ano 7. Número 71. Outubro.

Pimentel, B. (2001). Cell Phone Craze May Be Key For Philippines Future. San Francisco Chronicle. 11/02/2001;
http://www.sfgate.com/cgi-bin/article.cgi?file=/chronicle/archive/2001/02/11BU166485. DTL.

Rheingold, H. (2003). Smart Mobs. The Next Social Revolution. Cambridge: Perseus Publishing.

Schoen-Ferreira, T. E., Aznar-Farias, M., Silvares, E. F. M. (2003). A construção da identidade em adolescentes: um estudo exploratório. In: Estudos de Psicologia, 8, n°1. Natal. Janeiro-Abril.

Schultz, D. P, Schultz, S. E. (2002). Teorias da personalidade. Tradução: Eliane Fanner. São Paulo. Pioneira Thomson Learning.

Taylor, A. S.; Harper, R. (2001). Talking 'Activity': Young People and Mobile Phones. In: CHI 2001 Conference on Human Factors in Computing Systems. Seattle: 31/03 – 5/04/2001;
http://www.cs.colorado.edu/~palen/chi_ workshop/papers/TaylorHarper.pdf.

CHAPTER 40

THE INDIVIDUAL IN MANAGEMENT OF TECHNOLOGY

Marianne Hörlesberger

Technology Management; systems research,
Austrian Research Centers GmbH – ARC, Austria

Developing technology is the answer of the human being to the changing environment. Nowadays nobody doubts that innovation is the driving force for an enterprise, and a country. An important competence for innovation is creativity. Creativity is a core characteristic of a human being. The individual bears knowledge, abilities, and competences. Which role does an individual play in Management of Technology? What can we learn from artists? At the beginning of an innovation there is the idea, a WHAT. Bringing the idea to life is a question of HOW. At the end we get once more a WHAT but shaped by the HOW. The human being needs his individual creativity, enthusiasm, knowledge, methods and craft, social competence and the will to find the way to the goal through the whole process. This contribution will give one possible approach to the role of an individual in Management of Technology.

1. Introduction

Innovations keep companies and organisations alive. Creating new things is an important characteristic of the human being. The intellect strives for newness and does not care about the incorporation of this newness into our lives, environment and social structures. Hardly anybody, or rather better nobody, as some scientists of the theories of complex systems state, can estimate the consequences of the mutual influence of technology. Intelligence together with the intellect is able to incorporate newness into our environment, into our lives (Pearce, 1992). Developing the intellect and intelligence starts in the very early childhood. Among others telling fairy-tales and playing besides a mother-child bonding are very important for a healthy development. Based on such a development the individual is able to be creative as an adult. Brown and Eisenhardt illustrate the importance of being creative on the "edge of chaos", of a balance between chaos and structure in management (Brown and Eisenhardt, 1998). In this respect we can learn a lot from art.

The paper is structured in the following way. Firstly there is a short summary about handling technologies we are surrounded by, followed by a short insight into human behaviour. Then a brief survey about the human's brain prepares the reader for the part about "intellect – intelligence". The thoughts about "playing and the role of intuition" lead us to the chapter "the individual is creative on the edge of chaos". Finally the author finishes with a conclusion.

2. Shaping Technology

We are surrounded by a lot of different technologies. We use technologies 24 hours every day. Technology is inside and outside the human being. For instance there are implants inside. Outside you are surrounded by technologies rather than by nature especially if you live in a city. But even in the middle of nature you are confronted with technologies. Scientists do research on genes of all creatures. Genetic engineering and other biotechnologies mix with "unspoiled" nature. In the past the human being influenced nature directly. Today nature can be changed and shaped by the human being directly. This enormous influence affects not only the life in the industrialized world but also all aspects of life in developing countries hence globally. Furthemore the latest developed technologies have effects on the whole world much more quickly than the former ones.

"Technology shapes our future. We should shape technology. To do so we have to: understand technology well – what it is, how it is composed, how it evolves; know the people of influence – who they are and what they do; choose how to take responsibility – in our work, through our investments, through political action" (Van Wyk, 2004). The better you know all details and components of processes the better you can shape them. But is it possible to look through and to analyse especially the mutual effects of all technologies around us? Computer scientists for instance know or have found that not even the best expert is able to duplicate or to pursue all processes which are running in a working computer. Will the human being, will scientists and experts be able to understand and to follow how, where and why the latest developments in technology like in biotechnology, in nanotechnology, or in technologies which use different waves, and so on affect our lives?

Therefore different questions emerge. Which roll does an individual play in these processes with his creativity, morality, ethic laws, and skills? Who is responsible for these developments? How can the human being manage the phenomenon that nobody is able to analyse the mutual influence of all technologies on earth?

3. What is the Driving Force of Our Behaviour?

Behavioural researchers and researchers of the human brain have done a lot of work in the past few last years. They have tried to explain how our feeling, thinking, and acting come about. They say that the human being can only become aware of activities which are accompanied by the neo-cortex (Roth, 2003). On the one hand the processes of consciousness can be recorded quite well with modern methods, because they are bonded to neuronal processes and metabolism; for instance they need a lot of oxygen and sugar, and the chemical processes bounded on the consumption of chemical substances can be measured. But there is an understanding that some states of consciousness especially the forms of experience and attentiveness are embedded in a lot of unconscious processes. Our feelings influence the steering of our behaviour very strongly, both the conscious one and the unconscious one, researchers of the brain have found out. Strong feelings like stress, pain, fear, aggression, pleasure, happiness, amorousness and, love move us and we feel their power. The intellect and the reason arise quite late in the development of an individual and they can influence our emotions only to a certain extent.

Both the theory of sociology and the theory of neuroscience decline the traditional idea of the free will of the human being. All philosophers throughout the history have thought about the free will. The term "free will" in its traditional sense covers substantially the following aspects: creating actions through the will, the possibility of acting differently, if you want to act differently, and the responsibility of the action (Roth, 2003). We in the western world cannot imagine our society, our rules of law without a free will, at least partially. Neurologists and psychologists have avoided talking about the free will. This field has been in the competence of theologians, philosophers, and jurists. But the neurobiologist Benjamin Libet and his team published an investigation of the free will in the 1983 (Libet *et al.*, 1983). They found that your brain has already made a "decision" for instance in which direction you will do your next steps before you are aware of this decision (see also Libet, 1985). Since then a lot of articles and reports about different experiments on the brain have been published. The free will is the capability of wilful acting without a perceptible constraint, out of oneself. It is the feeling that you could act in a different way in a special situation if you merely want to act differently (Roth, 2003).

Gerhard Roth's core statement is: the social nature of the human being results from his biological nature and not the other way round. The human being is bounded to other human beings biologically, psychologically and for communication because of his innate mechanisms.

You can notice that there is a break with the strict anti-individualism and a revaluation of individualism in social science (Baecker, 1999). At least Gerhard Roth says in his book 2003 that the neurobiologists' claim of explaining the human being with his thinking, feeling, acting, has to be modest and does not explain the "soul" (Roth, 2003).

4. The Human Brain

Our brain can be divided into three different neural structures. Paul MacLean calls it "triune brain" with the Reptilian Brain, the Limbic System and the Neo-cortex (MacLean, 1990).

The reptilian brain is the archipallium brain, called by MacLean the "R-system". It includes the brain stem and the cerebellum. The R-system is responsible for the physical process in our body in muscles, balance and autonomic functions, such as breathing and heartbeat. The R-system is a means of acting in the physical world and "stores" our different learning-experiences about this physical world around us.

The middle part of the brain is called the limbic system (paleomammalian brain). (lat.: *limbus* means edge, border; the limbic system edges the R-system). There are many different neuron-modular systems with different functions in this area, but neurobiologists agree that this part of the brain is responsible for emotions. It is the place of our emotional bonding, and it takes also part in dreaming, visions of our inner world. The middle-emotional system connects the three parts of the brain to a unit. MacLean claims to have found a physical basis for the dogmatic and paranoid tendency, the biological basis for the tendency of thinking to be subordinated to feeling, to rationalize desires in the limbic system. He sees a great danger in all this limbic system power. As he understands it, this lowly mammalian brain of the limbic system tends to be the seat of our value judgements, instead of the more advanced neo-cortex. It decides whether our higher brain has a "good" idea or not, whether it feels true and right.

The neo-cortex, cerebrum, the cortex, or an alternative term, neopallium, also known as the superior or rational (neomammalian) brain, comprises almost the whole of the hemispheres (made up of a more recent type of cortex, called neo-cortex) and some sub-cortical neuronal groups. It corresponds to the brain of the primate mammals and, consequently, the human species. The neo-cortex is the seat of the intellect. It is the place for reflection based on the "reports" from the lower neighbours, the limbic system and the R-system.

As a limbic system is not able to exist without the R-system, the neo-cortex cannot work without the support of the lower systems, the limbic system and the R-system. The developed neo-cortex has access to causality, and there is a potentiality for changing the lower order of both prime systems.

The cortex is divided into the left and right hemispheres. The left half of the cortex controls the right side of the body and the right side of the brain the left side of the body. The feeling of a free will emerges in us after the limbic system has determined the functions, what we have to do. The intellect and the reason are important for questions of decisions.

But the limbic system determines how and with which arguments the neo-cortex is engaged.

5. Intelligence – Intellect

Intelligence is present in all creatures, in all forms of lives as Joseph Pearce says (Pearce, 1992). Intelligence strives after welfare state, well-being, and continuity. The term intelligence defined by Joseph Pearce corresponds more to the term intuition. Intuition is a person's capacity to obtain or have direct knowledge and/or immediate insight, without observation or reason. It is a holistic way, where we comprehend a fact intellectually-emotionally without analysing. The intellect is an important characteristic of the human being. It strives after newness, novelty, and possibilities. The intellect is the impulse in us for solving problems, which we have created before. The intellect is located in the brain, but the intelligence has to do with our heart, Joseph Pearce maintains (Pearce, 1992).

Both a creative discovery and a creative life emerge out of a connection of the intellect's passion with the unfathomable matrix of intelligence.

Excitements are transmitted with transmitters in our neuronal system in the brain. These transmitters have also been found in the heart and these transmitters in the brain are in some way connected to them. It has been noticed that processes in the heart precede the processes in the brain as well as in the body. The limbic system plays a core role in the dynamic between heart and brain. John and Beatrice Lacey of the National Institute of Mental Health found that our brain continuously sends information of our environment to our heart and the heart sends messages to our brain for an adequate answer of the environment's request (Lacey, 1978). Special hormones (so called "ANF") are produced in the auricle of the heart, which influence a lot of important organs of our body and theses hormones influence the limbic system enormously (Cantin and Genest, 1986). The human being has two poles of experiences, his individual self with its experience in the brain, and the universal and impersonal intelligence in the heart (Pearce, 1992).

Furthermore he maintains that a successful human life depends on the dialogue heart-brain.

How can a healthy relationship between brain and heart be developed? First the intellect has to be developed in the best possible way. Intelligence cannot be developed for itself. It needs something it can influence. Intelligence is the capability to work for the well-being (as ability and not as content). One cell in an organ works in this organ for a higher evolutionary possibility. Joseph Pearce takes this as an image like the intellect works for the intelligence. There are several steps for the development of the intellect (a lot of books illustrate this). And the different steps get developed without the intelligence. However there are "open windows" at special times of a human life, where the intelligence has the possibility for a development: birth, in the middle of adolescence, and in later life. Hardly any research has been done concerning adolescence and later life. The first step is described by different scientists and summarized by Joseph Pearce (Pearce, 1992). The imperative of nature and a stimulating environment are important for a successful development. As you can duplicate by watching different developments higher structures are able to be developed only if fundamental structures are available. Joseph Pearce calls the first important phase "mother-child bonding" and explains that the intelligence is not a sweet feeling, but it is a fundamental necessity and the basis for all bonding. Bonding develops in clear steps: mother-child, child-family, family-society, and bonding of a couple. It seems that the ability of bonding gets lost in our society (Pearce, 1992).

Carl Gustav Jung says that a child lives in the unconsciousness of its parents. Their convictions and expectances are decisive factors for world and self view of the child, even they are not talked about, and they do not become explicit.

6. Playing and the Role of Intuition

"If you would like your children to become intelligent, tell them fairy-tales. If you would like them to become highly intelligent, tell them even more fairy-tales" (Ascribed to Albert Einstein).

Playing is the fundament of creative intelligence. It is not given from the beginning, it has to be developed. One basis for the ability of playing is telling tales, as psychologists of development and the experiences of parents and educationists have found. The self-created pictures inside a child are very important for the later symbolical and metaphorical thinking as well as for the concrete and formal thinking (mathematics, philosophy and so on). Metaphors and symbols play a big role in our thinking and in the development of thinking in our every day life and in the high science too. It seems that the limbic structures in connection with the neo-cortex are responsible for the emergence of the metaphorical-symbolical abilities. All metaphor-symbolic actions have to be translated from the R-system via the limbic system to the neo-cortex. This is the reason for remembering our feelings, for what we have learnt.

Playing is the training of thinking in metaphors and symbols, and is the fundament of education and higher development.

Intuition means recognizing the essential. Intuitions give us information about the physical or emotional well-being, if such information is not available in the immediate surrounding. The intuition uses the dynamic between the limbic system and the neo-cortex, whereas the usual information of the environment comes from the dynamic between the limbic system and the R-system.

Joseph Pearce states that an intellect developed without the intuition is only for defence and manipulation, and this restrains the development of an intelligence which could lead us further than our ego. A well developed intuition harmonizes our accommodation with the environment with the right hemisphere of the neo-cortex. The nature of our neo-cortex is striving after newness. We need intuition for the integration of this newness in an adequate and conductive way.

The risk of the intellect lies in its possibility of cutting itself off the intelligence.

7. The Individual is Creative on the Edge of Chaos

"Each child is an artist. The problem is how a child can stay an artist while becoming adult" (Pablo Picasso).

Being creative makes us happy and healthy. Nowadays you can find many books and seminars where you can be trained in finding your own creativity and it has to do with your "core". The best case would be that we can be creative during our every day work, in our jobs as well as in leisure times in the sense how it is mentioned above, in the dynamic between intellect and intelligence. Would this be an adequate precondition for developing and managing technologies?

Human beings rather seek their convenient environments as they adapt the environment so it will be convenient for them, behaviour researchers found. Is this not an invitation for shaping and structuring our companies in such a way that the adequate employees are attracted? Just as creativity is important for a fulfilled human life innovation is the essence for the life of a company, but creativity in having ideas and the creativities in managing innovation companies depend upon creative, intuitive, and intelligent human beings.

"Limited structure combined with intense interaction creates enough flexibility for behaviour to be fresh, surprising, and adaptive, and provides just enough structure for a business to deliver products and services on target and on time", Brown and Eisenhardt found (Brown and Eisenhardt, 1998). In their book "Competing on the Edge" they cite the rules of Jazz improvisation. These rules give us a metaphor for convenient structures and rules in a company.

- At any time in a performance, know who the leader (soloist) is and where you are in the piece.
- The soloist should listen to and build off the work of other members of the band.
- Know the rules in order to know how and when to break them.
- Experiment as a group (e.g. by changing or eliminating structure) or as an individual (e.g. by overblowing or fiddling with your instrument).
- Expect occasional "trainwrecks". Recover and move on.
- Do not play the same solo over and over; practice new approaches and styles in familiar pieces. Incorporating the unexpected is the essence of great jazz.

We can find much stimulation in other fields of art. A further example is writing film scripts as Frank Daniels states. He gives clear rules what has to happen in the eight sequences of a film. First we think, especially in Europe, that we are restricted in our creativity by such rules, but screen-writers and film directors experience that the rules give them the necessary freedom for their creativity. Nevertheless it is a "walking on the edge", because Josef Mikl, a contemporary painter of Austria, says "the content creates the shape" (Interview in the radio, ORF oe1, October 2001).

How different is shaping and managing technology from creating and giving a performance, from writing a film script and directing a film? Successful people and managers know it. Books like "Competing on the Edge" show that the knowledge and experience of artists are very interesting and important for managing companies, for managing technologies, for managing business. Creativity is the content, the WHAT, the rules are the structure, the HOW. The shaped WHAT by the HOW is the product, the technology at the end of the process, the new WHAT. The secret is the well-balanced measure of freedom and structure. Finding this lies in the intellect and intelligence of an individual.

8. Conclusions

Shaping technologies, so that our future, the future of the earth, will be healthy, will be a home where the human being in all parts of the world is able to live his creativity, is a big challenge besides the danger of war and terrorism. The situation stimulates to be creative, to use the intellect and intelligence for creating technologies. As we saw above the education for a high intellect and intelligence starts very early, in the time of pregnancy. Carl Gustav Jung's argument stimulates parents to reflect their patterns of feeling, thinking and acting, because they influence the development of the children enormously. It seems playing in the sense of playing the scenes of tales a child has listened to before with simple things is getting lost. Where can they find a place and time to develop their own pictures? One of the biggest challenges is building and shaping places where this can happen. Education has to focus on the development of the dynamic between intellect and intelligence, the dynamic between brain and heart. Based on a well developed intellect and intelligence the rules of management of technology can be found, and therefore the conditions of being innovative in the whole process of management of technology.

References

Baecker, D. (1999). Organisation als System, Frankfurt am Main, Suhrkamp Taschenbuch Verlag.
Brown, Sh. L. and Eisenhardt, K. M. (1998). Competing on the Edge: Strategy as Structured Chaos, Harvard Business School Press, Boston, Massachusetts, ISBN 0-87584-754-4.
Cantin, M.; Genest, J. (1986). The heart as an endocrine gland. *Scientific American Feb.,* Vol 254, Nr. 2; 76
Gebert, D. (2004). Innovation durch Teamarbeit, eine kritische Bestandsaufnahme, Verlag W. Kohlhammer, Germany, ISBN 3-17-018095-9.
Jung, C. G. Über die Entwicklung der Persönlichkeit. 8. Auflage 1994, Walter-Verlag, AG Olten, 1972 , ISBN 3-530-40717.
Lacey, J. and B. (1978). Two-way communication between the heart and the brain: Significance of time within the cardiac cycle. *American Psychologist.,* 33(2):99-113.
Libet, B. (1985). Unconscious Cerebral Initiative and the Role of Conscious Will in Voluntary Action. The Behavioral and Brain Sciences VIII, 529-539.
Libet, B.; Gleason, C. A.; Wright, E., Pearl, D. K. (1983) Time of Conscious Intention to Act in Relation to Onset of Cerebral Activities (Readiness-Potential): The Unconscious Initiation of a Freely Voluntary Act. Brain 106, 623-642.
MacLean, P. D. (1990). The Triune Brain in Evolution. Plenum, New York.
Pearce, J. Ch. (1992). Evolution's End, Claiming the Potential of Our Intelligence. HarperCollins Publisher Inc., San Francisco, California, ISBN 0-06-250732-X.
Roth, G. (2003). Fühlen, Denken, Handeln; Wie das Gehirn unser Verhalten steuert, suhrkamp taschenbuch wissenschaft 1678, Germany, ISBN 3-518-29278-1.
Van Wyk, R. (2004). Technology: A Core Theory, 13th Conference of International Association for Management of Technology (IAMOT), Washington.

Author Index

Amyot, D. 183
Arif, M. 145
Baumgartner, R. J. 535, 553, 563
Bergman, J.-P. 101
Bochenek, G. 341
Boeck, H. 47
Bolisani, E. 119
Boly, V. 449
Brendle, B. 341
Chauvel, M. A. 589
Chiesa, V. 3
Corona-Armentats, J. R. 449
Corrêa Fleury, A. C. 577
Cunha, J. C. 515
Cunha, S. K. 515
Davenport, S. 195
De Massis, A. 3
Desmarteau, R. H. 17
Ebner, D. 535
Ebrahimi, M. 17
Egyedi, T. M. 427
Figueiredo, P. 241
Fraser, P. 33
Frattini, F. 3
Garnier, C. 17
Gutenberg, S. 241
Guth, W. 381
Hacklin, F. 211
Hörlesberger, M. 601
Iler, C. 341
Inganäs, M. 211
Jakobs, K. 427
Jolly, D. R. 465
Kasper, H. 279

Kato, M. 323
Kero, P. 59
Kim, M.-K. 87
Kohlbacher, F. 279
Kotnour, T. 341
Kruglianskas, I. 297
Lankila, M. 101
Larsson, M. 263, 485
Lefebvre, E. 47
Lefebvre, L.-A. 47
Lucas, P. C. 311
Marxt, C. 211
Milosevic, D. Z. 227, 381
Minshall, T. 33
Moraes, R. 297
Morel Guimaraes, L. 449
Nascimento, P. T. 311
Neemann, C. 397
Nilsson, C.-H. 485
Niwa, F. 323
Park, J.-H. 87
Patanakul, P. 381
Piovezan, L. H. 577
Probert, D. 33
Ragusa, J. 341
Sainio, L.-M. 101
Saives, A.- L. 17
Sbragia, R. 241
Scarso, E. 119
Schou, F. K. 135
Schuh, G. 397
Shenhar, A. J. 381
Sherif, M. H. 427
Srivannaboon, S. 227

Talbot, S. 499
Talhami, H. 145
Urness, M. 69
Valli, R. 33
Vasconcellos, E. 311
Wamba, S. F. 47

Watanabe, C. 355, 367
Weiss, M. 183
Yamada, A. 367
Yu, O. 409
Zilber, S. N. 163